IoT Security

IoT Security

Advances in Authentication

Edited by
Madhusanka Liyanage
School of Computer Science, University College Dublin, Ireland
Centre for Wireless Communications, University of Oulu, Finland

An Braeken
Industrial Engineering, Vrije Universiteit Brussels, Belgium

Pardeep Kumar
Department of Computer Science, Swansea University, UK

Mika Ylianttila
Centre for Wireless Communications, University of Oulu, Finland

Registered Offices
John Wiley & Sons, Inc., 111 River Street, Hoboken, NJ 07030, USA
John Wiley & Sons Ltd, The Atrium, Southern Gate, Chichester, West Sussex, PO19 8SQ, UK

Editorial Office
The Atrium, Southern Gate, Chichester, West Sussex, PO19 8SQ, UK

For details of our global editorial offices, customer services, and more information about Wiley products visit us at www.wiley.com.
Wiley also publishes its books in a variety of electronic formats and by print-on-demand. Some content that appears in standard print versions of this book may not be available in other formats.

Library of Congress Cataloging-in-Publication data applied for

Hardback ISBN – 9781119527923

Cover Design: Wiley
Cover Image: © jamesteohart/Shutterstock

Set in 10/12pt WarnockPro by SPi Global, Chennai, India
Printed and bound in Singapore by Markono Print Media Pte Ltd

10 9 8 7 6 5 4 3 2 1

Contents

About the Editors

Madhusanka Liyanage
School of Computer Science, University College, Ireland.
Centre for Wireless Communications, University of Oulu, Finland.

Madhusanka Liyanage received the BSc degree (First Class Honors) in electronics and telecommunication engineering from the University of Moratuwa, Moratuwa, Sri Lanka, in 2009, his ME degree from the Asian Institute of Technology, Bangkok, Thailand, in 2011 and an MSc degree from the University of Nice Sophia Antipolis, Nice, France in 2011. In 2016, Liyanage received a PhD in communication engineering from the University of Oulu, Oulu, Finland. He is currently an Assistant Professor/Ad Astra Fellow at the School of Computer Science, University College, Ireland. He is also an adjunct Professor at the Centre for Wireless Communications, University of Oulu, Finland. Moreover, he is a visiting lecturer at Yangon Technological University, Myanmar and University of Moratuwa, Sri Lanka. During 2018–2020, he was a Marie Curie Fellow at the School of Computer Science, University College, Ireland. In 2011–2012, he was a research scientist at I3S Laboratory and Inria, Shopia Antipolis, France. Also, he was a visiting research fellow at CSIRO-Australia, Lancaster University, The University of New South Wales, The University of Sydney, Sorbonne University and Oxford University from 2016 to 2019. His research interests include SDN, IoT, Blockchain, mobile and virtual network security. He is a Member of IEEE. Madhusanka is co-author of over 70 publications including two edited books with Wiley. He is also a management committee member of EU COST Action IC1301, IC1303, CA15107, CA15127, CA16116 and CA161226 projects. URL: http://madhusanka.com.

An Braeken
Industrial Engineering, Vrije Universiteit Brussel, *Belgium*.

An Braeken obtained her MSc Degree in Mathematics from the University of Gent in 2002. In 2006, she received her PhD in engineering sciences from the KU Leuven at the research group COSIC (Computer Security and Industrial Cryptography). She became professor in 2007 at the Erasmushogeschool Brussel (currently since 2013, Vrije Universiteit Brussel) in the Industrial Sciences Department. Prior to joining the Erasmushogeschool Brussel, she worked for almost 2 years at the management consulting company Boston Consulting Group (BCG). Her current interests include security and privacy protocols for IoT, cloud and fog, blockchain and 5G security. She is (co-)author of over 150 publications. She has been member of the program committee for numerous conferences and workshops (IOP2018, EUC 2018, ICNS 2018, etc.) and member of the editorial board for *Security and Communications* magazine. She has also been member of the organizing committee for the IEEE Cloudtech 2018 conference and the Blockchain in IoT workshop at Globecom 2018. In addition, since 2015 she is reviewer for several EU proposals and ongoing projects, submitted under the programs of H2020, Marie Curie and ITN. She has cooperated and coordinated more than 15 national and international projects. She has been STSM manager in the COST AAPELE project (2014–2017) and is currently in the management committee of the COST RECODIS project (2016–2019).

Pardeep Kumar
Department of Computer Science, Swansea University, UK.

Pardeep Kumar received his Bachelor of Engineering Degee, in Computer Science and Engineering in 2002 from the Institute of Technology and Management (now, NorthCap University), Haryana, India, his Masters of Technology in Computer Science and Technology in 2006 from Chaudhary Devi Lal University, Haryana, India, and his PhD in Computer Science in 2012 from Dongseo University, Busan, South Korea. Currently, he is working with the Department of Computer Science, Swansea University, United Kingdom. Previously, he worked with the Department of Computer Science, Oxford University, from 2016–2018, the Department of Computer Science, The Arctic University of Norway (UiT), Tromso, Norway (2015–2016), and the Centre for Wireless Communications (CWC), University of Oulu, Finland (2012–2015). His current research interests include security in sensor networks, smart environments, cyber physical systems, body area networks, Internet of Things, and cloud computing. URL: http://cs.swansea.ac.uk/~pkumar.

Mika Ylianttila
Centre for Wireless Communications, University of Oulu, Finland.

Mika Ylianttila (M.Sc, Dr.Sc, eMBA) is a full-time associate professor (tenure track) at the Centre for Wireless Communications (CWC), at the Faculty of Information Technology and Electrical Engineering (ITEE), University of Oulu, Finland. He is leading a research team and is the director of communications engineering doctoral degree program. Previously he was the director of Center for Internet Excellence (2012–2015), vice director of MediaTeam Oulu research group (2009–2011), and professor (pro tem) in computer science and engineering, and director of information networks study programme (2005–2010). He received his doctoral degree on Communications Engineering at the University of Oulu in 2005. He has coauthored more than 150 international peer-reviewed articles. His research interests include edge computing, network security, network virtualization and software-defined networking. He is a Senior Member of IEEE, and Editor in *Wireless Networks* journal.

List of Contributors

Alexander J. M. Milne received his degree in computer science from Swansea University, Wales, UK in 2018. Following his degree, he worked at Swansea University on a Cherish de funded project in collaboration with Oyster Bay Systems. He is currently a student at Swansea University doing a Chess II funded masters by research on Blockchain.

Anca Delia Jurcut received a Bachelor of Mathematics and Computer Science from West University of Timisoara, Romania (2007) and a PhD from University of Limerick, Ireland (2013). From 2008 to 2013, she was a Research Assistant with the Data Communication Security Laboratory at University of Limerick, and from 2013 to 2015 she was working as a postdoctoral researcher in the Department of Electronic and Computer Engineering at the University of Limerick and as a software engineer at IBM, Ireland. Since 2015, she has been an Assistant Professor with the School of Computer Science, University College Dublin, Ireland. Her research interests focuses on network and data security, security for internet of things (IoT), security protocols, formal verification techniques and applications of blockchain technologies in cybersecurity.

Anshuman Kalla is an Associate Professor at Department of Computer and Communication Engineering, School of Computing & Information Technology, Manipal University Jaipur, India. Dr. Kalla graduated as an engineer from Govt. Engineering College Bikaner in 2004. He received his Master of Science in Telecommunications and Wireless Networking from ISEP, France in 2008 and another Master from UNICE, France in 2011. He obtained Ph.D. degree in 2017. Dr. Kalla was recipient of Master's scholarships for pursuing both the Master programs. His area of interest is Future Networking – Information Centric Networking (ICN), Internet of Things (IoT), SDN and Blockchain.

Arnold Beckmann received his PhD and "Habilitation" in Mathematics from University of Münster. He gained post-doctoral experiences at University of Oxford, University of California in San Diego, and Vienna University of Technology. He is currently Professor of Computer Science at Swansea University, and Head of Department of Computer Science. Arnold conducts research in fundamentals of Computer Science, based on Mathematical Logic and Theoretical Computer Science. He is member of Council of the *Association Computability in Europe* (ACiE), and chair of *The Proof Society* (TPS). He is Editorial Board member of journals *Annals for Pure and Applied Logic, Archive for Mathematical Logic, Computability*, and Managing Editor of the book series Perspectives in Logic. He has edited 11 journal special issues and 5 proceedings volumes, and served as PC co-chair of 3 major conferences. Arnold has developed a profile for transferring his expertise to applications. He is founding member of the Swansea Blockchain Lab http://www.swanseablockchainlab.com/, and involved in several projects that explore the application of blockchain technology to real world problems.

Burkhard Stiller received the Diplom-Informatiker (M.Sc.) degree in Computer Science and the Dr. rer.-nat. (Ph. D.) degree from the University of Karlsruhe, Germany. He has been a Full Professor of the Communication Systems Group, Department of Informatics, University of Zurich since 2004. He held previous research positions with the Computer Laboratory, University of Cambridge, U.K., the Computer Engineering and Networks Laboratory, ETH Zurich, Switzerland, and the University of Federal Armed Forces, Munich, Germany. He did coordinate various Swiss and European industrial and research projects, such as BC4CC, Foodchains, AAMAIS, DAMMO, SmoothIT, SmartenIT, SESERV, and Econ@Tel, besides participating in others, such as M3I, Akogrimo, EC-GIN, EMANICS, FLAMINGO, symbIoTe, and ACROSS. His main interests are published in well over 250 research papers and include systems with a fully decentralized control (blockchains, clouds, peer-to-peer), network and service management (economic management), Internet-of-Things (security of constrained devices, LoRa), and telecommunication economics (charging and accounting).

Corinna Schmitt received the Diplom-Informatikerin (Bioinformatikin) from the Eberhard-Karls University of Tübingen (Germany) and the Dr. rer. nat. (Ph.D.) degree from the Technische Universität München (Germany). From spring 2013 to May 2018 she was employed at the University of Zurich (Switzerland) as "Head of Mobile and Trusted Communications" at the Communication Systems Group (CSG) of Prof. Dr. B. Stiller. Her focus was on constrained networks, security and privacy issues, as well as on Internet of Things (IoT) related issues. Now she is a researcher and laboratory supervisor at the Research Institute CODE at the Universität der Bundeswehr München (Germany). She continues her previous research with extension to the application area of military communication, smart environments, and critical infrastructures. Her work is documented in more than 30 publications, including several book chapters and journal articles, as well as the RFC 8272 on "TinyIPFIX for Smart Meters in Constrained Networks", and the ITU-T recommendation Y.3013 on "Socio-economic Assessment of Future Networks by Tussle Analysis". She contributed to several EU projects (e.g., AutHoNe, SmartenIT, FLAMINGO, symbIoTe) and different standardization organizations (IETF, ITU, ASUT) until now and continues with these activities and recruits research funds continuously. She is active in ACM and IEEE as TCP member, as well as reviewer for several journals and funding schemes, and organizer of conferences.

Dominik Bünzli received the Bachelor of Science in Informatics from the University of Zurich, Switzerland. He is currently completing his Master of Science in Informatics at the University of zurich with a focus on Data Science. He works part-time in a team dealing with data warehousing and event processing in a financial environment. His main interests include the Internet of Things, Blockchains and Big-Data.

Gaurav Sharma is currently working as a postdoc researcher at the Université libre de Bruxelles, Belgium. He received his Ph.D and M.E degree in Computer Science & Engineering from Thapar University, India. He received M.Sc. and B.Sc. Degree from CCS University, Meerut, India. His current research is based on exploring security threats in Multi-Processor System on-chips (MPSoCs). He is working on the project "Self-Organising circuits For Interconnected, Secure and Template computing (SOFIST)", funded through the ARC grant for Concerted Research Actions, Fédération Wallonie-Bruxelles. Dr. Sharma is a member of IEEE since 2008 and he has authored/coauthored more than 45 journal/conference articles and book chapters. He serves as reviewer of IEEE Systems Journal, IEEE Sensors Journal, Future Generation Computer Systems and Journal of Information Security and Applications. He was the TPC member of ICISSP'18, GlobeCom'18, IndiaCom'18, SPIN's18, CRIS'18. Dr. Sharma also edited a special issue on "Advanced Research in Privacy and Forensic Analytic of Web Engineering", International Journal of Information Technology and Web Engineering (IJITWE), IGI Global.

Gurjot Singh Gaba received the M.Tech (Electronics & Communication) degree from the Guru Nanak Dev Engg. College, Ludhiana of India in 2011. After completing his M.Tech, he joined Lovely Professional University, Jalandhar of India in July, 2011. His name is very well known in the field of Research. His research works are acknowledged by IEEE conference, national conference and international journals. He has been the executive member of RAECT national conference. His research interests are in the field of Wireless Sensor Networks. He is the live member of the Institute of Electrical and Electronics Engineers (IEEE-USA), life member of The Indian Science Congress Association, Kolkata (India), International Association of Engineers (IAENG-USA), International Association of Computer Science and Information Technology, Singapore.

Guy Van der Sande was born in Belgium in 1978. He received the Master's degree in electrotechnical engineering with a major in photonics from the Vrije Universiteit Brussel (VUB), Brussels, Belgium, in 2001. He received the title of Doctor in the applied sciences from the Department of Applied Physics and Photonics, VUB, in 2005. His Ph.D. program was awarded with the prize "Ignace Vanderschueren" for the best PhD thesis in Basic, Natural and Applied Science of the last 6 years at the Vrije Universiteit Brussel. In 2006, he was postdoctoral researcher at Optique Nonlinéaire Théorique (ULB) with prof. Thomas Erneux, where he did research on nonlinear dynamics vertical-cavity surface-emitting lasers, metamaterials and dynamics of spatially extended nonlinear optical systems. In 2007, he was a visiting scientist at the Institute for Cross-Disciplinary Physics and Complex Systems (IFISC) in the Universitat de les Illes Balears, Palma de Mallorca, Spain (under supervision of Prof. Dr. Claudio R. Mirasso and invited by Prof. Maxi San Miguel), where he worked on nonlinear dynamics of large networks of delay-coupled nonlinear oscillators. In 2013, he was granted a Research Professor fellowship at the Vrije Universiteit Brussel. Since 2012, he provides Physics and Photonics education to bachelor and master students in Sciences and Engineering. His current research interests include modeling and nonlinear dynamics of semiconductor lasers, synchronization phenomena and bio-inspired information processing. He is author of more than 60 journal papers and 80 conference papers of which 6 invited talks, with an h-index of 25.

Jose Razafindrakoto received his B.Sc. degree in Computer Science, Master of Research in Logic and Computation, and Ph.D. degree in Computer Science from Swansea University. He is currently an associate lecturer in Computer Science at Swansea University. His current research interests include Bounded Arithmetic, Propositional Proof Complexity and Computational Complexity Theory. Recently, he has developed an interest in blockchain technology, its underlying software development and related formal methods questions. He is also a founding member of the Swansea University Blockchain Lab.

Lars Keuninckx received a master of industrial engineering in electronics and telecommunications from Hogeschool Gent in 1996. He worked in the industry for several years designing electronics for automotive, industrial and medical applications. He received a bachelor in physics from the Vrije Universiteit Brussel in 2009. Subsequently, he earned his PhD in engineering at the Applied Physics Group (APHY) in 2016, where his main research topics were applications of complex dynamical behavior in electronic circuits and reservoir computing. Currently, he is at the Consciousness, Cognition and Computation (CO3) group of the faculty of psychology and education of the Université Libre de Bruxelles, where he works on computational modeling of the dynamical processes of consciousness.

Lina Xu is an Assistant Professor in the School of Computer Science in University College Dublin (UCD) since 2016. She has received her B.E. from Software Engineering school in Fudan University, China and a B.Sc from Computer Science in UCD through a joint Program. Then she started her PhD in Computer Science in UCD and received her doctoral degree in 2014. Then she worked in HP Labs as a research scientist from the end of 2014 till 2016. Her research interests include Internet of Things (IoT), 5G network, energy efficiency and smart networking, machine learning etc. Now she is working on several projects aiming to apply IoT technologies to assistant healthy ageing and smart transportation.

Mauri Honkanen M.Sc. with honors in Electrical Engineering from Helsinki University of Technology. He is currently working as a CEO of Vital Streams Oy that is a startup company productizing and commercializing a remote outpatient monitoring solution for cardiovascular diseases.

Mehrnoosh Monshizadeh Ph.D. in Telecommunication Networking at Electrical School of Aalto University, Finland. She is working as research security specialist at Nokia Bell Labs France. Her research interests include cloud security, mobile network security, IoT security and data analytics.

Michael Breach is Managing Director and co-founder of Oyster Bay Systems (https://www.oysterbaysystems.com/). With over 30 years' experience in the finance and lending sector, and over 40 years in the software development industry, he is responsible for the direction of the business. Having written several Oyster Bay's lending systems, including Fiscal, Michael is hands-on in the technical side of the business and is seen as an authority on compliant financial calculations. With a passion for discovery, Michael is also spearheading a blockchain technology project.

Norbert Preining got his PhD in Mathematics from the Vienna University of Technology in Austria. Several years of PostDoc at the same university were followed by a Marie Curie Fellowship in Siena, Italy, and then 7 years as Associate Professor at the Japan Advanced Institute for Science and Technology. He is currently working at the Research and Development Department of Accelia, Inc. (Tokyo). His main interests are Mathematical Logic, in particular intermediate logics and many-valued logics, formal/algebraic specification and verification, theory of computation. In addition, he is researcher and practitioner in machine learning/AI, as well as software developer.

Oskari Koskimies M.Sc. in Computer Science from Helsinki University, Finland. He is a Senior Security Architect at Elisa Corporation in Finland, working on the security and medical compliance of the Elisa Remote Patient Monitoring solution.

Pasika Ranaweera is currently pursuing his PhD studies in School of Computer Science, University College Dublin, Ireland. He obtained his Bachelor Degree in Electrical and Information Engineering in 2010 from University of Ruhuna, Sri Lanka and Master's Degree in Information and Communication Technology (ICT) in 2013 from University of Agder, Norway. Pasika is focused on enhancing the security measures in Multi-access Edge Computing (MEC) and Internet of Things (IoT) integration. His research directives extend to the areas lightweight security protocols, 5G and MEC integration technologies, Privacy preservation techniques and IoT security.

Pawani Porambage is Postdoctoral Researcher at the Centre for Wireless Communications, University of Oulu, Finland. She obtained her Bachelor Degree in Electronics and Telecommunication Engineering in 2010 from University of Moratuwa, Sri Lanka and her Master's Degree in Ubiquitous Networking and Computer Networking in 2012 from University of Nice Sophia-Anipolis, France. In 2014 she was a visiting researcher at CSG, University of Zurich and Vrije Universiteit Brussel. Her main research interests include lightweight security protocols, security and privacy in IoT, MEC, Network Slicing, and Wireless Sensor Networks. She has co-authored more than 30 peer reviewed scientific articles.

Veronika Kuchta obtained her Diploma degree (equiv. to Master degree) in Mathematics from Karls-Ruprecht University of Heidelberg (Germany) and her PhD degree in applied cryptography form University of Surrey (United Kingdom) under the supervision of Dr. Mark Manulis. The focus of her PhD thesis was on distributed cryptographic protocols. From December 2015 to March 2018 she was a postdoctoral researcher with the Université libre de Bruxelles (ULB, Belgium). In this time she was a member of the QualSec (quality and security of information systems) group under the supervision of professor Olivier Markowitch and she was working on a project called SeCloud, focusing on security-driven engineering of cloud-based applications. Since March 2018 she has been Research Fellow at Monash University, Melbourne, Australia under the supervision of Dr. Joseph K. Liu. Her research interests lie in the areas of post-quantum cryptography, zero-knowledge proofs, blockchain. She designed several lattice-based cryptographic protocols which were pubished and presented at the international conferences. Furthermore, she is interested in different aspects of code-based cryptography and their relation to lattice-based cryptoschemes and application to the research area of blockchain.

Vikramajeet Khatri graduated with M.Sc. IT from Tampere University of Technology, Finland. He is working as security specialist at Nokia Bell Labs Finland. His research interests include intrusion detection, malware detection, IoT security and cloud security.

Preface

The Internet of things (IoT) is the network of physical devices such as vehicles, home appliances sensors, actuators and other electronic devices. The development of the Internet offers the possibility for these objects to connect and exchange data. Since the IoT will play a major role in human life, it is important to secure the IoT ecosystem in order for its value to be realized. Among the various security requirements, authentication to the IoT is important as it is the first step in prevention of any negative impact of possible attackers.

This book provides reference material on authentication in IoT. It offers an insight into the development of various authentication mechanisms to provide IoT authentication across various levels such as user level, device level and network level. This book offers reference material which will be important for all relative stakeholders of mobile networks, for example, network operators, cloud operators, IoT device manufacturers, IoT device users, wireless users, IoT standardization organizations and security solution developers.

IoT

Over the last four decades, the Internet has evolved from peer-to-peer networking, world-wide-web, and mobile-Internet to the IoT. The IoT is a network consisting of animals, people, objects, physical devices, e.g., home appliance sensors, actuators, vehicles, digital machines, and other electronic devices that can collect and exchange data with each other without human intervention. The communication in IoT is either between people, between people and devices, and between devices themselves, also called machine-to-machine (M2M). Many benefits are realized through these interactions using a variety of technologies including sensors, actuators, controls, mobile devices and cloud servers as now people and things can be connected any time, any place, with anything and anyone, ideally using any path or network and any service.

Following analysis of Statista in 2019, it is estimated that there will be approximately 31 billion connected IoT devices worldwide in 2020, which may even be doubled by 2024. From 2017 onwards, the overall market of IoT has become worth more than one billion US dollars annually. According to the same study, it turns out that the largest IoT market is represented by the consumer electronics industry. The highest IoT-related investments have been seen, for the moment, in travel, transportation and hospitality industries. A very promising market for realizing large growth in IoT is considered to be the automotive industry. Other markets are retail, logistics, construction and agriculture. Consequently, it can be concluded that IoT will have a vital impact on a lot of industries. It will, in many cases, enable a smarter decision-making process based on context-aware information. The main motivations to integrate IoT are the increase in both efficiency and convenience.

The Need for Security

One of the most remarkable attacks that made people more aware of the importance of security in IoT is the Mirai Botnet attack in 2016. In this attack, the hackers simply scanned the Internet for open Telnet ports. Once found, they logged into the devices by trying a set of username/password combinations, that were often used as default by the manufacturer and that had never changed. This allowed them to collect an army of compromised devices, including baby monitors, home routers, air-quality monitors, and personal surveillance cameras, ready to perform powerful distributed denial of service attacks (DDoS). At its peak, approximately 600 000 IoT devices were infected. The most well-known DDoS attack performed by the Mirai Botnet was on the Domain Name System provider Dyn in October 2016. This attack caused large numbers of users from Europe and North America with the non-availability of many major Internet services like Amazon, BBC, Spotify, etc., for several hours.

As the IoT is continuously growing, it becomes an even more interesting point of attack for the hackers. According to a threat report from the security firm Symantec, the number of attacks on IoT devices increased in only one year by 600% from 2016 to 2017, corresponding to 6000 and 50 000 reported attacks respectively. Besides the DDoS attacks, mining of cryptocurrencies has also been reported as a popular activity for hackers. Another important threat is ransomware, with WannaCry and Petya/Not Petya as the most well-known examples, resulting in a large and worldwide take down of systems.

Consequently, we can conclude that much more attention needs to be given to security when including IoT into any business. As the IoT is a kind of galaxy of devices, technologies, and concepts of information, it is hard to understand what happens if malicious interventions of attackers jeopardize the security and privacy of IoT users, devices, and networks when these IoT galaxies are implemented with poor or no security. It is, therefore, important that all elements of the chain, being the device, user and network operator, integrate the required security mechanisms to guarantee end-to-end security in the communication between these devices over the Internet and on local networks. This will not only result in the gaining of trust and acceptance with the end-users, but will also avoid direct physical harm to humans, perhaps even loss of life.

The Need for Authentication

By secure communication, we mainly consider the security features of confidentiality, integrity of the transmitted messages, and authentication of the sending and receiving devices. Confidentiality and integrity are well studied security features and can rather easily be realized through lightweight symmetric primitives following the establishment of a secret shared key. However, to establish such a secret shared key, authentication of the devices, being the verification of the identities while sending and receiving messages, is required. Achieving authentication in a robust way is far harder.

Authentication mechanisms should be considered at a number of different levels, going from user, device to network. Each level has its own particularities, resulting in different types of solutions to offer efficient authentication. For instance, at the user level, biometrics play a very important role in the integration of authentication schemes. For the device, variants such as physical unclonable functions (PUFs), but also tamper resistant memory are important aspects to be considered. Finally, at the network level, different architectures should be considered.

One common feature for all the proposed solutions is efficiency, both from a communication as well as a computation point of view. Additional security features, such as anonymity, unforgeability and non-repudiation, are also required in some cases. Therefore, mechanisms to be considered will be based on symmetric key (aiming for efficiency) and public key cryptography (aiming for additional security features).

Moreover, the choice of the authentication mechanism is also largely dependent on the specific use case since each use case has a different type of architecture, resulting in different requirements with respect to security features and efficiency.

In any case, one of the main goals is to keep the computational, communication and storage overhead as low as possible at the side of the IoT device, which is typically the most constrained device.

Intended Audience

One of the major challenges for IoT adoption is (robust) authentication, which is a basic security process and is sorely needed at first place in the IoT. Although authentication is one of the paramount requirements of IoT networks, many of the authentication related techniques and standards are still under development. Therefore, there are only a limited number of books, which partly address the authentication in IoT. However, such rapid adaptation of IoT networks will soon raise the requirement of a complete handbook of Authentication in IoT.

This book will be of key interest for:

Consumer Internet of things (CIoT). As the consumer adoption of the IoT is evolving, it is important to understand the typical authentication mechanisms to keep illegal entities away from the IoT networks. This book will offer the required guidelines for authentication and its techniques to protect the IoT from unauthorized entities.

Service providers. Service providers are currently actively looking to adopt IoT technology to offer new and state-of-the-art secure services to IoT customers. This book will be a great source of security material that can provide insights for the authentication mechanisms in IoT networks.

Network operators (NOs). Network operators try equally to reach large customer bases who will switch to IoT networks. Security is the key requirement while connecting the IoT devices with the core networks of large operators.

IoT device manufacturers. Security is one of the key areas of interest for IoT device manufacturers as security challenges outpace the traditional tools available to the market.

This book will offer a single source of all the authentication-related topics for the device manufacturer.

Academics. IoT security has already been an area of research and study for major educational institutions across the world. With IoT evolvement as the future of humans, there is no such reference and book available (particularly on authentication in IoT) that academics can use for teaching this as an area of interest.

Technology architects. IoT is going to cross the traditional mobility borders and is going to have an equal impact on all enterprises and organizations who plan to transform into digital businesses. It would be critical for architects to start aligning their technology and security architectures to the future needs of IoT standards. This book offers resources to design and build an authentication architecture and maintain it.

IoT organizations and digital organizations. IoT is going to change the way industrial networks are built, 5G is going to provide the underlying platform for IoT networks. Security has remained the top priority for industries due to criticality and sensitivity of the data and information flows in their networks. A beforehand, knowledge of 5G security principles, components and domains is going to help industries lay out a foundation of IoT security. This book will provide the guidelines and best practices for 5G based IoT security and authentication.

Book Organization

The book is divided into five parts covering various aspects of IoT authentication: IoT Overview, IoT network level, IoT user level, IoT device level and IoT use cases.

Part I consists of an introduction to IoT and an introduction to the corresponding security threads. Chapter 1 introduces IoT in a pedagogical manner by presenting its evolution, the taxonomy and the proposed architectures and standardization efforts. It also illustrates some of the popular applications of IoT. In Chapter 2, security challenges at every layer are addressed in detail by considering both the technologies and the architecture used. A thorough survey is provided, together with a classification of the existing vulnerabilities, exploitable attacks, possible countermeasures and the access control mechanisms including authentication and authorization. Additionally, solutions for remediation of the compromised security, as well as methods for risk mitigation, with prevention and improvement suggestions are discussed.

Part II contains the chapters related to the protection at network level. In Chapter 3, different methods to provide key establishment and authentication using symmetric key based mechanisms limited to hashing, xoring and encryption/decryption operation are discussed. A new key management protocol for wireless sensor networks with hierarchical architecture, using solely symmetric key based operations, is proposed. Chapter 4 describes the utilization of Elliptic Curve Cryptography (ECC) for designing security protocols in terms of authentication, key establishment, signcryption, and secure group communication. Chapter 5 provides a general overview of a post-quantum security primitive, being the lattice-based primitive. The chapter summarizes how this primitive can be applied to IoT and gives a review on the state-of-the-art of proposed applications in literature.

Part III is about the user-level authentication and consists of four chapters. Chapter 6 deals with the anonymous mutual authentication scheme in multi-access edge computing environments (MEC). It will utilize the password-based approach for the user authentication. Chapter 7 proposes a biometric-based access control model in industrial IoT applications. The model will perform a robust authentication and establish a session key between the user and smart devices IIoT. In Chapter 8, authentication is discussed in case the user can experience IoT enabled services without carrying any gadget, also called the naked approach. A use case from the medical and healthcare sector has been worked out in order to enable the patient an ambient Internet of Everything experience. Chapter 9 discusses a user-friendly Web-based framework for handling user requests automatically by addressing user concerns for mobility support, ownership support, and immediate privilege updates having the goal of limiting the involvement of any third-parties in the process chain and also to inform all involved parties immediately about any status changes.

Part IV of the book contains two chapters related to device-level authentication. In Chapter 10, an authentication mechanism is discussed in case the IoT nodes contain a PUF, which is a low-cost primitive exploiting the unique random patterns in the device allowing it to generate a unique response for a given challenge. The advantage of a PUF at the IoT is that even when the key material is extracted, an attacker cannot take over the identity of the tampered device. However, in practical applications, the verifier, orchestrating the authentication between the two IoT nodes, represents a cluster node in the field, who might be vulnerable for corruption or attacks. In the proposed authentication mechanism, additional protection has been provided for this. Chapter 11 presents an encryption and authentication scheme suitable for ASIC or Field-Programmable Gate Array (FPGA) hardware implementation, which is based on the generalized synchronization of systems showing chaotic dynamical behavior. The strength of the system relies on the unobservability of the internal states of a strongly nonlinear system having a high-dimensional phase space.

Part V contains three chapters dedicated to a use case in the healthcare, smart grid, connected cyber physical system. Chapter 12 introduces a remote patient monitoring platform that consists of three main parts, patient monitoring devices, cloud backend and the hospital's clinician application. The system has been implemented for a pilot project and in a joint research with neuro and cardiology departments of Helsinki University Hospital (HUS).

Chapter 13 proposes a secure and efficient privacy-preserving scheme in a connected smart grid network. The scheme is based on ECC, outperforming both in communication and communication costs. Chapter 14 first discusses the overlapping in cyber physical system and IoT, and then proposes a cyber physical trust system that utilizes the blockchain as a security tool. The security strength is shown in terms of data authenticity, integrity and identity.

Acknowledgments

This book focuses on IoT authentication, which has been created through the joint efforts of many people. First of all, we would like to give thanks to all of the chapter authors for doing a great job!

This book would not have been possible without the help of many people. The initial idea for this book originated during our joint research work in 6Genesis Flagship (grant no. 318927) project, two COST Action projects (i.e. CA15127 RECORDIS and CA16126 SHELD-ON) and RESPONSE 5G (Grant No: 789658) Marie Skłodowska-Curie Actions (MSCA) project. We thank the European Union and MSCA Research Fellowship Programme who funded the above projects. We thank all the reviewers for helping us select suitable chapters for the book. Moreover, we thank anonymous reviewers who evaluated our proposal and provided many useful suggestions for improvement. We also thank Sandra Grayson from John Wiley and Sons for her help and support in getting this book published.

Also, the authors are grateful to the School of Computer Science at University College Dublin, Centre for Wireless Communication (CWC) at University of Oulu, Department of Computer Science at Swansea University and Department of Industrial Engineering at Vrije Universiteit Brussel for hosting the IoT Security related research projects which helped us gain the fundamental knowledge for this book. Last but not least, we would like to thank our core and extended families and friends for their love and support in getting the book completed.

Madhusanka Liyanage, An Braeken, Pardeep Kumar and Mika Ylianttila

Part I

IoT Overview

1

Introduction to IoT

Anshuman Kalla, Pawani Prombage, and Madhusanka Liyanage

Abstract

The successful existence of the Internet, its proven potential to cater to day-to-day needs of people from all walks of life and its indispensability to society at large, together have propelled the evolution of the current Internet to the next level termed as the Internet of Things (IoT). As a witness to the dawn of IoT revolution, what we are experiencing (and will continue in to do so in the near future at an exponential and astonishing rate) is the intelligent presence and communication of the physical objects or things around us with themselves (M2M) and/or with humans (M2H). Emergence of such a kind of pervasive inter-networking ecosystem has enormous scope in terms of market growth and applications which have (to some extent) and will prove with greater force its efficacy to improve quality of life. Though it is bit early to precisely define the depth of coverage and the long-term impact of IoT applications, nevertheless particularly in domains like healthcare, agriculture, city and home/office automation, industrial and energy management, etc. the immediate applications of IoT are easily conceivable. For realization and rapid development of such IoT applications, formal establishment of IoT architecture and standardization of related protocol suites are vital as they ensure co-existence and co-operation of cross-vendor devices as well as applications. Nevertheless, as with any other hyped research area, IoT has also become victim of its own success and hitherto no one architecture is globally accepted with a common consensus.

In the midst of this, this chapter intends to introduce IoT in a pedagogical manner to the readers. More specifically, the chapter guides the reader through the evolution of IoT, discusses the pertinent taxonomy and proposed architectures, probes the various efforts for standardization of IoT and illustrates some of the popular applications of IoT. While dealing with promising IoT applications, the chapter presents a comprehensive view comprised of the constituent components and major stakeholders to fit-in, characteristics and key factors to focus, enabling technologies to leverage and categorize each application to understand the various viewpoints.

IoT Security: Advances in Authentication, First Edition.
Edited by Madhusanka Liyanage, An Braeken, Pardeep Kumar, and Mika Ylianttila.
© 2020 John Wiley & Sons Ltd. Published 2020 by John Wiley & Sons Ltd.

1.1 Introduction

The evolution towards 5G is widely characterized by exponential growth in the number of computing devices embedded in everyday objects and interconnected over the Internet. Over 50 billion devices are expected on the cellular networks by the year 2020, compared to 12.5 billion devices in 2010 [41] and about 28 billion devices estimated in 2017 [6]. This massive interconnection of proliferating heterogeneous physical objects is technically termed as the Internet of Things (IoT). Such a kind of networking ecosystem enables communication-capable resource-constrained heterogeneous objects or devices to be connected over the Internet, in addition to the interconnections of computationally resourceful devices like computers, smartphones, PDA, etc. Thus, IoT renders the entire Internet space as the working area for such devices. In other words, the IoT paradigm begins to facilitate devices to acquire smartness by performing all sorts of operations (monitor, exchange, process, compute, make decisions indigenously or collaboratively) and accordingly take the required actions, based on the information being sensed anywhere across the globe. IoT system is poised to generate a significant surge in demand for data, computing resources, as well as networking infrastructures in order to accommodate these myriads of interconnected devices. Meeting these stringent demands necessitates appropriate improvisations to existing network infrastructures as well as computing technologies; one of such alterations is Multi-Access Edge Computing (MEC) formerly know as Mobile Edge Computing [55]. Analogically, IoT can be viewed as the sensory and nervous system of the future Information and Communication Technology (ICT) whereas the brain's inherent capabilities to store, process and take decisions would be furnished by technologies like cloud computing, mobile edge computing, parallel computing as well as the sciences of big data analytics, artificial intelligence, machine learning, etc. Ensuring synergy between these technologies is the key to success.

1.1.1 Evolution of IoT

Initially, computer networking began with the aim of economic and efficient sharing (or accessing) of scarce and expensive (computing) resources. Soon, the development of TCP/IP protocol suites fueled the growth and lead to the advent of the global networking facility known as the Internet. Since then, the Internet has evolved tremendously and has achieved several decades of a successful existence. The years of maturity of the current Internet and the advancements in relevant underlying technologies have paved the way for the emergence of IoT. As shown in [63], the evolution of the Internet consists of five phases (Figure 1.1). Initial phase dealt with connecting together computers followed with the second phase that gave rise to the World Wide Web which connected a large number of computers as a web. Then, the mobile-Internet came into picture which enabled mobile devices to be connected to the Internet and later peoples' identities also stepped in and joined the Internet by the means of social networks. Finally, the present phase nurtures the advent of IoT that envisions the connection of day-to-day physical objects to the Internet.

Similar to that of the Internet, IoT also has its own journey; it is the culmination of convergence of different visions like Things oriented, Internet oriented, and Semantic oriented [31, 32]. According to the definition in [45], IoT allows people and things to be connected anytime, anyplace, with anything and anyone, ideally using any path or network and any service. Continuing the momentum, one recent proposition named

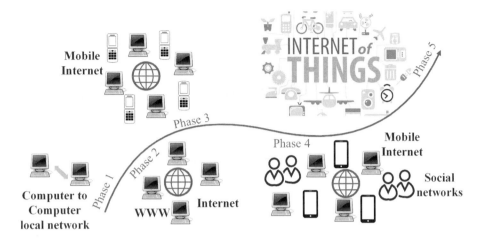

Figure 1.1 Evolution of IoT.

as Social Internet of Things (SIoT) aims to interconnect the IoT to human social networks [33]. SIoT explains how the objects are capable of establishing social relationships in an autonomous way with respect to their owners. Another prominent facet of IoT is Industrial Internet of Things (IIoT) which intends to transform the entire existing industrial manufacturing and maintenance system to a smart enterprise automation system provisioned with higher levels of intelligence and cognitive computing. It is realized by securely interconnecting industrial assets over the Internet and leveraging relevant technologies (for e.g. cloud computing) which leads to precise supervision in industrial environments and an increase in the return on investment.

1.2 IoT Architecture and Taxonomy

IoT has set the stage for interconnecting billions to trillions of objects through the Internet [31] and this number is expected to grow at an unprecedented rate. IoT devices are optimistically estimated to reach 75.44 billion by the end of 2025 and, moreover, 10 smart devices per capita is expected by 2025 as compared to two smart devices in 2015 [30]. Triggered by the requirement of seamless connectivity of such an enormously large number of heterogeneous objects, IoT entails a flexible layered architecture. In this direction, although an increasing number of architectures have been proposed for IoT with close collaboration between research and industry, hitherto, none has received common consensus and thus no reference model is yet firmly established [29, 85]. The architectural modeling of IoT is based on modifying the OSI (Open Systems Interconnection) standard with appropriate adjustments to the data link, network, and transport layers.

As discussed in much literature [33, 53, 87] IoT operates on three basic layers termed as Perception, Network, and Application as shown in Figure 1.2. The perception layer or the 'Device Layer' interacts with physical objects and components using the smart devices like RFID tags, sensors, actuators, etc. The key responsibilities include data acquisitions, processing the state information associated with smart objects and transmitting the raw data or processed information to the upper layers. The network layer enables optimal routing and data transmission through integration

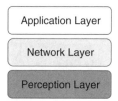

Figure 1.2 Three layer architecture [87].

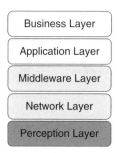

Figure 1.3 Five layer architecture [50].

of heterogeneous and disparate networks using various connecting devices (hub, switch, gateways, routers), communication technologies (Bluetooth, WiFi, optical fibers, Long-Term Evolution (LTE)) and protocols (IEEE 802.15.4, 6LoWPAN, Zigbee, Z-Wave, CoAP, MQTT, XMPP, DDS). The application layer provides the essential services or operations to the users through the analyzed and processed perception data. This fulfills a combination of social and industrial demands in numerous domains including smart grid, smart transportation, smart cities, e-health, data services, etc.

More recent literature highlights the use of the five-layer model (Figure 1.3) to represent the architectural frameworks of IoT [31, 50, 58, 81, 87]. When viewed together, the 5-layer architectures have many features in common, however, they introduce multiple intermediate layers between the perception and application layers. As per the 5-layer architecture discussed in [50] the middleware layer supports the service management, receives information from the network layer, processes the information, performs ubiquitous computation and provides link to the database. The top most business layer manages the overall IoT system by determining the release of and charging of various IoT applications and by building the business models, graphs, flow charts, etc.

In addition to the layered IoT model, in [85], the authors have surveyed the industry oriented reference IoT architectures. Some renowned architectures are identified as Reference Architecture Model Industrie 4.0 (RAMI 4.0) [18], Industrial Internet Reference Architecture (IIRA) [24] and Internet of Things-Architecture (IoT-A) [13] (Figure 1.4). The IoT-A concentrates on the generic aspects of informatics instead of the application facets of semantics whereas the IIRA focuses on the functionality of the industry domain like business, operations (prognostics, monitoring, optimization), information (analytics and data), and application (UIs, APIs, logic, and rules). RAMI 4.0 is domain specific and extends the view of the IIRA toward the life cycle and value streams of manufacturing applications. Furthermore, there have been a number of architectures designed for industrial IoT frameworks including SENSEI [82], ASPIRE [2], SmartSantander [73], iCore [3] and FIESTA-IoT [37].

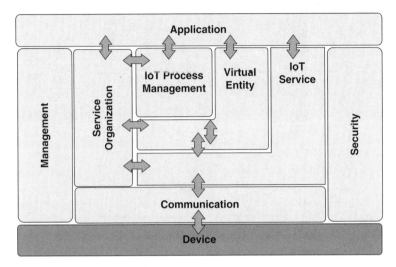

Figure 1.4 IoT-A architecture [13].

1.3 Standardization Efforts

Over the past couple of years, standardization of IoT has gained momentum since many organizations have stepped in and have boosted their contribution to develop a suite of protocols as well as open standards for IoT deployment that support inter-operable communication [43, 48, 75]. Among them Internet Engineering Task Force (IETF) has taken the lead in standardizing communication protocols for resource-constrained devices such as Routing Protocol for Low Power and Lossy Networks (RPL), Constrained Application Protocol (CoAP), Low-Power Wireless Personal Area Networks (6LoWPAN), etc. [75]. Moreover, many other organizations and communities including the International Telecommunication Union (ITU), European Telecommunication Standards Institute (ETSI), 3rd Generation Partnership Project (3GPP), World Wide Web Consortium (W3C), EPCglobal, Object Management Group (OMG), Organization for the Advancement of Structured Information Standards (OASIS), and Institute of Electrical and Electronics Engineers (IEEE) have made a noteworthy contribution to consolidate IoT standardization activities. ETSI introduced Machine-to-Machine standards relevant to IoT communication, whereas ITU coordinated activities on aspects of identification systems for M2M. Figure 1.5 presents an overall summary of the most prominent standards and protocols used for the realization of IoT.

In the application layer, CoAP is the most widely used protocol which defines a constrained web protocol based on REpresentational State Transfer (REST) on top of well known HTTP functionalities [74]. Although CoAP is not a compressed version of HTTP, nevertheless, a subset of HTTP functions with small header (low overhead) and reduced complexity parsing in an optimized way to equip constrained devices with low-memory footprint and less computational capability. CoAP supports UDP transport with application layer reliable unicast and best-effort multicast, proxy caching capabilities and resource discovery. Message Queue Telemetry Transport (MQTT) provides embedded connectivity between applications and middleware on one side and networks and communications on the other side [15]. MQTT-SN is specifically

Applications protocols

	RESTful	Transport	Publish/ Subscribe	Request/ Response	Security	QoS	Header Size (Byte)	Organization
COAP	√	UDP	√	√	DTLS	√	4	IETF
MQTT	X	TCP	√	X	SSL	√	2	OASIS
MQTT-SN	X	UDP	√	X	SSL	√	2	OASIS
XMPP	X	TCP	√	√	SSL	X	-	IETF
AMQP	X	TCP	√	X	SSL	√	8	OASIS
DDS	X	TCP UDP	√	X	SSL DTLS	√	-	OMG
HTTP	√	TCP	X	√	SSL	X	-	IETF W3C

Network layer & middlewear protocols

	Functionality	Characteristics	Organization
mDNS	Service discovery	Zero configuration. Uses IP multicast UDP packets. Run with no or less infrastructure	IETF
DNS-SD	Service discovery	Utilizes mDNS. Zero configuration. Wide area service discovery by clients.	IETF
RPL	Routing	Distance vector protocol. DODAG topology.	IETF
6LoWPAN	Encapsulation	Adaptation layer for IPv6 . Header compression. Fragmentation.	IETF
6TiSCH	Encapsulation	Low-power operations for IPv6 over IEEE802.15.4e TSCH mode. Industrial IoT.	IETF

Physical layer protocols

	Spreading Technique	Radio Band (MHz)	MAC Access	Data Rate (bps)	Scalability	Organization	Range
IEEE 802.15.4	DSSS	868/915/2400	TDMA, CSMA/CA	20/40/250K	65K nodes	IEEE	10 m-100m
BLE	FHSS	2400	TDMA	1024K	5971slaves	Bluetooth group	< 100m
EPCglobal	DS-CDMA	860~960	ALOHA	Varies5~640K	-	EPCglobal	< 50 m
LTE-A	Multiple CC	varies	OFDMA	1G (up), 500M (down)	-	3GPP	< 100km
Z-Wave	-	868/908/2400	CSMA/CA	40K	232nodes	Sigma Designs	30 m (Indoor) & 100m (Outdoor)
LoRa	LoRa (CSS)	0.125-0.25	Unslotted ALOHA	0.3K to 50K	Upto millions	LoRa Alliance	< 15 km
Sigfox	BPSK	0.1	Unslotted ALOHA	100or 600	Upto millions	SNO	30 km – 50 km (Rural) & 10 km (Urban)
NB-IoT	QPSK	0.18	FDMA/OFDMA	20K(up) 250K(down)	50K	3GPP	< 35 km
Insteon	-	904	TDMA + simulcas	38.4k	256	Smartlabs	< 45 m

Figure 1.5 Summary of standardization efforts available for IoT.

defined for sensor networks and is tailored to adapt to the peculiarities (and dynamics) of the wireless communication environment [77]. Extensible Messaging and Presence Protocol (XMPP) was designed originally for chatting and message exchange applications and later was reused in both IoT and SDN (Software Defined Networking) [72]. Advanced Message Queuing Protocol (AMQP) is an open standard application layer protocol for IoT and supports message-oriented environments [10]. Some of the key features of AMQP include message orientation, queuing, routing (including point-to-point and publish-and-subscribe), reliability and security. OMG introduces another publish/subscribe protocol, named as Data Distribution Service (DDS), which suits IoT and M2M communication due to the excellent Quality-of-Service (QoS) levels and the broker less architecture that guarantees reliability [17].

Due to the high scalability of IoT, it requires a standard Domain Name System (DNS) type resource management mechanism to register and discover resources in a self-configured, efficient, and dynamic way. Multicast DND (mDNS) and DNS Service Discovery (DNS-SD) can browse the network for discovering resources and services offered by IoT devices [38, 71]. IETF has designed RPL as a link-independent distance-vector routing protocol which is based on IPv6 for resource-constrained nodes [83]. In RPL the nodes construct a Destination Oriented Directed Acyclic graph (DODAG) by exchanging distance vectors and root with a controller. The 6LowPAN protocol is an adaptation layer allowing to transport IPv6 packets over IEEE 802.15.4 networks with a maximum packet size of 127 bytes [62]. The standard provides compression of IPv6 and UDP/ICMP headers and fragmentation for reassembling of IPv6 packets. A new working croup called 6TiSCH is recently developed by IETF for standardizing IPv6 to pass through Time-Slotted Channel Hopping (TSCH) mode of IEEE 802.15.4e datalinks [40].

The IEEE 802.15.4 protocol specifies a sub-layer for Medium Access Control (MAC) and physical (PHY). It defines a frame format and headers (including source and destination addresses) and also explains how nodes can communicate with each other [9]. This standard is used by IoT, M2M and WSN due its salient features which are the low-power consumption, low data rate, low cost, interoperability, reliable communication and high message throughput. Bluetooth Low-Energy (BLE) is another good candidate for IoT applications as it offers wider range, lower latency, and minimal amount of power over the classic Bluetooth [19]. RFID technology uses Electronic Product Code (EPC) unique identification numbers while EPCGlobal has become a universal standard [84]. Long-Term Evolution Advanced (LTE-A) is a scalable and lower-cost protocol which fits well for M2M communication and IoT applications in cellular networks [14, 47]. Z-Wave is yet another low-power protocol which is originally designed for automation networks in smart home applications and later was developed for small commercial domains [7].

In addition to those standards that define the operational framework of IoT, there exist many other protocols for security, interoperability and management purposes. Since the conventional security protocols on the Internet are not always compatible with the resource-constrained IoT devices, the customized protocols have been defined (e.g., IPsec [49], Datagram Transport Layer Security (DTLS) [70], Host Identity Protocol Diet-exchange (HIP-DEX) [61]). Furthermore, some other management protocols are available such as IEEE 1905.1 [12] for interoperability, Long Range Wide-Area Network (LoRaWAN) [4], Wireless Smart Ubiquitous Network (Wi-SUN) [1], Narrow Band IoT (NB-IoT), Sigfox and Zigbee [11] for low-power wide area networks (LPWANs). The unlicensed spectrum for LPWAN, LoRa radio, defines Physical and Data Link layers of LPWANs whereas LoRaWAN is analogous to Network and Transport layers of OSI communication stack. LoRaWAN is an open network protocol that manages communication between gateways and end-devices [4].

According to the latest forecast report [22] from Rethink research, most of the growth of LPWAN technologies will be supported by NB-IoT, LTE-M and Wi-SUN from 2017 to 2023 period. Moreover, they anticipate that LoRa and Sigfox respectively have slightly increasing and constant growth rates during the next seven years.

1.4 IoT Applications

Let's now divert our attention toward different types of application scenarios that can benefit from IoT revolution. A panorama of typical and potential applications including healthcare (e.g., patient monitoring and surgeries), smart energy, smart automotive (e.g. autonomous vehicles), industrial automation, etc. is shown in figure 1.6.

IoT applications have been categorized in different ways based on the scope of functionalities, number of devices required for deployment v/s reliability, level of scope of usage, etc. [44], [20], [56]. Four-category levels of IoT applications [44], in the increasing order of scope of offered functionalities, are Identity-related services, Information Aggregation services, Collaborative-Aware services and Ubiquitous services. Identity-related services map every physical object into the network space (thereby making them addressable) as well as implement network resolution services. Some sort of such services is intrinsic to all IoT applications. Next, the services that readily collect appropriately aggregate data from heterogeneous objects in order to send it for processing come under the category of information aggregation services.

Figure 1.6 Different IoT Applications.

Table 1.1 Categorization of Different IoT Applications.

IoT Application	Categories as per [44]	Categories as per [20]	Categories as per [56]
Smart Home	Collaborative aware	Massive IoT	Individual level
Smart City	Collaborative aware	Massive IoT	Infrastructural level
Smart Energy	Information aggregation	Massive IoT	Infrastructural level
IoT Automotive	Information aggregation	Critical IoT	All-inclusive level
Healthcare	Information aggregation	Massive IoT	All-inclusive level
Gaming, AR, VR	Information aggregation	Massive IoT	Individual level
Retail	Collaborative aware	Massive IoT	Organizational level
Wearable	Information aggregation	Massive IoT	Individual level
Smart Agriculture	Collaborative aware	Massive IoT	Organizational level
Industrial Internet	Collaborative aware	Critical IoT	Organizational level

Based on the received aggregated data, collaborative-aware services take decisions and accordingly the responses/actions are coordinated to the point of actuation. Finally, the Ubiquitous services ensure the network-wide pervasive presence and anytime availability of underlying collaborative-aware services. IoT applications, when designed to rise up to the level of ubiquitous services, will yield maximum benefit, however, it requires smooth integration of technologies, protocols and standards.

Among the wide range of IoT use cases, the market is heading towards two key categorized areas namely massive IoT and mission critical IoT [20]. In massive IoT, large numbers of low-cost low-powered devices typically emit a low volume of non-delay sensitive data. The devices need to report to the cloud on a regular basis and therefore require seamless connectivity and good coverage. The application areas of massive IoT comprise smart home, smart agriculture, asset management and smart metering. By contrast, the critical IoT applications have very high demands of reliability, availability, and low latency.

Based on the scope of the usage and adaptation, IoT applications are categorized into four levels of applications [56]; infrastructural level, organizational level, individual level and all-inclusive level. At infrastructural level, applications like smart city, smart energy, smart tourism etc. are placed where they have potential, in turn, to create the next level of the ecosystem. Industrial Internet, smart agriculture, retail etc. come under organizational level since such applications aim to automate the working of an organization. Quite obviously, the applications that fall under the category of individual level are smart home, gaming, wearable etc. However, there are few applications that have wider scope and can span through all the levels such as medical and healthcare, automotive, education etc.

Table 1.1 exhibits one-way of categorizing different IoT applications whereas Table 1.2 reveals the characteristics of those IoT applications.

1.4.1 Smart Home

The concept of smart home is not new and has been around long before even the birth of IoT. The idea of smart home is to monitor, manage, optimize, access remotely, in

Table 1.2 Characteristics of Different IoT application [5, 8, 23, 64, 66].

IoT Application	Data type	Backhaul Connectivity	Expected latency
Smart Home	Stream / Historical data	Realtime	1 ms -1000 s
Smart City	Stream / Massive data	Realtime	≤1ms
Smart Energy	Stream / Massive data	Realtime/ Intermittent	1ms - 10 mins
IoT Automotive	Stream / Massive data	Realtime	≤1ms
Remote surgery	Stream data	Realtime	≤200 ms
Remote consultancy	Stream data	Realtime	1 ms-100 s
Gaming, AR, VR	Stream / Massive data	Realtime	≤1ms
Retail	Stream / Historical data	Realtime/ Intermittent	≤1 ms
Wearable	Stream data	Intermittent	Several Hours
Smart Agriculture	Historical data	Intermittent	Several hours
Industrial Internet	Stream / Massive data	Realtime	≤1 ms
Tactile Internet	Stream	Realtime	≤1 ms

short, fully automate the home environment comprising household devices and home appliances while minimizing human effort. IoT vision intrinsically promises to furnish the much needed underpinning ecosystem that supports the easy accomplishment of smart home application.

IoT-based smart home makes use of both local (but limited) storage and processing units (example gateway or hub) as well as cloud infrastructure [78, 88]. With augmentation of edge computing performance is expected to be improved significantly as the operations are not computational intensive. Apparently, the achieved gain would be in terms of latency, load balancing, traffic reduction and progressive-resource utilization.

Smart home is sometimes seen as an extension of smart grid concept [78]. From that perspective, the primary intent of smart home is to optimize the energy consumption taking into account various inputs like usage pattern and real-time presence of residents, the external environment (e.g. weather condition), time of the day, balance units of pre-paid electricity account etc.

Prominent stakeholders of smart home are shown in the figure 1.7. In addition, the *major components* constituting the smart home application are smart security & surveillance systems, smart HVAC (Heating, Ventilation and Air Conditioning), self-adjustable and smart customization of the environment based on the user's profile, smart energy management and smart object traceability via IoT powered GIS. The key *factors* i.e. challenges pertinent to the smart home are high privacy and security, high reliability, high interoperability, strong adaptation to multipath error prone wireless environment etc. Various contending technologies and standards for IoT driven smart home are ZigBee, 6LoWPAN, low power WiFi, BLE, RFID, Insteon, cloud computing etc.

Undoubtedly, full-flexed realization of IoT enabled smart home application has enormous potential to enhance the experience of personal living.

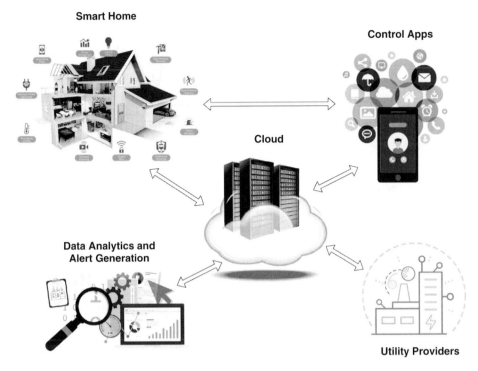

Figure 1.7 Stakeholders of Smart Home Environment.

1.4.2 Smart City

Many countries have already embarked on their plan of smart city projects, including Germany, USA, Belgium, Brazil, Italy, Saudi Arabia, Spain, Serbia, UK, Finland, Sweden, China and India [27, 30]. This trend indicates the rise of IoT from its infancy to blossoming state. As depicted in figure 1.8, some of the *major components* of an IoT driven smart city are smart hygiene, smart mobility & traffic management, smart governance,

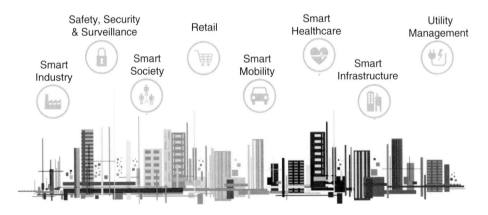

Figure 1.8 Use of IoT in Smart Cities.

smart development of infrastructure, smart commutation, smart surveillance and smart utility management.

The *factors* that comparatively deserve more attention while developing an IoT-based smart city are high scalability, low latency, high reliability, high availability and high security. Depending on the specific nature of the service offered via smart city projects a subset of technologies can be exploited from figure 1.5. In general, for small range connectivity RFID, ZigBee, Bluetooth, Wi-Fi, etc. technologies are used, however, to provide extensive long range backhaul connectivity to massive IoT devices spread across a city, technologies like GSM, WCDMA, 3G, LTE and 5G (in near future) could be employed.

1.4.3 Smart Energy

In a nutshell, the tasks involved in the energy sector are power generation at the source, transmission of power over high-voltage lines from source up to substations (generally located in the vicinity of the point of high demand), distribution to the end consumers, billing at predefined time cycle, and 24x7 monitoring & maintenance comprising of fault detection and rectification. In this sphere, the early usage of IoT is quite conspicuous since multipurpose smart meters and smart thermostats are already deployed [68]. Continuing the momentum, IoT has an extensive role to play both from a utility provider's perspective as well as from the consumer's perspective.

Major components of IoT-driven smart energy, as depicted in the figure 1.9, are smart energy generation including renewable energies (can be thought as an instance of IIoT application), smart maintenance involving prediction-based early diagnosis of failure and subsequently proactive rectification, smart access to real-time usage to optimize consumption and for billing purposes, smart capacity building based on the conclusions drawn from data analytics applied over huge data collected from the users. Moreover the prime *factors* to keep in mind in context of IoT powered smart energy application are high (self) sustainability, high resiliency, high safety and high level of optimization. Underlying promising technologies for IoT that have potential to drive smart energy application are LoRa, SigFox, Power-Line Communication (PLC), IEEE 802.15.4, Z-Wave, etc [69].

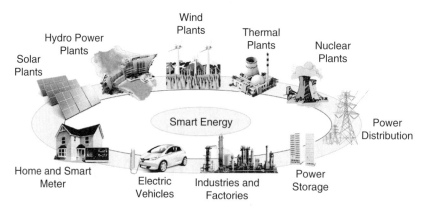

Figure 1.9 Components of Smart Energy System.

1.4.4 Healthcare

IoT driven healthcare applications are envisioned to roll-out in a massive way and it is envisioned to capture the biggest chunk of the future IoT market by 2025 [29]. IoT has built-in capabilities to support well all sorts of medical healthcare; preventive, diagnostic, therapeutic and rehabilitation healthcare [36]. Interestingly, on one side, the healthcare sector demands IoT paradigm to bestow living beings with solutions that can monitor various physiological parameters, detect symptoms and thereby (early) diagnose, suggest preventive measures, and progressively adapt treatment based on AI and ML approaches. On the other side, medical IoT can guide pharmaceutical companies to develop and design new medicines based on data analytics of IoT generated big data of recent patients and take further appropriate measures when required.

Figure 1.10 gives a glimpse into the role of IoT in the healthcare domain. Various entities which can directly or indirectly benefit from healthcare application of IoT are patients, doctors, supporting staff (i.e. nurse and technicians), hospitals, medical insurance companies and pharmaceutical industries. Medical IoT devices such as smart watch, bands, shoes, clothes, etc. can sense and communicate in real-time the vital signs of an admitted patient remotely to a doctor who, if required, can instruct the attending nurse to take action on an urgent basis. One can imagine numerous *major components* of IoT healthcare applications, some of them are smart remote health monitoring, smart asset management for hospitals, smart medical inventory optimization based on

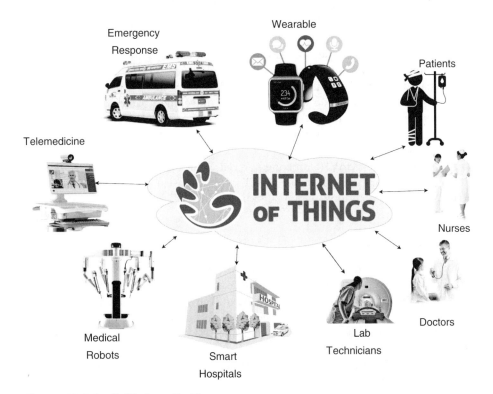

Figure 1.10 Role of IoT in Smart Healthcare.

real-time healthcare data analytics, smart patient-doctor rapport, smart augmented treatment and surgeries, etc.

It is worth noting, the *factors* that are of paramount importance among others while developing IoT solutions for healthcare application are ultra-high safety, ultra-high precision, high trustworthiness, high privacy and low energy consumption [52]. Various technologies that have the potential to play significant role in this area are BLE, WBAN (IEEE 802.15.6), LR-WPAN (IEEE 802.15.4) and NB-IoT [65].

1.4.5 IoT Automotive

By 2040, it is optimistically estimated that 90% of the overall sales of vehicles will be either highly automated (level - 4) vehicles or fully automated (level - 5) vehicles [59]. Many companies, for example Tesla, Google, BMW, Ford, Uber etc. are working in this direction [54]. IoT in conjunction with cloud computing and Mobile Edge Computing (MEC) [51] would play a significant role in the realization of connected and autonomous vehicles (aka self-driving or driverless or robotic vehicles) in the coming future. Vehicles gradually reaching higher levels of autonomy can have by and large following as their *major components*; smart self-optimization and maintenance to be carried out by the vehicle itself or remotely by the owner, smart security and safety systems, smart customization of ambience based user's profile, smart navigation, etc.

Today, there are approximately 60 to 100 sensors embedded in a single vehicle and soon the number will rise to 200 [21]. This trend falls in line with the fundamental prerequisite of IoT-enabled vehicles, which is the presence of all sorts of sensors in abundance. Thus, the adaptation of IoT in the automobile industry is well anticipated. Various underlying technologies for vehicular communications are Bluetooth, ZigBee, Dedicated Short Range Communication (DSRC), WiFi and 4G cellular technology [51]. However, based on [26] the vital *factors* that need to be tackled are seamless real-time last mile connectivity when the vehicle is on the move, high bandwidth, low power, high privacy and minimization of roaming issue which in-turn minimizes the associated impacts like an increase in latency and fluctuations in the price of home v/s foreign network providers.

Moreover, drones or UAV (Unmanned Aerial Vehicles) acting as sensor devices open-up avenues for a large number of IoT applications in many domains such as agriculture, mining operations, public safety and industrial inspection services [67].

1.4.6 Gaming, AR and VR

Yet another exciting application of IoT is in the realm of gaming, AR and VR, since IoT phenomenon has innate potential to enhance and uplift the perceived experience of users. Basically, Virtual Reality (VR), Augmented Reality (AR) and Mixed Reality (MR) provide an experience of the computer-generated illusory world in simulated environments, however, they vary in the degree of immersive presence and interactivity. AR adds digital elements to a live view often by using the camera on a smartphone and VR implies a complete immersion experience that shuts out the physical world. An MR experience combines elements of both AR and VR, where real-world and digital objects interact.

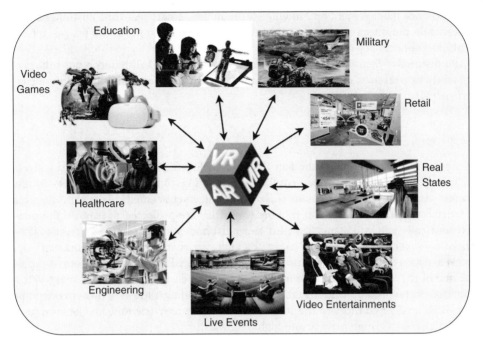

Figure 1.11 Selected AR and VR transformational use cases.

Interestingly, the immersive nature of available virtual reality devices backed-up by IoT ecosystem has set the foundation for the next level of gaming experience that can be supplemented by Brain-Control Interface (BCI) control signals [39]. Over the years, the purpose behind gaming has transformed and is no more confined merely to entertainment. As a result, gamification is yet another fascinating area which has been explored in various sectors like education & training, health management, market research, etc. [60]. A multitude of other VR, AR and MR centric propitious use cases of IoT are shown in figure 1.11. Real-time immersive viewing of remote live events will enrich the experience of television several fold. Moreover, a conglomeration of IoT, VR, MR & AR, in conjunction with cloud computing, mobile edge computing, artificial intelligence and data analytics will set a new path for the way we perceive and work in this world [76].

The prominent *factors* to consider for IoT-energized Gaming, AR and VR applications are real-time connectivity, high bandwidth, low power consumption, and most importantly ultra-low latency and interactive response time that essentially bridges the gap between virtuality and reality in order to minimize cyber-sickness.

1.4.7 Retail

Among others, retail is yet another captivating area that aspires to harness the capabilities of IoT. Entire new sets of real-time services can be introduced that at one end enhances the customer experience to the next level and on the other hand alleviates the way the retail sector (i.e. business-to-consumers (B2C)) is managed and maintained. *Major components* of retail applications of IoT are smart supply chain and logistics, smart finance management and intelligent prediction, smart real-time customer assistance while purchasing, smart complainy management system, smart post-purchase relations and feedback system.

The distinctive *factors* to bear in mind while developing IoT-based retail solutions are aesthetic presence, customer friendliness, context awareness and high QoE of customers [34].

1.4.8 Wearable

By now it is difficult to deny the fact that Wearable IoT (WIoT) devices and gadgets are closely interwoven in our present life style. Ranging from trendy devices like fitness tracker and smart watches, to fancy smart attire and essential medical devices, wearable IoT devices have hit the market in a big way. They are projected to capture the market drastically and will stand just next to smartphones in consumer electronics [79]. *Major components* of WIoT are smart tracking, smart infotainment, smart clothing, smart assistance, smart medical monitoring and personal security alarms, etc.

Some of the key *factors* to take into account while designing and developing WIoT (devices, protocols, applications etc.) are ultra-high safety, ultra-low power consumption, high level of comfort to the human body, highly user-friendly, low latency, high context awareness, high privacy and high bandwidth.

1.4.9 Smart Agriculture

Agriculture forms the very base of human civilization and also serves as means to provide a livelihood for farmers. As per [25] the percentage of employment in agriculture to the total employment has dramatically reduced from 43.24 % in 1991 to 25.95 % in 2018. This is alarming as the outcome of agriculture is lifesaving and meets a primary need of humans. In this context, IoT is anticipated to play a remarkable role to alleviate the overall situation.

Various tasks involved in IoT-driven smart agriculture are monitoring and acquisition of ambient data which is of a large variety and of huge volume, followed by its aggregation and exchange over the network, making short-term and long-term decisions based on data analytics and artificial intelligence and, at times, remotely actuating the decisions with help of field robots. It can also help in the prediction of yield to ensure the economic value to be gained, as well as early detection of diseases spread in crops, to enable timely preventive measures. Various technologies used are Bluetooth, RFID, Zigbee, GPS as well as other technologies gaining popularity in this application which are SigFox, LoRa, NB-IoT, edge computing, and cloud computing [80].

Some of the *major components* of IoT-based smart agriculture are smart plowing, smart seedbed preparation and planting, smart irrigation, smart fertilization, smart harvesting, smart stock maintenance, smart livestock management which deals with smart animal tracking, smart health monitoring, smart feed and fodder management, etc. An IoT solution for smart agriculture has different demands and so takes into account the following key *factors*; low-cost (both CAPEX and OPEX), low-power, highly reusable, cross-operational, highly efficient (big) data management, resource efficient, scalable and progressively extendable solutions.

1.4.10 Industrial Internet

An industrial application of IoT is referred as IIoT and at times is also referred to as the Industrial Internet [35, 46]. IIoT securely interconnects industrial assets (taken in a broader sense), over the Internet with the use of wide varieties of technologies shown in figure 1.5 and also by leveraging relevant technologies and concepts like cloud computing, edge or fog computing, big data analytics, Artificial Intelligence (AI) and Machine Learning (ML). Thus, IIoT has huge potential to facilitate precise supervision in an industrial environment, optimize value of productivity at low cost of operations and maintenance and finally, leads to an increase in the return on investment for stakeholders by offering ameliorated QoS [24].

Major components of IIoT are smart manufacturing, smart customer relationships, smart supply chain, smart budgeting, smart asset (all inclusive) and resource management. Some of the key *factors* to emphasizes are low latency, high precision, massive data management and analytics and perpetual connectivity.

1.4.11 Tactile Internet

The IoT networks started from the mobile Internet which interconnects billions of smart phones, portable devices and laptops. Mobile Internet plays a vital role in many domains such as health, energy, transport, education, logistics and many consumer industries. Currently, we are making use of the present generation of mobile networks which interconnect billions of IoT devices. Technologies such as NB-IoT have developed to provide reliable and efficient connectivity for these IoT devices. Tactile Internet is considered as the next generation of IoT networks [16, 28]. It is an advance version of the IoT networks with ultra-reliable, ultra-responsive and ultra-security network connectivity with extremely low latency. Thus, tactile Internet can deliver real-time control and physical tactile experiences remotely. These features will open up new domains of tactile Internet services such as remote surgery, robotics, autonomous vehicles and many more. Figure 1.12 shows the evolution of tactile Internet.

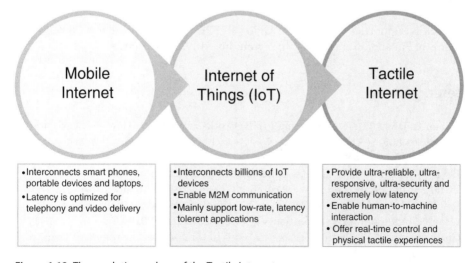

Figure 1.12 The revolutionary leap of the Tactile Internet.

The very first telesurgical operation was carried out as early as 2001 [57]. However, telesurgery is still not in the mainstream due to the technical limitations, especially in underlying communication networks. With the characteristics of tactile Internet, telesurgery types of procedure are realistic [86]. In industrial environments, Tactile Internet can be used for remote mining in high-risk areas, efficient manufacturing of highly customized products and remote inspection, maintenance and repair [42].

1.4.12 Conclusion

The chapter presents a lucid and compact summary of the IoT world and several pertinent facets of IoT, especially, the plethora of exciting IoT applications that open doors to a new world of services and users' experience. The IoT vision embraces the existence of all sorts of resource-constrained communication-capable targeted or generic purpose heterogeneous smart objects. Thus, in addition to the search for new technologies, IoT has apparently revived numerous forgotten technologies and has also effectuated the resurgence of research activities in these respective domains.

For IoT to transform into a profitable venture and find its fully flexed pragmatic realization, numerous factors have been discussed specific to each diversified application. In essence, IoT's global acceptance primarily depends on a few key factors like *reliability* with the bounded response time, *economic and incremental approach* which takes into account existing infrastructure thereby incentivizing the stakeholders, *seamless integration over heterogeneity* in terms of objects, technologies, protocols, platforms, applications, etc., *scalability* in terms of enormous objects distributed over a fleet of locations to be connected over the Internet, *availability* supporting infrastructure to gather, exchange and process big data, and finally *ability to draw intelligent conclusions* by performing data mining, data analytics, machine learning which in-turn helps with decision making.

Acknowledgement

This work is supported by European Union RESPONSE 5G (Grant No: 789658) and Academy of Finland 6Genesis Flagship (grant no. 318927) projects.

References

1 WiSun Alliance (2017). Comparing IoT Networks at a Glance: How Wi-SUN compares with LoRaWAN and NB-IoT. WiSUN technical White paper. www.wi-sun.org/wp-content/uploads/Wi-SUN-Comparing-IoT-Networks.pdf (accessed 16 July 2019).
2 EU Fp7 ASPIRE project. https://aspire-fp7.eu/ (accessed 16 July 2019).
3 EU FP7 iCore project. http://www.coreproject.eu (accessed 16 July 2019).
4 LoRAWAN specifications, LoRa Alliance Technology. https://lora-alliance.org/resource-hub/lorawanr-specification-v11 (accessed 16 July 2019).
5 Chadwick, S. (2018) Putting Sensors to Work in the Factory Environment: Data to Information to Wisdom. https://itpeernetwork.intel.com/putting-sensors-to-work-in-factory-environment/#gs.q2drfl (accessed 16 July 2019).

6 Sage Business Researchermore than 28 billion devices connect via internet of things. http://businessresearcher.sagepub.com/sbr-1863-102197-2772812/20170306/more-than-28-billion-devices-connect-via-internet-of-things. (accessed 17 July 2019).

7 Smart Home products with Z-Ware. Z-wave Technology. https://www.z-wave.com/ (accessed 16 July 2019).

8 GSMA. (2016) Unlocking Commercial Opportunities From 4G Evolution to 5G. GSMA Network Technical Report. https://www.gsma.com/futurenetworks/wp-content/uploads/2017/03/704_GSMA_unlocking_comm_opp_report_v5.pdf (accessed 16 July 2019).

9 IEEE. (2012). IEEE standard for local and metropolitan area networks-part 15.4: Low-rate wireless personal area networks (lr-wpans) amendment 1: Mac sublayer. *IEEE Std 02.15.4e-2012* (Amendment to IEEE Std 802.15.4-2011), 1–225.

10 OASIS. (2012). OASIS Advanced Message Queuing Protocol (AMQP) Version 1.0. *OASIS Standard*. http://docs.oasis-open.org/amqp/core/v1.0/os/amqp-core-overview-v1.0-os.html (accessed 16 July 2019).

11 Zigbee Alliance. (2012). ZigBee Specification. 1–622. ZigBee Alliance. http://www.zigbee.org/wp-content/uploads/2014/11/docs-05-3474-20-0csg-zigbee-specification.pdf (accessed 16 July 2019).

12 IEEE. (2013). 1905.1-2013 –IEEE Standard for a Convergent Digital Home Network for Heterogeneous Technologies. *IEEE Standard.*

13 Bauer, M., Boussard M., Bui, N. et al. (2013). Deliverable D1.5 Final architectural reference model for the IoT v3.0. IoT-A Consortium. https://www.researchgate.net/publication/272814818_Internet_of_Things_-_Architecture_IoT-A_Deliverable_D15_-_Final_architectural_reference_model_for_the_IoT_v30 (accessed 16 July 2019).

14 Wannstrom, J. (2013). LTE-Advanced Specifications. 3GPP Org. https://www.3gpp.org/technologies/keywords-acronyms/97-lte-advanced (accessed 16 July 2019).

15 OASIS. (2014). MQTT v3.1.1. OASIS Standard. https://mqtt.org/ (accessed 16 July 2019).

16 ITU-T. (2014). The tactile internet. ITU-T Technology Watch Report. https://www.itu.int/dms_pub/itu-t/oth/23/01/T23010000230001PDFE.pdf (accessed 16 July 2019).

17 OMG. (2015). Data Distribution Service (DDS) Specification Version 1.4. OMG Standard. https://www.omg.org/spec/DDS/About-DDS/ (accessed 16 July 2019).

18 Platform Industrie. (2015). Reference Architecture Model Industrie 4.0 (RAMI 4.0). Platform Industrie. https://www.plattform-i40.de/PI40/Redaktion/EN/Downloads/Publikation/rami40-an-introduction.pdf?__blob (accessed 16 July 2019).

19 Bluetooth SIG Working Group. (2019). Bluetooth Core Specification Version 5.1. Bluetooth SIG. https://www.bluetooth.com/specifications/bluetooth-core-specification/ (accessed 16 July 2019).

20 Ericsson Working Group. (2016). Cellular Networks for Massive IoT. Ericsson White Paper. https://www.ericsson.com/en/white-papers/cellular-networks-for-massive-iot–enabling-low-power-wide-area-applications (accessed 16 July 2019).

21 Automotive Sensors and Electronics Expo, 2017. Detroit, USA (14–15 June 2017).

22 Rethink Technology Research (2017). LPWAN Revenue Forecast, 2017 to 2023. Rethink Technology Research. https://rethinkresearch.biz/report/lpwan-market-forecast-2017-2023/ (accessed 16 July 2019).

23 iGillot Research. (2017). The Business Case for MEC in Retail: A TCO Analysis and its Implications in the 5G Era. Intel technical White Paper, IGR. https://www.intel

.com/content/www/us/en/communications/multi-access-edge-computing-brief.html (accessed 16 July 2019).

24 Shin, S.-W., Crawford, M., and Mellor, S. (eds) (2017). The Industrial Internet of Things Volume G1: Reference Architecture. 1–58. Industrial Internet Consortium (IIC). https://www.iiconsortium.org/IIC_PUB_G1_V1.80_2017-01-31.pdf (accessed 16 July 2019).

25 Employment in agriculture (% of total employment) (modeled ILO estimate). The World Bank, (2018). https://data.worldbank.org/indicator/SL.AGR.EMPL.ZS (accessed 16 July 2019).

26 IoT in the Automotive Industry. Tata Communications (2018). https://www.tatacommunications.com/wp-content/uploads/2018/02/IOT-IN-THE-AUTOMOTIVE-INDUSTRY.pdf (accessed 16 July 2019).

27 Adapa, S. (2018). Indian smart cities and cleaner production initiatives–integrated framework and recommendations. *Journal of Cleaner Production* 172: 3351–3366.

28 Aijaz, A., Dohler, M., Aghvami, A.H., et al. (2017). Realizing the tactile internet: Haptic communications over next generation 5g cellular networks. *IEEE Wireless Communications* 24 (2): 82–89.

29 Al-Fuqaha, A., Guizani, A.M., Mohammadi, M. et al. (2015). Internet of things: A survey on enabling technologies, protocols, and applications. *IEEE Communications Surveys Tutorials* 17 (4): 2347–2376.

30 Alavi, A.H., Jiao, P., Buttlar, W.G. and Lajnef, N. (2018). Internet of things-enabled smart cities: State-of-the-art and future trends. *Measurement* 129: 589–606.

31 Atzori, L., Iera, A. and Morabito, G. (2010). The internet of things: A survey. *Computer Networks* 54 (15): 2787–2805.

32 Atzori, L., Iera, A. and Morabito, G. (2017). Understanding the internet of things: definition, potentials, and societal role of a fast-evolving paradigm. *Ad Hoc Networks* 56: 122–140.

33 Atzori, L., Iera, A., Morabito, G. and Nitti, N. (2012). The social internet of things (SIoT): When social networks meet the Internet of Things: Concept, architecture and network characterization. *Computer Networks* 56 (16): 3594–3608.

34 Balaji, M. and Roy, S.K. (2017). Value co-creation with internet of things technology in the retail industry. *Journal of Marketing Management* 33 (1–2): 7–31.

35 Boyes, H., Hallaq, B., Cunningham, J. and Watson, T. (2018). The industrial internet of things (IIoT): An analysis framework. *Computers in Industry* 101: 1–12.

36 Brucher, L. and Moujahid, S. (2017). The Internet-of-Things A revolutionary digital tool for the healthcare industry. Deloitte. https://www2.deloitte.com/content/dam/Deloitte/lu/Documents/life-sciences-health-care/lu_digital-tool-healthcare-industry_062017.pdf (accessed 16 July 2019).

37 Carrez, F., Elsaleh, T., Gomez, D. et al. (2017). A reference architecture for federating IoT infrastructures supporting semantic interoperability. *Networks and Communications (EuCNC)*, Oulu, Finland (12–15 June 2017). IEEE.

38 Cheshire, S. and Krochmal, M. (2013). DNS-Based Service Discovery. IETF RFC 6763. https://tools.ietf.org/html/rfc6763 (accessed 16 July 2019).

39 Coogan, C.G. and He, B. (2018). Brain-computer interface control in a virtual reality environment and applications for the internet of things. *IEEE Access* 6: 10840–10849.

40 Dujovne, D., Watteyne, T., Vilajosana, X. and Thubert, P. (2014). 6tisch: deterministic IP-enabled industrial Internet (of Things). *IEEE Communications Magazine* 52 (12): 36–41.

41 Evans, D. (2011). The internet of things: How the next evolution of the internet is changing everything. *CISCO white paper* 1 (2011): 1–11.

42 Fettweis, G.P. (2014). The tactile internet: Applications and challenges. *IEEE Vehicular Technology Magazine* 9 (1): 64–70.

43 Gazis, V. (2017). A survey of standards for machine-to-machine and the internet of things. *IEEE Communications Surveys Tutorials* 19 (1): 482–511.

44 Gigli, M. and Koo, S.G. (2011). Internet of things: Services and applications categorization. *Advanced Internet of Things* 1 (2): 27–31.

45 Guillemin, P. and Friess, P. (2009). The Industrial Internet of Things Volume G1: Reference Architecture. The Cluster of European Research Projects, Technical Report.

46 Gurtov, A., Liyanage, M. and Korzun, D. (2016). Secure communication and data processing challenges in the industrial internet. *Baltic Journal of Modern Computing* 4 (4): 1058–1073.

47 Hasan, M., Hossain, E. and Niyato D. (2013). Random access for machine-to-machine communication in LTE-advanced networks: issues and approaches. *IEEE Communications Magazine* 51 (6): 86–93.

48 Kae, V.P., Fukushima Y. and Harai, H. (2016). Internet of things standardization ITU and prospective networking technologies. *IEEE Communications Magazine* 54 (9): 43–49.

49 Kent, S. and Seo, K. (2005). Security Architecture for the Internet Protocol. IETF RFC 4301.

50 Khan, R., Khan, S.U., Zaheer, R. and Khan, S. (2012). Future internet: The internet of things architecture, possible applications and key challenges. *2012 10th International Conference on Frontiers of Information Technology*, Islamabad, Pakistan (17–19 December 2012). IEEE.

51 Kong, L., Khan, M.K., Wu, F. et al. (2017). Millimeter-wave wireless communications for IoT-cloud supported autonomous vehicles: Overview, design, and challenges. *IEEE Communications Magazine* 55 (1): 62–68.

52 Kumar, T., Braeken, A., Liyanage, M. and Ylianttila, M. (2017). Identity privacy preserving biometric based authentication scheme for naked healthcare environment. *2017 IEEE International Conference on Communications (ICC)*, Paris, France (21–25 May 2017). IEEE.

53 Lin, J., Yu, W., Zhang, N. et al. (2017). A survey on internet of things: Architecture, enabling technologies, security and privacy, and applications. *IEEE Internet of Things Journal* (99): 1–1.

54 Litman, T. (2017). *Autonomous vehicle implementation predictions*. Victoria Transport Policy Institute Victoria, Canada.

55 Liyanage, M., Ahmad, I., Abro, A.B. et al. (2018). *A Comprehensive Guide to 5G Security*. New York: John Wiley & Sons.

56 Lu, Y., Papagiannidis, S. and Alamanos, E. (2018). Internet of things: A systematic review of the business literature from the user and organisational perspectives. *Technological Forecasting and Social Change* 136: 285–297.

57 Marescaux, J., Leroy, J., Gagner, M. et al. (2001). Transatlantic robot-assisted telesurgery. *Nature* 413 (6854): 379.

58 Miorandi, D., Sicari S., Pellegrini, F.D. and Chlamtac, I. (2012). Internet of things: Vision, applications and research challenges. *Ad Hoc Networks* 10 (7): 1497–1516.

59 Munster, G. and Bohlig, A. (2017). Auto Outlook 2040: The Rise of Fully Autonomous Vehicles. Loupventures. https://loupventures.com/auto-outlook-2040-the-rise-of-fully-autonomous-vehicles/ (accessed 16 July 2019).

60 Nacke, L.E. and Deterding, C.S. (2017). The maturing of gamification research. *Computers in Human Behaviour* 450–454.

61 Nie, P., Vähä-Herttua, J., Aura, T. and Gurtov, A. (2011). Performance analysis of HIP diet exchange for WSN security establishment. 51–56. ACM.

62 Thubert, P. (2011). Compression Format for IPv6 Datagrams over IEEE 802.15.4-Based Networks. IETF RFC 6282.

63 Perera, C., Zaslavsky, A., Christen, P. and Georgakopoulos, G. (2014). Context aware computing for the Internet of Things: A survey. *IEEE Communications Surveys Tutorials* 16 (1): 414–454.

64 Perez, M., Xu, S., Chauhan, S. et al. (2016). Impact of Delay on Telesurgical Performance: Study on the Robotic Simulator dV-Trainer. *International Journal of Computer Assisted Radiology and Surgery* 11 (4): 581–587.

65 Porambage, P., Manzoor, A., Liyanage, M., et al. (2019). *Managing mobile relays for secure e2e connectivity of low-power IoT devices. IEEE Consumer Communications & Networking Conference, 2019*, Las Vegas, USA (11–14 January 2019). IEEE.

66 Porambage, P., Okwuibe, J., Liyanage, M. et al. (2018). Survey on multi-access edge computing for internet of things realization. *IEEE Communications Surveys & Tutorials* 20 (4): 2961–2991.

67 Rajakaruna, A., Manzoor, A., Porambage, P. et al. (2018). Lightweight dew computing paradigm to manage heterogeneous wireless sensor networks with UAVs. arXiv preprint arXiv:1811.04283. https://arxiv.org/pdf/1811.04283.pdf (accessed 16 July 2019).

68 Ramamurthy, A. and Jain, P. (2017). The internet of things in the power sector opportunities in Asia and the Pacific. Asian Development Bank https://www.adb.org/publications/internet-of-things-power-sector-opportunities (accessed 25 June 2019).

69 Reka, S.S. and Dragicevic, T. (2018). Future effectual role of energy delivery: A comprehensive review of internet of things and smart grid. *Renewable and Sustainable Energy Reviews* 91: 90–108.

70 Rescorla, E. and Modadugu, N. (2012). Datagram Transport Layer Security Version 1.2. IETF RFC 6347.

71 Cheshire, S. and Krochmal, M. (2013). Multicast DNS. IETF RFC 6762.

72 Saint-Andre, P. (2011). Extensible Messaging and Presence Protocol (XMPP): Core. IETF RFC 6120.

73 Sanchez, L., Muñoz, L., Galache, J.A. et al. (2014). Smartsantander: Iot experimentation over a smart city testbed. *Computer Networks* 61: 217–238.

74 Shelby, Z., Hartke, K., and Bormann, C. (2014). The Constrained Application Protocol (CoAP). IETF RFC 7552.

75 Sheng Z., Yang, S., Yu, Y. et al. (2013). A survey on the ietf protocol suite for the internet of things: standards, challenges, and opportunities. *IEEE Wireless Communications* 20 (6): 91–98.

76 Shoikova, E., Nikolov, R. and Kovatcheva, E. (2018). Smart digital education enhanced by ar and iot data. *INTED2018 Proceedings, 12th International Technology, Education and Development Conference*, Valencia, Spain (5–7 March 2018). INTED.

77 Stanford-Clark, A. and Truong, H.L. (2013). MQTT For Sensor Networks (MQTT-SN) Protocol Specification Version 1.2. OASIS Standard.

78 Stojkoska, B.L.R. and Trivodaliev, K.V. (2017). A review of internet of things for smart home: Challenges and solutions. *Journal of Cleaner Production* 140: 1454–1464.

79 Sun, H., Zhang, Z., Hu, R.Q. and Qian, Y. (2018). Wearable communications in 5g: challenges and enabling technologies. *IEEE Vehicular Technology Magazine* 13 (3): 100–109.

80 Talavera, J.M., Tobon, L.E., Gomez, J.A. et al. (2017). Review of IoT applications in agro-industrial and environmental fields. *Computers and Electronics in Agriculture* 142: 283–297.

81 Tan, L. and Wang, N. (2010). Future internet: The internet of things. *2010 3rd International Conference on Advanced Computer Theory and Engineering (ICACTE)*, Chengdu, China (20–22 August 2010). ICACTE.

82 Tsiatsis, V., Gluhak, A., Bauge, T. et al. (2010). The SENSEI Real World Internet Architecture. *Towards the Future Internet* 247–256.

83 Winter, T., Thubert, P., Brandt, A. et.al. (2012). RPL: IPv6 Routing Protocol for Low-Power and Lossy Networks. IETF RFC 6550.

84 Uckelmann, D., Harrison, M. and Michahelles, F. (2011). *An Architectural Approach Towards the Future Internet of Things*, 1–24. Berlin, Heidelberg: Springer Berlin Heidelberg.

85 Weyrich, M. and Ebert, C. (2016). Reference architectures for the internet of things. *IEEE Software* 33 (1): 112–116.

86 Wong, E., Dias, M.P.I. and Ruan, L. (2016). Tactile internet capable passive optical lan for healthcare. *OptoElectronics and Communications Conference (OECC) held jointly with 2016 International Conference on Photonics in Switching (PS)*, Niigata, Japan (3–7 July 2016). IEEE.

87 Wu, M., Lu, T.-J., Ling, F.-Y. et al. (2010). Research on the architecture of internet of things. In *2010 3rd International Conference on Advanced Computer Theory and Engineering (ICACTE)*, Chengdu, China (20–22 August 2010). IEEE.

88 Zaidan, A.A., Zaidan, B.B., Qahtan, M. et al. (2018). A survey on communication components for iot-based technologies in smart homes. *Telecommunication Systems* 69 (1): 1–25.

2

Introduction to IoT Security

Anca D. Jurcut, Pasika Ranaweera, and Lina Xu

Abstract

In a world with "things" and devices interconnected at every level, from wearables to home and building automation, to smart cities and infrastructure, to smart industries, and to smart-everything, the Internet of Things (IoT) security plays a central role with no margin for error or shortage on supply. Securing, including authentication of these devices, will become everyone's priority, from manufacturers to silicon vendors (or IP developers), to software and application developers, and to the final consumer, the beneficiaries of the security "recipe" that will accompany these IoT products. Together, all consumers of these products need to adapt to the market demands, innovate, and improve processes, grasp new skills and learn new methods, raise awareness and embrace new training and curricula programs.

In this chapter, we provide a thorough survey and classification of the existing vulnerabilities, exploitable attacks, possible countermeasures as well as access control mechanisms including authentication and authorization. These challenges are addressed in detail considering both the technologies and the architecture used. Furthermore, this work also focuses on IoT intrinsic vulnerabilities as well as the security challenges at every layer. In addition, solutions for remediation of the compromised security, as well as methods for risk mitigation, with prevention and suggestions for improvement are discussed.

Keywords *internet of things; security; attacks; countermeasures; authentication; authorization*

2.1 Introduction

The rapid proliferation of the Internet of Things (IoT) into diverse application areas such as building and home automation, smart transportation systems, wearable technologies for healthcare, industrial process control, and infrastructure monitoring and control is changing the fundamental way in which the physical world is perceived and managed. It is estimated that there will be approximately 30 billion IoT devices by 2020. Most of these IoT devices are expected to be of low-cost and wireless communication technology based, with limited capabilities in terms of computation and storage. As IoT systems are increasingly being entrusted with sensing and managing highly complex eco-systems,

IoT Security: Advances in Authentication, First Edition.
Edited by Madhusanka Liyanage, An Braeken, Pardeep Kumar, and Mika Ylianttila.

questions about the security and reliability of the data being transmitted to and from these IoT devices are rapidly becoming a major concern.

It has been reported in several studies that IoT networks are facing several security challenges [1–7] including authentication, authorization, information leakage, privacy, verification, tampering, jamming, eavesdropping, etc. IoT provides a network infrastructure with interoperable communication protocols and software tools to enable the connectivity to the internet for handheld smart devices (smart phones, personal digital assistants [PDAs] and tablets), smart household apparatus (smart TV, AC, intelligent lighting systems, smart fridge, etc.), automobiles, and sensory acquisition systems [1]. However, the improved connectivity and accessibility of devices present major security concerns for all the parties connected to the network regardless of whether they are humans or machines. The infiltration launched by the Mirai malware on the Domain Name System (DNS) provider Dyn in 2016 through a botnet-based Distributed Denial of Service (DDoS) attack to compromise IoT devices such as printers, IP cameras, residential gateways, and baby monitors represents the fertile ground for cyber threats in the IoT domain [82]. Moreover, the cyber-attack launched at the Ukrainian power grid in 2015 targeting the Supervisory Control and Data Acquisition (SCADA) system caused a blackout for several hours and is a prime example of the gravity of resulting devastation possible through modern day attacks [2]. The main reasons for the security challenges of current information-centric automated systems is their insecure unlimited connectivity with the internet and the non-existent access control mechanisms for providing secure and trustworthy communication. Furthermore, the problem of vulnerabilities in IoT systems arises because of the physical limitations of resource-constrained IoT devices (in terms of computing power, on-board storage and battery-life), lack of consensus/standardization in security protocols for IoT, and widespread use of third-party hardware, firmware, and software. These systems are often not sufficiently secure; especially when deployed in environments that cannot be secured/isolated by other means. The resource constraints on typical IoT devices make it impractical to use very complex and time-consuming encryption/decryption algorithms for secure message communication. This makes IoT systems highly susceptible to various types of attacks [1, 3–7]. Furthermore, addressing the security vulnerabilities in the protocols designed for communication is critical to the success of IoT [8–12, 97, 98].

This chapter focuses on security threats, attacks, and authentication in the context of the IoT and the state-of-the-art IoT security. It presents the results of an exhaustive survey of security attacks and access control mechanisms including authentication and authorization issues existing in IoT systems, its enabling technologies and protocols, while addressing all levels of the IoT architecture. We surveyed a wide range of existing works in the area of IoT security that use a number of different techniques. We classify the IoT security attacks and proposed countermeasures based on the current security threats, considering all three layers: perception, network, and application. This study aims to serve as a useful manual of existing security threats and vulnerabilities within the IoT heterogeneous environment and proposes possible solutions for improving the IoT security architecture. State-of-the-art IoT security threats and vulnerabilities have also been investigated, in terms of application deployments such as smart utilities, consumer wearables, intelligent transportation, smart agriculture, industrial IoT, and smart city have been studied. The insights presented on authentication and authorization aspects for the comprehensive IoT architecture are the prime contributions of this chapter.

The remainder of this chapter is organized as follows. Section 2.2 provides the IoT classification of attacks and their countermeasures according to the IoT applications and different layers of the IoT infrastructure. Section 2.3 addresses the importance of authentication with respect to security in IoT and presents in detail the existing authentication and authorization issues at all layers. Section 2.4 introduces other security features and their related issues. Additionally, solutions for remediation of the compromised security, as well as methods for risk mitigation, with prevention and suggestions for improvement discussed in the same section. A discussion on the authentication mechanisms in the IoT domain, considering the most recent methodologies, has been presented in Section 2.5. Section 2.6 introduces future research directives such as blockchain, 5G, fog and edge computing, quantum, AI and network slicing. Finally, Section 2.7 concludes this study.

2.2 Attacks and Countermeasures

Security is defined as a process to protect a resource against physical damage, unauthorized access, or theft, by maintaining a high confidentiality and integrity of the asset's information and making information about that object available whenever needed. The IoT security is the area of endeavor concerned with safeguarding connected devices and networks in the IoT environment. IoT enables the improvement of several applications in various fields, such as, smart cities, smart homes, healthcare, smarts grids, as well as other industrial applications. However, introducing constrained IoT devices and IoT technologies in such sensitive applications leads to new security challenges.

IoT is relying on connectivity of myriads of devices for its operation. Hence, the possibility of being exposed to a security attack is highly probable. In Information Technology (IT), an attack is an attempt to destroy, expose, alter, disable, steal, or gain unauthorized access to an asset. For example, cryptographic security protocols are key components in providing security services for communication over networks [10]. These services include data confidentiality, message integrity, authentication, availability, nonrepudiation, and privacy [3]. The proof of a protocol flaw is commonly known as an "attack" on a protocol and it is generally regarded as a sequence of actions performed by a dishonest principal, by means of any hardware or software tool, in order to subvert the protocol security goals. An IoT attack is not much different from an assault against an IT asset. What is new is the scale and relative simplicity of attacks in the IoT – millions and billions of devices are a potential victim of traditional style cyber-attacks, but on a far greater scale and often with limited or no protection.

The most prevalent devices connected to serving IoT applications for infotainment purposes are smart TVs, webcams, and printers. A vulnerability analysis has been conducted in [81] on these devices using Nessus[1] tool to observe that approximately 13% of the devices out of 156 680 were vulnerable which were further classified as critical, high, medium, and low. The vulnerabilities that exist , for example, MiniUPnP, Network Address Translation Port Mapping Protocol (NAT-PMP) detection, unencrypted telnet, Simple Network Management Protocol (SNMP) agents, Secure Shell (SSH) weak

---

1 https://www.tenable.com/products/nessus/nessus-professional

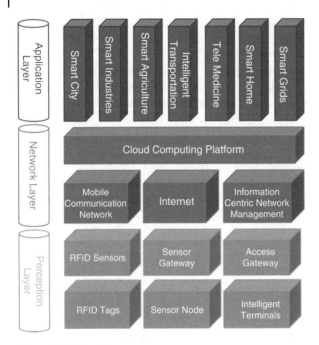

Figure 2.1 IoT architecture.

algorithms and File Transfer Protocol (FTP) inherited by webcams, smart TVs and printers are further identified based on manufacturer models.

In this section, we present the results of our study on the existing vulnerabilities, exploitable attacks and possible countermeasures in the context of the IoT and the state-of-the-art IoT security. We surveyed a wide range of existing work in the area of IoT security that use different techniques. We classified the IoT security attacks and the proposed countermeasures based on the current security threats, considering all three layers: Perception, Network, and Application. Figure 2.1 illustrates the typical architecture of IoT and entities which are considered under each layer. Table 2.1 summarizes the taxonomy of attacks and viable solutions of IoT categorized under each layer. These attacks and their corresponding solutions will be further discussed below.

2.2.1 Perception Layer

The devices belonging to the perception layer are typically deployed in Low-power and Lossy networks (LLNs), where energy, memory, and processing power are constricted compared to the localization of network nodes in conventional internet platforms [1]. Therefore, including secure public key encryption-based authentication schemes would not be feasible because they require high computational power and storage capacity. Hence, developing a lightweight cryptographic protocol would be a challenging task when scalability, context-awareness, and ease of deployment must also be considered [2].

There are several problems and attacks to be considered for the perception layer. We will be addressing these as shown in Table 2.1 and also by discussing the existing problems and attacks for perception nodes, sensor nodes and sensor gateways.

Table 2.1 Taxonomy of attacks and solutions in IoT layers.

Layer/component	Attacks	Solutions
a. Perception Layer		
Perception Nodes RFID	Tracking, DoS, repudiation, spoofing, eavesdropping, data newness, accessibility, self-organization, time management, secure localization, tractability, robustness, privacy protection, survivability, and counterfeiting [13].	Access control, data encryption which includes non-linear key algorithms, IPSec protocol utilization, cryptography techniques to protect against side channel attack [9], [14], Hashed-based access control [15], Ciphertext re-encryption to hide communication [16], New lightweight implementation using SHA-3 appointed function Keccak-f (200) and Keccak-f (400) [17]
Sensor nodes	Node subversion, node failure, node authentication, node outage, passive information gathering, false node message corruption, exhaustion, unfairness, sybil, jamming, tampering, and collisions [18, 19]	Node authentication, Sensor Privacy
Sensor Gateways	Misconfiguration, hacking, signal lost, DoS, war dialing, protocol tunneling, man-in-the-middle attack, interruption, interception, and modification fabrication [20]	Message Security, Device Onboard Security, Integrations Security [21]
b. Network layer		
Mobile Communication	Tracking, eavesdropping, DoS, bluesnarfing, bluejacking, bluebugging alteration, corruption, and deletion [1], [5], [38]	Developing secure access control mechanisms to mitigate the threats by employing biometrics, public-key crypto primitives and time changing session keys.
Cloud Computing	Identity management, heterogeneity which is inaccessible to an authentic node, data access controls, system complexity, physical security, encryption, infrastructure security and misconfiguration of software [22]	**Identity privacy** – Pseudonym [23–25], group signature [24], connection anonymization [26, 30] **Location privacy** – Pseudonym [23–25], one-way trapdoor permutation [25, 27] **Node compromise attack** – Secret sharing [27–29], game theory [26], population dynamic model [27] **Layer removing/adding attack** – Packet transmitting witness [25, 27, 30], aggregated transmission evidence [27] **Forward and backward security** – Cryptographic one-way hash chain [23, 24] **Semi-trusted/malicious cloud security** – (Fully) homomorphic encryption [31], zero knowledge proof [32]
Internet	Confidentiality, encryption, viruses, cyberbullying, hacking, identity theft, reliability, integrity, and consent [33]	Identity Management for confidentiality [34], Encryption schemes for confidentiality of communication channels [35], Cloud based solutions to establish secure channels based on PKI for data and communication confidentiality [35]
c. Application Layer	Data privacy, Tampering Privacy, Access control, disclosure of information [18]	Authentication, key agreement and protection of user privacy across heterogeneous networks [1], Datagram Transport Layer Security (DTLS) for end-to-end security [36], Information Flow Control [28]

2.2.1.1 Perception Nodes

Radio frequency identification (RFID) nodes and tags are typically used as perception nodes. RFID tags could be subjected to Denial of Service (DoS – from radio frequency interference), repudiation, spoofing, and eavesdropping attacks in the communication RF channel [1, 6, 13]. Moreover, reverse engineering, cloning, viruses (the SQL injection attack in 2006), tracking, killing tag (using a pre-defined kill command to disable a tag), block tag (employing a jammer such as a Faradays' cage) and side-channel attacks through power analysis are attacks which could compromise the RFID physical systems [84]. These attacks are feasible because of the low resources of RFID devices and comparatively weaker encryption/encoding schemes. Solutions to overcome these vulnerabilities and the corresponding exploitable attacks include access control, data encryption which includes non-linear key algorithms, IPSec protocol utilization, cryptography techniques to protect against side-channel attacks [9, 14], hashed-based access control [15], ciphertext re-encryption to hide communication [16], new lightweight implementation using SHA-3 appointed function Keccak-f (200) and Keccak-f (400) [17].

2.2.1.2 Sensor Nodes

Sensor nodes, such as ZigBee, possess additional resources compared to RFID devices with a controller for data processing and interoperability of sensor components, a Radio Frequency (RF) transceiver, a memory, the power source and the sensing element [1]. Even though the sensor nodes follow a fairly secure encryption scheme due to the elevated resources, attacks such as node tampering, node jamming, malicious node injection, Sybil, and collisions [18, 19] could exploit the vulnerabilities due to the nature of transmission technology and remote/distributed localization of them. A malware exploiting a flaw in the radio protocol of ZigBee caused a Save Our Souls (SOS) code illumination in smart Philips light bulbs as a demonstration of weakness in sensor node systems in 2016 [82]. Additionally, GPS sensors are vulnerable to jamming or data-level and signal-level spoofing which results in Time Synchronization Attacks (TSAs) targeted on Phasor Measurement Units (PMUs) of various IoT deployments that rely on GPS for locating or navigation-based services [85]. Possible countermeasures for such attacks are node authentication and sensor privacy techniques.

2.2.1.3 Gateways

Sensory gateways are responsible for checking and recording various properties such as temperature, humidity, pressure, speed, and functions of distributed sensor nodes. User access, network expansion, mobility, and collaboration are provided using sensor gateways.

These channels are also vulnerable to several attacks such as misconfiguration, hacking, signal lost, DoS, war dialing, protocol tunneling, man-in-the-middle attack, interruption, interception, and modification fabrication [20]. Moreover, perception layer devices could be subjected to Side-Channel Attacks (SCA) such as Differential Power Analysis (DPA), Simple Power Analysis (SPA), timing, and acoustic cryptanalysis [6]. To ensure security with respect to sensory gateways; message security, device on board security and integrations security are suitable proposed solutions [21].

2.2.2 Network Layer

Network Layer facilitates the data connectivity to perception layer devices for accomplishing the functionality of various applications in the Application layer. Because this layer is a connectivity provider for other layers, there are probable security flaws which would compromise the operations of the entire IoT architecture.

2.2.2.1 Mobile Communication

Mobile devices are the main interfaces of human interaction for IoT technology which range from smart phones, PDAs to mini-PCs. The state-of-the-art for mobile devices are extensively resourceful with their location services, biometric sensors, accelerometer/gyroscope, extended memory allocations, etc. The connectivity options range from RF, Low Rate Wireless Personal Area Networks (LR-WPAN/IEEE 802.15.4), Near Field Communication (NFC), Wireless Fidelity (Wi-Fi) to Bluetooth. However, these devices are vulnerable to DoS, sinkhole, bluesnarfing, bluejacking, blue bugging, alteration, corruption, deletion of data, and traffic analysis attacks [1, 5, 6, 38]. In addition, mobile devices are also vulnerable for phenomena such as cloning, spoofing, and various battery draining attacks explained in [83]. Even the technologies LR-WPAN, Bluetooth, and Wi-Fi are vulnerable to data transit attacks [82]. However, current standards of mobile devices have the means to improve the security through development of secure access control mechanisms to mitigate threats by employing biometrics, public-key crypto primitives, and time-changing session keys.

2.2.2.2 Cloud Computing

A cloud computing platform is the prime entity in IoT for centralized processing and storage facilitation for IoT applications [37]. Through cloud computing, IoT applications can enable higher computing power with unlimited storage capacity for a low cost, while maintaining versatile accessibility. Reliance on standalone dedicated server-based services is superseded by remote cloud-based server farms with outsourced services. However, outsourcing information to be stored in a remote location could raise security concerns. Privacy preservation is the most inevitable issue with cloud computing among other flaws such as physical security, anonymity, data access control failure, identity management, and direct tampering of the cloud servers [1, 22]. Several security solutions have been proposed in different areas for clouds including: (i) Identity privacy – Pseudonym [23], [24], [25], group signature [24], connection anonymization [26, 30]; (ii) Location privacy – Pseudonym [23–25], one-way trapdoor permutation [25, 27]; (iii) Node compromise attack – Secret sharing [27–29], game theory [26], population dynamic model [27]; (iv) Layer removing/adding attack – Packet transmitting witness [25, 27, 30], aggregated transmission evidence [27] (v) Forward and backward security – Cryptographic one-way hash chain [23, 24]; (vi) Semi-trusted/malicious cloud security – (Fully) homomorphic encryption [31], zero knowledge proof [32].

2.2.2.3 Internet

The term Internet stands for the holistic global networking infrastructure which scopes from private, public, academic, cooperate networks to government networks [1]. The connectivity through the Internet is formulated by Transmission Control

Protocol/Internet Protocol (TCP/IP) and secured through various protocols such as Secure Socket Layer (SSL)/Transmission Layer Security (TLS), IPSec, and SSH. In IoT, however, Datagram Transport Layer Security (DTLS) is used as the communication protocol [1, 37]. Since the Internet is accessible for everyone, the amount and nature of vulnerabilities outweigh the effectiveness of existing secure communication protocols [3–5, 7, 8, 10, 11] due to its implosive access capacity. Probable attacks are viruses, worms, hacking, cyber bullying, identity theft, consent, and DDoS [1, 33]. Countermeasures to overcome these attacks include Identity Management for confidentiality [34], Encryption schemes for confidentiality of communication channels [35], Cloud-based solutions to establish secure channels based on Public Key Infrastructure (PKI) for data and communication confidentiality [35].

2.2.3 Application Layer

As illustrated in Figure 2.2, possible applications for IoT are expanded into every industry available in the current era, in addition to myriads of non-industrial applications developed for automation purposes. In general, feasible attacks on the IoT application layer could be represented in two forms. They are software-based and encryption-based attacks. In the software attacks, most attacks are based on malicious software agents, apart from the phishing attacks, where the attacker reveals the authentication credentials of the user by impersonating as a trusted authority. Malware, worms, adware, spyware, and Trojans are highly probable occurrences with the heterogeneity of IoT applications and their broader services [6]. Encryption-based

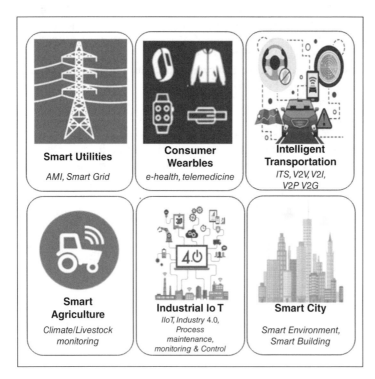

Figure 2.2 IoT applications.

attacks are the approaches taken to exploit the procedural nature of the cryptographic protocols and their mathematical model through extensive analysis. Cryptanalytic attacks, ciphertext only attacks, known plaintext attacks, and chosen plaintext attacks exemplify such possible threats [18].

There are several solutions proposed in the literature for the security of IoT applications such as authentication, key agreement and protection of user privacy across heterogeneous networks [1], DTLS for end-to-end security [36], and information flow control [28]. The countermeasures for software-based authentication should be taken for mitigating attacks such as phishing attacks; through the verification of the identity of malicious adversaries before proceeding.

2.2.3.1 Smart Utilities – Smart Grids and Smart Metering

Smart Grids are the future of energy distribution for all industrial and residential sectors. IoT plays a major role in smart grids for establishing the communication and monitoring protocols with consumers of energy. Smart grid is a decentralized energy grid with the ability to coordinate the electricity production in relation to the consumption or consumption patterns of the consumer. These systems are featuring monitoring technology called as Advanced Metering Infrastructure (AMI)/smart metering/net metering; which can measure and update the power consumption parameters to both entities in real time [39]. Additionally, smart grids are incorporating renewable energy sources commissioned in the vicinity of the consumer to cater the bidirectional energy flow for mitigating energy deficiencies [1].

Figure 2.3 illustrates a Smart Grid Architectural Model (SGAM) proposed by the coordinated group of European Committee for Standardization – European Committee for

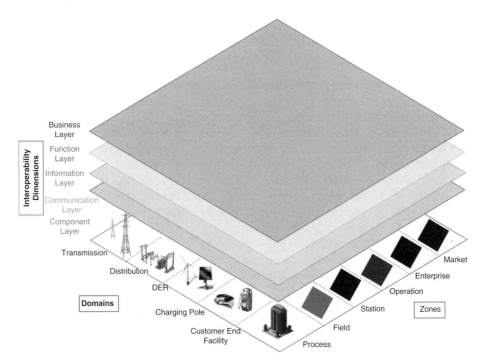

Figure 2.3 Smart grid architectural model.

Electrotechnical Standardization – European Telecommunications Standards Institute (CEN-CENELEC-ETSI), which offers a framework for smart grid use cases [76]. This architecture formulates three dimensions which amalgamate five functional interoperability layers with energy sector domains and zones that account for power system management [77]. This holistic framework is capable of reinforcing the design stages of the smart energy systems. The IoT technologies could be amalgamated with the SGAM framework to establish the bi-directional communication.

All the monitoring applications are developed with IoT infrastructure, with grid controlling access granted to the grid, controlling officers for pursuing configurations while the consumers can only visualize the consumption details via a mobile device. The information circulated through the AMI may pose a privacy concern for consumers for disseminating information regarding their habits and activities, where the impact could be severe for industries. Due to the heterogeneous nature of communication equipment deployed with IoT, and rapidly increasing population and industries, it would cause scalability issues for security. Smart grids are distributed across the power serving area and are, therefore, exposed to adversaries.

As the energy distribution system is the most critical infrastructure that exists in an urban area, the tendency to convert prevailing wired power-line communication (sending data over existing power cables) based controlling and monitoring channels to the wireless medium, with the introduction of IoT technologies, would expose the entire system to unintended security vulnerabilities. The intruders, using the proper techniques, could perpetuate AMI interfaces stationed at every household or industrial plant. Once access is granted to the hostile operators, potential outcomes can be devastating from disrupting the level of energy flow from a local grid substation to overloading the nuclear reactor of a power station. The availability of the grid could be compromised from IP spoofing, injection, and DoS/DDoS attacks [39]. Thus, access controlling for devices used in AMI and grid-controlling systems should be secured with extra countermeasures.

2.2.3.2 Consumer Wearable IoT (WIoT) Devices for Healthcare and Telemedicine

IoT-based healthcare systems are the most profitable and funded projects in the entire world. This is mainly due to the higher aggregate of aging people and the fact that health is the most concerning aspect of human life. A sensory system embedded with actuators is provided for individuals to use as a wearable device (i.e. wearable Internet of Things [WIoT] device), illustrated in Figure 2.4. A WIoT device can be used for tracking and recording vitals such as blood pressure, body temperature, heart rate, blood sugar, etc., [39]. This data can be conveyed and stored in a cloud as a Personal Health Record (PHR) to be accessed by the user and the assigned physicians.

Since the data handled in IoT-based healthcare is personal, privacy is the most demanding security issue. Hence, the access control mechanism for wearable devices as well as for PHRs must be well secured. However, employing strong crypto primitives for enhancing the authentication protocols of PHRs is possible as they are also stored in cloud environments. Hence, the same privacy concerns presented in Section 2.3.2.2 under cloud computing apply. Moreover, a method for assuring anonymity of patients should be developed in case the PHRs are exposed to external parties, because they are stored in Cloud Service Providers (CSPs). Wearable devices also face the resource scarcity issues for battery power, memory, and processing level [39]. Thus, a lightweight

Figure 2.4 WIoT devices.

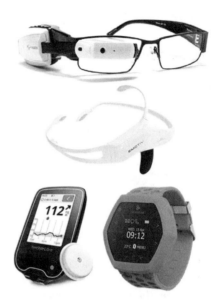

protocol for authentication and access control should be employed [101]. Similar to all other IoT applications, heterogeneous wearable devices produced by different manufacturers would employ diverse technologies for developing communication protocols. Thus, developing a generic access control policy would be extremely challenging.

2.2.3.3 Intelligent Transportation

Intelligent Transportation Systems (ITS) are introduced to improve transportation safety and decrease traffic congestions while minimizing environmental pollution. In an ITS system, there are four main components; vehicles, roadside stations, ITS monitoring centers and security systems [39]. All information extracted from vehicular nodes and roadside stations are conveyed to the ITS monitoring center for further processing; while the security subsystem is responsible for maintaining overall security. The entire system could be considered as a vehicular network, where the communication channels are established between Vehicle-to-Vehicle (V2V), Vehicle-to-Infrastructure (V2I), Vehicle-to-Pedestrian (V2P), and Vehicle-to-Grid (V2G) [39]. These communication links are implemented using technologies like RFID and Dedicated Short Range Communication (DSRC) for launching a large Wireless Sensor Network (WSN) [1]. The vehicular nodes and the entire data storing and monitoring infrastructure form a viable IoT deployment.

Figure 2.5 illustrates an ITS model, which enables communication among vehicular entities traveling through different mediums (airborne, land, and marine) with various technologies such as satellite, mobile, Wireless Local Area Network (WLAN), etc. Such a system would enable services like real-time updated navigation, roadside assistance, automated vehicular diagnostics, accident alerting system and self-driving cars [78]. Thus, massive divergence in the applicability of ITS deployment raises the requirement for a ubiquitous wireless connectivity with access points.

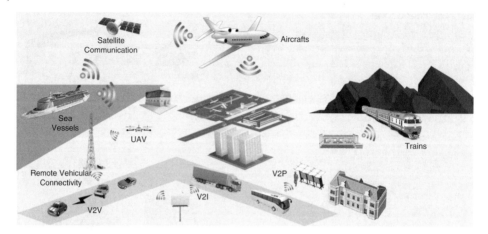

Figure 2.5 Intelligent transportation system.

As mentioned above, a larger number of entry points to a vehicular network makes it vulnerable to diverse attacks which can be targeted toward many sources [39]. At the same time, the privacy of drivers should be ensured from external observers, though drivers are not participating in any authentication activity. Authentication mechanisms are initiated between V2V interfaces where they can be exploited by an invader impersonating another vehicle or a roadside station. Therefore, a mechanism to verify the identity of the vehicles or roadside stations should be developed with a proper authentication mechanism employing a Trusted Third Party (TTP).

In some V2V communication systems, an On Board Diagnostic (OBD) unit is utilized to extract information directly from the Engine Control Unit (ECU) [1]. The OBD port could be used to manipulate the engine controls of a vehicle and could then be remotely accessed via the systems being developed. Thus, securing the access to the OBD port is vital.

2.2.3.4 Smart Agriculture

Agriculture is the most crucial industry in the world as it produces food and beverages by planting crops such as corn, rice, wheat, tea, potatoes, oats, etc. With the rapid worldwide population growth accounting for resource depletion, pollution, as well as the scarcity of human labor; agriculture is becoming an demanding industry. Automation is the most probable alternative for improving the effectiveness of the agriculture industry. Thus, IoT could play a vital role in such automation. IoT infrastructure could be deployed to perform climate/atmospheric, crop-status monitoring and livestock tracking. Climatic sensors, water/moisture level sensors and chemical concentration/acidity sensors along with visual sensors could be deployed for crop-status monitoring, while automated water and fertilizer dispersing mechanisms are in place within the bounds of the plantation. Additionally, livestock tracking is another aspect of smart farming implemented through the deployment of Local Positioning Systems (LPSs) on farm animals.

This type of a smart system would provide benefits such as the ability to utilize the fertilizer and water usage while maximizing crop production through mitigated effects of climatic deficiencies. The fruition science and "Hostabee" are two live cases of smart agriculture solutions used currently by the plantation industry [78].

Because of the diverse nature of sensor devices used in smart agriculture applications, integrating them into a holistic system may raise concerns about the compatibility of technologies among the variety of manufacturers and those protocols in which communication is established. As the plantations or fields are extending to larger areas, the number of IoT-enabled sensory systems to be deployed will be immense. Handling the data flow of such a large number of individual sensors with different data representations dispersed throughout a broad geographical region exerts the requirement for a communication technology with a higher coverage and moderate data rates which could not be satisfied by low-range communication technologies such as Bluetooth or NFC. However, DSRC would be a suitable technology to create a WSN with smart agriculture sensors, as it is compatible with ITSs.

As the IoT devices are disseminated across a larger geographical extent, the probability of any IoT device being compromised is high as they are exposed. Perception level attacks are probable with these devices as they are sensory nodes and would have limited resources for both processing and storing information. The spoofing, impersonation, replay and Man-in-the-Middle (MiM) attacks are probable with this application [80]. This raises the requirement for a proper authentication scheme as all perception level attacks could be mitigated using such a countermeasure.

2.2.3.5 Industrial IoT (IIoT)

M2M-based automation systems are quite common for industries such as oil and gas manufacturers. These industries are vast and the machinery employed is massive, expensive, and poses a significant risk to machine operators. Functions such as oil exploration by drilling, refining, and distributing are all conducted using automated machinery controlled through Programmable Logic Controllers (PLCs) based on SCADA systems. Although, the current M2M infrastructure is ideal for controlling the machinery, remote monitoring and accessibility is limited while a proper data storage and processing mechanism for decision making is not yet available. Thus, the requirement for IoT arises to improve operational efficiency by optimizing control of robots, reducing downtime through predictive and preventive maintenance, increasing productivity and safety through real- time remote monitoring of assets [78]. IoT sensor nodes could be deployed at the machinery while monitoring tools could be integrated without affecting the operation of SCADA systems. Hence, SCADA systems could be optimized to enhance productivity.

The Smart Factories term is an adaptation of Industrial Internet of things (IIoT), introduced as "Industry 4.0" to represent the Fourth Industrial Revolution (4IR) [79]. This standard signifies a trend of automation and data exchange in manufacturing industries which integrate Cyber-Physical Systems (CPS), IoT, and Cloud Computing based Data analytics [78, 79]. The interoperability, information transparency, technical assistance and decentralized decision making are the design principles of Industry 4.0 standard. BOSCH has developed connected hand-held tools which could monitor location, current user, and task-at-hand; analyzed and utilized for improving the efficiency in industrial labor [78]. Thus, the deployment of IoT across industry is imminent.

The security of industrial applications is a major concern, as any hostile intrusion could result in a catastrophic occurrence for both machinery and human operators. The SCADA systems are no longer secure (e.g. Considering the recent events [2]) due to their isolated localization and operation. However, main controlling functions are maneuvered within the control station located inside the industrial facility, while limited egress connectivity is maintained via satellite links with VSAT (Very Small Aperture Terminal) or microwave in the case of offshore or any other industrialized plants of such nature.

Due to their offline nature, the probability of any online intrusion is minimal. Though, any malicious entity such as a worm or a virus injected to the internal SCADA network could compromise the entire factory. Once inserted into the system, the intention of the malicious entity would be to disrupt the operations of the facility and its machinery. Thus, limiting the possibility for any malicious insurgence from the internal network and employing effective Intrusion Detection System (IDS) to detect malicious entities, would be the most suitable countermeasure for this application.

2.2.3.6   Smart Buildings, Environments, and Cities

Smart city is a holistically expanded inclusion of smart buildings and smart environments along with other smart automation systems formed to improve the quality of life for residents in a city. This is, in fact, the most expandable version of any IoT application in terms of cost for infrastructure deployment and geographical extent. In this concept, as shown in Figure 2.6, sensors are deployed throughout the building, environment, or the city for the purpose of extracting data of varied parameters such as temperature, humidity, atmospheric pressure, air density/air quality, noise level, seismic detection, flood detection, and radiation level. CCTV streams and LPSs would be a valuable input for smart building and smart cities to detect intrusions, monitor

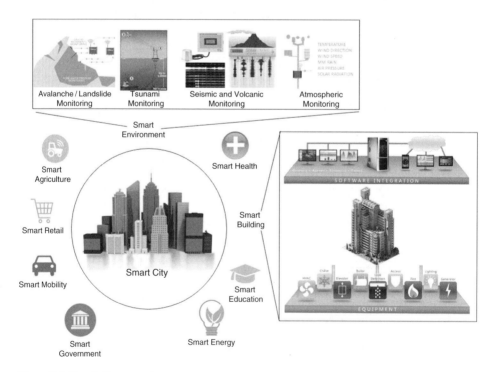

Figure 2.6 Smart city concept.

traffic and emergencies. All other smart systems explained in the previous sections are, in fact, subsystems of a functional smart city.

Due to various parameters to be gathered from the sensory acquisitions, heterogeneity is immense and the implementation is arduous [39]. At the same time, management of the gathered Big Data content is not scalable. Thus, providing security for all the applications in smart cities would be extremely challenging. Most of the Big Data content extracted from the sensors is forwarded to clouds through M2M authentication. Because of large data transmissions, cryptographic schemes should be lightweight and the authentication mechanism should be dynamic. DoS or DDoS attacks are most probable and could be mitigated with a strong authentication mechanism [1]. Individual sensors could be compromised resulting in the initiation of fake emergencies and access control methods should be improved to avoid such inconsistencies at the sensor level.

The paper [40] introduces applications of IoT with specific focus on smart homes. The study presented in [40] claims that although smart homes are offering comfortable services, security of data and context-oriented privacy are also a major concern of these applications. The security and privacy issues in IoT applications have also been studied in [41].

2.3 Authentication and Authorization

Authentication and access control mechanisms hold a great deal of significance in IoT. Without a proper mechanism for access control, entire IoT architecture could be compromised, as IoT devices are highly reliant on the trustworthiness of the other components to which they are connected. Thus, a proper access control mechanism is paramount to mitigate the flaws in the current IoT infrastructure.

Access control mechanisms are comprised of two stages (Figure 2.7) [1]: (i) Authentication and (ii) Authorization.

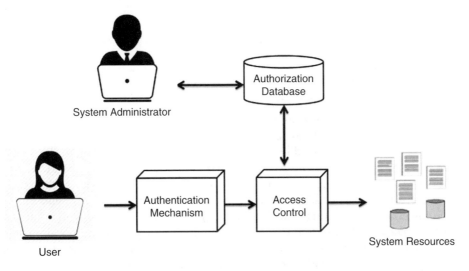

Figure 2.7 Typical access control System.

2.3.1 Authentication

Authentication is the process of verifying the identity of an entity [2]. The entity to be verified could either be human or a machine. Authentication is the first phase of any access control mechanism which can determine the exact identity of the accessing party in order to establish the trust of the system. In most cases, authentication is initiated between a human and a machine in a process to log into the internet banking portal by entering the credentials. However, in this scenario, the access-seeking entity does not have a guarantee regarding the identity of the access granting entity. In order to overcome this concern, mutual-authentication should be established between the entities, by verifying the identity of the access-granting entity with the involvement of a TTP, such as a Certificate Authority (CA) [2]. CAs are globally recognized institutions which are responsible for issuing and maintaining secure digital certificates of web entities registered under them. These certificates are imperative for the operation of all modern day authentication protocols such as SSL/TLS, IPSec, and HTTPS.

The process of authentication is merely facilitating credentials of an entity to the access granting system, which are unique to that entity and could only be possessed by them. This mechanism could be enabled with or without a TTP. The credentials used are often categorized as factors. The authentication schemes' accuracy and efficiency depend on the number of factors engaged in the mechanism. The types of factors are listed below.

- Knowledge factor – passwords, keys, PINs, patterns
- Possession factor – Random Number Generators (RNG), ATM card, ID card
- Inherence factor – Biometrics such as fingerprint, palm print, iris, etc.

Recent innovations in embedding biometric sensors to smart handheld devices have enabled the possibility of using multi-factor multi-mode (if more than one bio metric is used for verification) Human-to-Machine (H2M) authentication protocols for IoT devices. However, Machine-to-Machine (M2M) authentication can only be conducted using cryptographic primitives. Moreover, including strong cryptographic primitives (PKI, Hashing, Timestamps, etc.) for the authentication protocols involved is crucial in order to ensure data confidentiality, integrity, and availability; as the credentials being conveyed are highly sensitive and unique for the authenticating entity.

2.3.2 Authorization

Authorization is the process of enforcing limits and granting privileges to the authenticated entities [42]. In simple terms, this is determining the capabilities of an entity in the system. In order for an entity to be authorized for performing any action, the identity of that entity should be verified first through authentication. According to Figure 2.7, an administrator usually configures the authorization database for granting access and rights to system resources. Each resource is assigned with different rights such as read, write, and execute. Depending on the level of authorization (clearance) being set by the administrator, each authenticated entity can perform different actions on resources. A typical access control system has a policy for granting rights. These policies could vary from Discretionary Access Control (DAC), Mandatory Access Control (MAC) or a Multi-Level Security (MLS) model such as Role Based Access Control (RBAC) [42]. In DAC, the administrator specifies the rights, while in MAC there are rules set by the system for assigning rights for subjects. Clearances are granted according to the role of the authenticated entity (Roles: course coordinator, lecturer, or student in a university) in RBAC.

2.3.3 Authentication at IoT Layers

Authentication is the most critical security requirement in IoT for preserving user identity and mitigating the threats mentioned in the previous sections. With each IoT application, more hardware devices are introduced to be integrated to the IoT network. The authentication is the mechanism used to ensure the connectivity of those components to the existing ones. Authentication mechanisms involve cryptographic primitives for transmitting credentials securely. The strength of the scheme is entirely dependent on the crypto primitives being used. However, developing a generic solution would not be feasible, because differing layers attribute different requirements in IoT and the resources available for processing, memory, and energy are diverse. Therefore, we will discuss the authentication requirements for each layer.

2.3.3.1 Perception Layer

The Perception layer includes all the hardware devices or the machines to extract data from IoT environments. In most cases, the authentication initiates as M2M connections. Thus, in this layer, authentication could be conducted either as peer authentication or origin authentication [1]. In peer authentication, validation occurs between IoT routing peers, preliminary to the routing information exchanging phase; while validating the route information by the connected peer IoT devices with its source is origin authentication. This method enhances the security in M2M communication. Though as mentioned previously, devices in Perception layer are inheriting inadequate resources for generating strong cryptographic primitives.

2.3.3.1.1 Perception Nodes These nodes are distributed across the IoT environment. Mostly, they are RFID tags and RFID readers/sensors, where few RFID tags are connected to a RFID reader. The connection establishment between RFID tags and the reader does not involve an authentication mechanism and would be vulnerable if the RFID tags could be cloned. An Identity Based Encryption (IBE) scheme was proposed by [39] for establishing secure communication channels between RFID tags. Due to resource scarcity, an authentication protocol might be implemented using techniques such as Elliptic Curve Cryptography (ECC) based Diffie-Hellman (DH) key generation mechanism [1]. The generated keys, once they are transmitted to each end, could be used as the shared symmetric key for information transferring via the medium securely [42]. However, MiM attacks are still feasible and could be solved by employing the ephemeral DH method; changing the ECC DH exponents for each connection establishment as a session key.

2.3.3.1.2 Sensor Nodes and Gateways Sensor nodes face the similar security flaws as the perception nodes. Thus, deploying a proper authentication scheme could eliminate the possibility of being exposed at a very low level. However, sensors are much more intelligent and resourceful than perception nodes. Hence, M2M authentication could be established as peer authentications and the origin authentication could be established via the sensor gateway. Similarly, to the perception nodes, ECC-based DH key exchange would be ideal for sensor nodes, where the ephemeral exponents are facilitated by the sensor gateway acting as a TTP. Identity validation of the sensor gateway should be conducted prior to any data transfer. Even though using certificates for identity determination is not practical, a similar parameter such as a serial number could be used when registering the sensor node in the IoT environment and all the identities are stored in the sensor gateway for validation. Sensor gateway should also possess a unique identity for mutual authentication to be established between the sensor node and the gateway. Moreover, countermeasures such as integrity violation detection (using Hashed Message Authentication Code – HMAC or Cipher Block Chaining Mandatory Access Control – CBC-MAC) and timestamps should be employed with the authentication protocols involved.

2.3.3.2 Network Layer

IoT network layer is integrated on top of the existing TCP/IP internet protocols. In this section, we discuss the significance of the authentication for the components of the network layer.

2.3.3.2.1 Mobile Communication

Security for mobile communication at the network layer was not a critical necessity until the inception of IoT, as most of the mobile applications were relying on the inbuilt security protocols of the corresponding mobile technology (such as Global System for Mobile Communication – GSM, Wireless Code Division Multiple Access – WCDMA, High Speed Packet Access – HSPA or Long Term Evolution – LTE). With IoT, inbuilt authentication schemes are no longer foolproof, considering the potentiality for integrating technologies embedded in addition to the mobile technologies. Current security level and comprised resources (such as processor, memory, and operating system) in mobile devices are adequate for designing tamper resistance authentication protocols at the network layer [5]. However, the existing key generation algorithms used in TCP/IP protocols for generating large and costly asymmetric keys (Rivest Shamir Adleman (RSA), ElGamal, or Paillier), are still not feasible to be used with mobile devices. Thus, generating unbreachable and lightweight keys would be the most challenging task in mobile communication.

Yao et al. [96] proposed a lightweight no-pairing Attribute Based Encryption (ABE) scheme based on ECC that is designed for handheld devices. Even though the improved mathematical complexity and linear relationship of the number of attributes with computational overhead are improving the robustness of the proposed ABE scheme, scalability of the scheme would be highly questionable with enormous amount of IoT devices. IBE schemes are also adoptable, if taking the identity parameter as the mobile number or the user Social Security Number (SSN) for developing the authentication scheme integrating with ECC [39].

Current mobile devices include different biometric sensors for extracting biometrics such as fingerprint, iris, facial, and voice imprints. Biometrics can be used as unique keys that could be used for authentication and can be employed with H2M authentication. As the majority of the mobile devices at operation in an IoT environment are handled by a human user, the authentication design and key generation could be based on biometrics. The security of the biometrics schemes could be enhanced using several biometrics (multi-mode) integrated into multi-factor authentication schemes. These biometrically generated keys could be used as the signatures of each mobile entity, for the verification of their identities and, for conveying a secure session key among the communicating parties with proper encryption schemes. Additionally, authentication credentials should be checked for probable integrity violations in order to avoid MiM attacks.

2.3.3.2.2 Cloud Computing Clouds are the storage facility of IoT architecture and they are quite resourceful in terms of memory and processing [37]. Thus, authentication should employ strong keys that are generated using public-key algorithms such as RSA or ElGamal, which are inviolable cryptographic primitives if the executing authentication mechanism are computationally feasible with the available resources. A symmetric key (Advanced Encryption Standard [AES], Triple Data Encryption Standard (TDES), etc.) to be used in data transferring between the IoT devices and the cloud could be generated and shared among the entities engaged in communication. Existing CAs could be used to validate the identity of the parties involved in communication via mutual authentication schemes for establishing the trust.

However, the main concern in cloud computing is privacy of the user data. A strong authentication scheme does not ensure the misuse of information by the CSP. Thus, approaches such as blockchain and homomorphism should be considered for enhancing the privacy. The authentication schemes would be more secure in these schemes, as blockchain support pseudonymity (the nodes are identified from hashes or public keys – CA not required and simplify the authentication scheme) and the homomorphism facilitates an additional layer of encryption to secure the communication [39].

Authorization techniques in clouds should be also be considered, as accessing the information in the clouds is vital for the IoT design. Existing access control mechanisms such as RBAC and MAC are no longer scalable or interoperable. Thus, a novel method called Capability-Based Access Control (CapBAC), which uses capability-based authority tokens to grant privileges to entities was proposed by Kouicem et al. [39].

2.3.3.2.3 The Internet Even though authentication in most applications on the Internet is pursued by either SSL or IPSec protocols; IoT uses the DTLS as its communication protocol. However, the dependability of CAs for validating authentication parties still exists. Chinese CA WoSign was issuing certificates for false subjects in 2016, leaving an easier access to systems through wrongfully validated certificates for the attackers [2]. This happens when the trust of the system is centralized into a single entity. Thus, distributed access control schemes such as OpenPGP (widely used for email encryption) have formidable odds in succeeding in IoT infrastructure. Hokeun et al. in [2] introduces a locally centralized and globally distributed network architecture called Auth. Auth is to be deployed in edge devices for providing authorization services for locally registered entities, by storing their credentials and access policies in its database. Since the other instances of Auth are being distributed globally in the network, this maintains the trust relationships among them for granting authorizations for IoT devices acting as a gateway. Providing a solution to the trust issue of CAs is the main concern for the Internet, as the security level in existing protocols is quite adequate.

2.3.3.3 Application Layer

Heterodyne nature of the IoT predicates the requirement for different approaches of access control mechanisms for different applications. Most of the existing application layer H2M authentication schemes are two-factor authentication schemes, while the M2M ones are web based such as in SSL. The applicability and effectiveness of existing schemes are evaluated for each IoT application, since a generic solution is infeasible.

2.3.3.3.1 Smart Utilities – Smart Grids and Smart Metering When using proper techniques, the intruders could perpetuate AMI interfaces stationed at every household or industrial plant. Once the access is granted to the hostile operators, potential outcomes could be devastating, from disrupting the level of energy flow from a local grid substation to overloading the nuclear reactor of a power station. Thus, access to the smart grids should only be granted to the local grid operator and the monitoring center, avoiding any interfacing through the AMI access points. Local-grid operator authentication mechanism could be employed with a two-factor authentication scheme with a username, password, and RNG. A biometric scheme could be used depending on the availability of biometric extraction devices. As the controlling access is given to the operator, an authorization scheme such as RBAC should be employed, as scalability concern does not exist due to the limited number operators available for a smart grid. An M2M authentication interface is executed between the smart grid and monitoring center for information access. Existing security protocols such as SSLs could be used for authentication.

The access to AMI meter could be given to the residential consumer for the purpose of monitoring statistics. This access could also be based on two-factor authentication or biometrics as access is only given to read the data and not to manipulate it. Smart Grid has the ability to access the AMI meter through M2M authentication and should be secured with strong crypto primitives for preventing any MiM information extraction. Certificates should be issued to all the smart grids by a CA and identities should be validated preferably via a mutual authentication scheme when establishing a grid-to-grid communication channel. A mechanism should be embedded with an authentication protocol to validate the AMI units for detecting possible tampering scenarios.

2.3.3.3.2 Consumer Wearable IoT (WIoT) Devices for Healthcare and Telemedicine In a telemedicine system, the parties to grant access are solely the patients and their physicians. Thus, access should be limited. Authentication protocols should be always H2M when accessing the information, while M2M authentication operates when updating sensory information from wearable devices to the server. Access to the patient should be granted in a two-factor authentication scheme if a PC is being used for access. If the patient is using a mobile device to access the server, three-factor authentication scheme could be employed by integrating biometrics. Though, storing all the credentials including biometric templates at the authentication database would not be scalable with expanded healthcare services. Still, authentication should be thorough because accessing PHRs is private and confidential. Cloud servers' access to physicians could be granted from a two-factor authentication scheme. Storing and accessing PHRs at the cloud could be secured using the blockchain concept to counter any obvious privacy concerns with CSPs. An IBE scheme could be adopted to enhance the message transferring in the authentication protocol.

2.3.3.3.3 Intelligent Transportation and Logistics Since the vehicles attribute high mobility, the connectivity of an established wireless link across vehicular entities may vary rapidly. Hence, the availability of a consecutive/fixed inter-link would be uncertain.

Thus, dynamic handover mechanisms should be adopted between vehicular nodes for maintaining a consistent connection with each communicating vehicular node. Hence, those handover-based connections might require a lightweight approach for authentication as they are highly dynamic.

Each vehicle should have an Identity-based private key (embedded with its credentials – chassis no., registration no., manufacturer, model, etc.). However, the keys should be generated from an IBE or ABE lightweight mechanism unlike public-key encryption schemes which require costly resources to generate. Authentication protocols are more likely to be M2M mechanisms, where the machines are the vehicles. Therefore, verifying the identity of each vehicular node engaged in communication is paramount to avoid malicious node invasions through a TTP-based identity verification. An ECC-based ephemeral DH scheme could be employed for establishing a shared symmetric session key once the authentication phase is concluded following validation of vehicle identities. All V2V, V2I, V2G, and Vehicle to Cloud (V2C) connections could be implemented in the same manner.

Additional to the approaches discussed earlier, Software Defined Networks (SDN) and Blockchain concepts are highly recommended to ensure the security requirements in the Application layer [1, 39].

2.3.3.3.4 *Smart Agriculture* As mentioned in the previous section regarding attacks, agriculture IoT devices intrinsically require a lightweight authentication protocol as they are vulnerable to external intervention and sparse resources with perception level nodes. With a lesser resourced platform, implementing a mutual-authentication scheme would be questionable. In [80], a logic based on the Burrows-Abadi-Needham (BAN) modal logic was proposed and tested using Automated Validation Information Security Protocol Application (AVISPA) for verification, which was validated for MiM and replay attacks. However, a frequently changing session key usage is a vital necessity to prevent perception level attacks. This session key establishment could be employed with a technique such as ephemeral DH or ECC for lesser resource utilization.

2.3.3.3.5 *IIoT* Most IIoT processes are M2M due to their automated platforms. Further, IIoT process operations are continuous as their work cycles might extend to hours. With the amount of controlling data flowing through the communication channels, simultaneous authentication of each sensory node might lessen the efficiency of the entire smart factory. Thus, a methodology for a scheduled authentication scheme, which does not affect industrial performance, should be established. However, the authentication at each sensory node could be evaded, as there could be hundreds of minor sensors connected to massive machines, which would not be feasible for authentication of each node frequently. Only the control information transfer of machines that is subject to authentication, as a single controlling command, could continuously last for hours. These authentication phases could employ heavy cryptographic primitives as there is no scarcity of computational resources.

2.3.3.3.6 *Smart Buildings, Environments, and Cities* Designing a generic authentication protocol for smart cities is not practically feasible. This concept is formed from an entity such as a smart home, inadequate security measures could compromise the privacy of users at any level of use [81]. However, this application could be visualized from the perspective of the three layers in IoT. Similar methods proposed for access control in the perception layer could be adopted for the sensory system in smart environments. The Network layer accompanies all the internet integrated data connections and routing devices along with severs (clouds), in addition to the mobile devices. Mobile devices could use three-factor authentication schemes incorporated with web-based SSL or DTLS protocols, while cloud servers and routing nodes could be authenticated with cryptographically generated keys. Authentication protocols in smart cities are likely to change with the requirements and applications, as all other applications mentioned under this section are sub-applications of a smart city.

2.4 Other Security Features and Related Issues

IoT systems have their own generalized features and requirements regardless of the diversified nature of its applications such as heterogeneity, scalability, Quality of Service (QoS)-aware, cost minimization due to large-scale deployment, self-management including self-configuration, self-adaptation, self-discovery, etc. The last, but not least, general feature/requirement of an IoT system is to provide a secure environment to gain robustness against communication attacks, authentication, authorization, data-transfer confidentiality, data/device integrity, privacy, and to form a trusted secure environment [43]. IoT systems are fundamentally different from other transitional WSN systems [44] in many ways. (i) The diversity of the types of applications, the capabilities, and attributes of the IoT devices and deployed environments (ii) The holistic design of the system is mostly driven by the applications and it is essential to consider who the users are, what are the purposes and expected outcomes of the applications, etc. An IoT system is required to manage a large variety of devices, technologies, and service environments as the system itself is highly heterogeneous, where the connected IoT devices or equipment can range from simple temperature sensors to high-resolution smart cameras. The communication, computing, and power capability of each device can be unique and unique from others. These resource and interoperability constraints limit the feasibility for a standard security solution.

2.4.1 The Simplified Layer Structure

The traditional Open Systems Interconnection (OSI) has Seven layers: (i) The Physical Layer (Layer 1) is responsible for the transmission and reception of wire level data. (ii) The Data Link Layer (Layer 2) is responsible for link establishment and termination, frame traffic control, sequencing, acknowledgment, error checking, and media access management. (iii) The Network Layer (Layer 3) is implemented for routing of network traffic. (iv) The Transport Layer (Layer 4) is responsible for message seg-mentation, acknowledgement, traffic control, and session multiplexing. (v) The Session Layer (Layer 5) is responsible for session establishment, maintenance, and termination. (vi) The Presentation Layer (Layer 6) is responsible for character code translation, data conversion, compression, and encryption. (vii) The Application Layer (Layer 7) includes resource sharing, remote file access, remote printer access, network management, and electronic messaging (email). Since IoT systems normally have a huge variety, ranging from the choice of the hardware to the type of applications, the traditional seven network layers are simplified to three layers: perception layer, networking layer and application layer, as shown in Figure 2.1. The perception layer can be seen as the combination of the traditional physical layer and the MAC layer. It can include 2D bar code labels and readers, RFID tags and reader-writers, camera, GPS, sensors, terminals, and sensor network. It is the foundation for the IoT system [45]. The networking layer is responsible for the data transmission and communication inside the system and with the external Internet. It should be aware of the different underlying networks no matter whether it is wired, wireless, or cellular. It can provide support for different communication modes including base station, access point based or Machine-to Machine type based. The application layer provides services to the end users and collects data from different scenarios. IoT has high potential to implement smart and intelligent application for any scenario in nearly every field. This is mainly because IoT can offer both (i) data collection through sensing over natural phenomena, medical parameters, or user habits and (ii) data analysis and predictive modeling for tailored services. Such applications will cover aspects including personal, social, societal, medical, environmental, and logistics, having a profound impact on both the economy and society [43]. The perception and network layer together are considered the foundation for the whole IoT system. Together, these two layers provide the backbone and fundamental infrastructure of an IoT system. However, the architecture design and detailed implementation can normally only be confirmed after knowing the application layer design. Where the system will be deployed, what size the field will be and what kind of data will be collected are all issues involved in the applications, but highly affect the decision making on the perception layer and network layer.

2.4.2 The Idea of Middleware

Researchers from academia and industry are exploring solutions to enhance the development of IoT from three main perspectives: scientific theory, engineering design, and user experience [46]. These activities can enrich the technologies for IoT, but also increase the complexities, when implementing such a system in the real world. For this reason, the concept of IoT middleware has been introduced and many systems are already available [47–51]. However, when describing the formal definition for IoT middleware, researchers have different understandings. In some circumstance, IoT middleware is equivalent to IoT Operating System (OS). In general, middleware can simplify and accelerate a development process by integrating heterogeneous computing and communication devices, as well as supporting interoperability within the diverse applications and services [52]. Most existing implementations for middleware are designed for WSN and not for a service-oriented IoT system. Though, certain IoT-Specific middleware exists [53, 54]. In reality, middleware is often used to bridge the design gap between the application layer and the lower infrastructure layers. The requirements for middleware service for the IoT can be categorized into functional and non-functional groups. Functional requirements capture the services or functions such as abstractions and resource management [55]. Non-functional requirements capture QoS support or performance issues such as energy efficiency and security [56].

The Internet of Everything (IoE) aims to connect the objects, buildings, roads and cities and also to make the platform accessible. However, this feature will significantly increase the vulnerabilities of the system and, the inherent complexity of the IoT further complicates the design and deployment of efficient, interoperable, and scalable security mechanisms. It has been clearly stated that all typical security issues (authentication, privacy, nonrepudiation, availability, confidentiality, integrity) exist across all layers and the entire function box to a certain degree. However, when implementing security solutions, different layers of a variety of systems will have specialized priorities [78].

An essential task of the middleware is to provide secure data transmission between the upper and lower layers. For inner system communication, it should guarantee that the data passed to the application layer from the infrastructure is safe and reliable to use – integrity. Integrity in this scenario involves maintaining the consistency, accuracy, and trustworthiness of data over the transmission. Conversely, the middleware should also ensure that the control comments and queries from the applications/end users are verified and it is harmless for the system to take actions – non-repudiation. Non-repudiation features ensure that users cannot deny the authenticity of their signature for their documents and footprints for their activities. In addition, the middleware must protect the data transmission and information exchange between the upper and lower layers from illegal external access by any arbitrary user. The data must not be disclosed to any unauthorized entities – confidentiality.

2.4.3 Cross-Layer Security Problem

It has been frequently argued that although layered architectures have been a great success for wired networks, they are not always the best choice for wireless networks. To address this problem, a concept of cross-layer design is proposed and it is becoming popular. This concept is based on an architecture where different layers can exchange information in order to improve the overall network performance. A substantial amount of work has been carried out on state-of-the-art cross-layer protocols in the literature recently [57]. Security can be considered as one of the most critical QoS features in IoT systems. Wireless broadcast communication is suffering security risks more than others while multi-hop wireless communication is in a worse situation, as there is no centralized trusted authority to distribute a public key in a multi-hop network because of the nature of its distribution. Current proposed security approaches may be effective in a particular security issue in a specific layer. However, there still exists a strong need for a comprehensive mechanism to prevent security problems in all layers [58]. Security issues like availability need to be addressed not only at each layer, but a good cross-layer design and communication is encouraged. IoT systems are generally large and complex systems with many interconnections and dependencies, such as in smart cities [59].

If the availability of any of the three layers (perception, network, and application) fails, the availability of the whole system collapses. The lower layer infrastructure must protect itself from malicious behavioral patterns and harmful control from unauthorized users. The application layer should be available for all authorized users continuously without any service overloading-type interruption from unauthorized users.

2.4.4 Privacy

As the new European General Data Protection Regulation (GDPR)[2] has become enforceable on the 25 May 2018, protecting user data and securing user privacy are urgent

---

2 https://www.eugdpr.org

and predominant issues to be solved for any IoT application. Users' data can neither be captured nor used without their awareness. Privacy has the highest priority for all existing and future application development, including IoT systems [100]. User identities must not be identifiable nor traceable. Under the new legislation, data processing must involve:

1) Lawful, fair, and transparent processing – emphasizing transparency for data subjects.
2) Purpose limitation – having a lawful and legitimate purpose for processing the information in the first place.
3) Data minimization – making sure data is adequate, relevant and limited, and organizations are sufficiently capturing the minimum amount of data needed to fulfill the specified purpose.
4) Accurate and up-to-date processing – requiring data controllers to make sure information remains accurate, valid, and fit for purpose.
5) Limitation of storage in a form that permits identification – discouraging unnecessary data redundancy and replication.
6) Confidential and secure – protecting the integrity and privacy of data by making sure it is secure (which extends to IT systems, paper records and physical security)
7) Accountability and liability – the demonstrating compliance. As a well-known statement in security, there are security issues at all perception, network, and application layers.

Some other security problems can be addressed effectively and efficiently on a certain layer level, such as implementing privacy components on the application layer. In a healthcare system, patients should be totally aware who is collecting and using their data. They also should have control over the data and who they want to share it with, as well as how and where their data is being used. The applications should provide services and interface to allow users to manage their data. Users must have tools that allow them to retain their anonymity in this super-connected world. The same scenario can be applied to systems such as smart home, smart transportation, etc. IoT applications may collect users' personal information and data from their daily activities. Many people would consider that data or information predicted from the data as private. Exposure of this information could have an unwanted or negative impact on their life. The use of the IoT system should not cause problems of privacy leaking. Any IoT applications which do not meet with these privacy requirements could be prohibited by law. The IoT system must seriously consider the implementation of privacy by the 7 data protection principles, providing user-centric support for security and privacy from its very own foundations [60].

2.4.5 Risk Mitigation

Mitigating the risk of an intrusion attempt or attack against an IoT device is not an easy thing to do. Having a higher degree of security protection at every level will discourage the attacker to pursue his/her goals further, by causing a higher amount of effort and time needed versus the benefits. Mitigation needs to start with prevention, by involving every actor in the market, from manufacturers to consumers and lawmakers, and to make them understand the impact of the IoT security threats in a connected world. Another way to mitigate risk is to keep abreast of the times by improving and innovating, from the ground up, and by finding new methods and designs to outgrow the shortcomings of the market.

2.5 Discussion

Authentication for IoT is a paramount necessity for securing and ensuring the privacy of users, simply due to the fact that an impregnable access control scheme would be impervious for any attack vector originating outside of the considered trust domain, as explained in the previous sections of this chapter. Authentication schemes in IoT applications are generally implemented at the software level, where it exposes unintentional hardware and design vulnerabilities [82]. This fact constitutes the requirement of a holistic approach for securing access to the systems via the employment of impregnable authentication schemes. However, developing a generic authentication scheme to counter all possible attack scenarios would be improbable and an arduous attempt due to the heterogeneity of the IoT paradigm. A layered approach that identifies the distinct authentication requirement is desired to formalize a holistic trust domain.

For perception level entities, IBE or ECC would be ideal authentication schemes to generate commendable cryptographic credentials with available resources. The mobile entities, where actual users are interfacing to IoT systems are storing personalized credentials such as photos, medical stats, access to CCTV systems, GPS location (GPS), daily routines, financial statistics, banking credentials, emergency service status and online account statistics, are emphasizing the need for privacy preservation at this level. As proposed in Section 2.3.3.1.1, adopting IBE, ABE, ECC, or biometric-based mechanisms should ensure security. Novel mechanisms such as CapBAC could be employed to launch a scalable access control scheme for cloud computing platforms for IoT applications. However, the potential for deploying edge computing paradigms in the edge of the network indemnifies the cloud computing services from external direct access, as the access control would be migrated to the edge along with the service platform. The internet technologies of IoT-enabled systems are more secured than the perception level and mobile level entities with the deployed protocols such as DTLS, SSL, and IPSec. Due to the dependency of a CA or TTP for employing such strong and secure protocols, the future of Internet security enhancements would be focused on developing distributed access control schemes to eliminate the single point of failure. Each IoT application composes different devices and systems to accomplish the intended outcome which attributes diverse protocols in hardware and software. Thus, the authentication schemes should be application specific and context aware of resource constraints associated with the diversified deployments. As privacy is the main concern on IoT to be ensured through impregnable access control schemes, the GDPR initiative is a timely solution established to constrict the IoT service providers (both software and hardware) from developing and marketing products with vulnerabilities.

Current researches have focused on developing novel methods for authentication in the IoT domain. We are briefly introducing a few of these recent approaches to demonstrate the state-of-the-art technologies.

In [86], Ning et al. has proposed an aggregated proof-based hierarchical authentication (APHA) scheme to be deployed on existing Unit IoT and Ubiquitous IoT (U2IoT) architecture. Their scheme employs two cryptographic primitives; homomorphic functions and Chebyshev polynomials. The proposed scheme has been verified formally using Burrows-Abadi-Needham (BAN) logic. However, the scalability of the scheme with the extent of multiple units has not been verified with a physical prototype.

There are various initiatives on Physical Unclonable Functions (PUF) to be used for IoT device authentication. A PUF is an expression of an inherent and unclonable instance-specific unique feature of a physical object which serves as a biometric for non-human entities, such as IoT devices [89]. Hao et al. are proposing a Physical Layer (PHY) End-to-End (E2E) authentication scheme which generates an IBE-based PHY-ID which acts as a PUF with unclonable PHY features RF Carrier Frequency Offset (CFO) and In-phase/Quadrature-phase Imbalance (IQI) extracted from collaborative nodes in a Device to Device (D2D) IoT deployment [87]. This mechanism is ideal for perception-level nodes to be impervious to impersonation or malicious node injection attacks, as it is using physical measurements which are unique for each entity and for its location of operation in generating an identity for devices. However, the proposed scheme relies on a TTP called Key Generation Center (KGC). KGC generates the asymmetric key credentials for the nodes in its contact. The reachability of a certain KGC is limited due to the low power D2D connectivity. Thus, multiple KGCs deployed to accomplish the coverage should be managed with a centralized control entity. This enables the attack vectors on decentralized KGC entities. Moreover, the reliance on CFO and IQI features require the nodes to be stationary. This would be an issue considering most IoT devices are mobile and their RF-based characteristics vary in a timely manner. Aman et al. proposed a PUF-based authentication protocol for scenarios when an IoT device is connecting with a server and a D2D connectivity focused on its applicability in vehicular networks. Authentication is based on a Challenge Response Pair (CRP), where the outcome of the CRP is correlated with the physical microscopic structure of the IoT device, which emphasizes its unique PUF attributes with the inherent variability of the fabrication process in Integrated Circuits (ICs). The proposed protocol was analyzed using Mao and Boyd logic, while Finite State Machine (FSM) and reachability analysis techniques have been adopted for formal verification. Even though the performance of the protocol has been analyzed in terms of computational complexity, communication overhead and storage requirement, its scalability with simultaneous multiple IoT device connections to the server have not been addressed. However, this approach would be a feasible solution for V2E applications as the PUF could be successfully integrated with vehicles.

A human-gait pattern based on the biometric extraction scheme WifiU has been proposed in [88] as a case study that uses Channel State Information (CSI) of the received Wi-Fi signals for determining the gait pattern of the person carrying the transmitter. The gait patterns are becoming a novel biometric mode and this solution is a cost-effective approach which does not employ any floor sensors or human wearables. However, the applicability of WifiU for IoT devices raises concerns over scalability, accuracy of the gait-pattern extraction from CSI, reliability of CSI measurement and Wi-Fi interference. Chauhan et al. in [90] proposes a Recurrent Neural Network (RNN) based on human breath print authentication system for mobile, wearable, and IoT platforms employing a derived breath print as a biometric through acoustic analysis. Even though this approach depicts a viable biometric solution for human interfacing

IoT applications, the breath print extraction would be dependent on the health, climatic circumstances and physical stability of the user.

If the proposed authentication schemes are not fully holistically applicable for IoT deployments, optimum solutions at different layers and specific applications could be aggregated to form an impregnable access control system, where the interconnectivity across them should be maintained by decentralized trust domain managers. However, the access control mechanism optimum for each application should be investigated for each case in order to ensure robustness.

2.6 Future Research Directions

This section proposes several new research approaches and directions that could have a high impact for the future of the IoT security.

2.6.1 Blockchain

The blockchain is a distributed database of online records. Typically used in financial transactions for the Bitcoin cryptocurrency, the peer-to-peer blockchain technology records transactions without exception, in exchange, to form an online ledger system. Blockchain technologies are immutable, transparent, trustworthy, fast, decentralized, and autonomic, providing solutions that can be public, consortium, or private. Due to the success of Bitcoin, people are now starting to apply blockchain technologies in many other fields, such as financial markets, supply chain, voting, medical treatment and security for IoT [61, 99]. There are expectations that blockchain will revolutionize industry and commerce and drive economic change on a global scale [62].

Blockchain technology leads to the creation of secure mesh networks, where IoT devices will interconnect while avoiding threats such as impersonation or device spoofing. As more legitimate nodes register on the blockchain network, devices will identify and authenticate each other without a need for central brokers and certification authorities. The network will scale to support more and more devices without the need for additional resources [63].

Smart contracts open the way to defining a new concept, a decentralized autonomous organization (DAO), sometimes labeled as a Decentralized Autonomous Corporation, an organization that runs through rules maintained on a blockchain. The legal status of this new brand of business organization is rather seen as a general partnership, meaning that its participants could bear unlimited legal liability. Ethereum blockchain, for example, is a public blockchain network optimized for smart contracts that use its cryptocurrency, called Ether (ETH). There is a huge interest in Ethereum, as a blockchain technology for the future. In 2017, Enterprise Ethereum Alliance was formed and already counts close to 250 members, like Samsung, Microsoft, J.P.Morgan, Toyota, ING, Consensys, BP, Accenture and many others. Etherum has become the second highest traded cryptocurrency in 2017, after Bitcoin, with a volume of transactions for over half of million euros in a single 24-hour period.

As with each disruptive concept that turns into an effective offering, the blockchain model is not perfect and has its flaws and shortcomings. Scalability is one of the main issues, considering the tendency toward centralization with a growing blockchain. As the blockchain grows, the nodes in the network require more storage, bandwidth, and computational power to be able to process a block, which currently leads to only a handful of the nodes being able to process a block. Computing power and processing time is another challenge, as the IoT ecosystem is very diverse and not every device will be able to compute the same encryption algorithms at the desired speed. Storage of a continuously increasing ledger database across a broad range of smart devices with small storage capabilities, such as sensors, is yet another hurdle. The lack of skilled people to understand and develop the IoT-blockchain technologies together is also a challenge. The lack of laws and a compliance code to follow by manufacturers and service providers is not helping both the IoT and blockchain to take off as expected.

IOTA solves some problems that the blockchain does not. One of them is centralization of control. As history shows, small miners create big groups to reduce the variation of the reward. This activity leads to concentration of power, computational, and political, in possession of just a handful of pool operators and gives them the ability to apply a broad spectrum of policies, like filtering on or postponing certain transactions.

2.6.2 5G

For the first time in history, LTE has brought the entire mobile industry to a single technology footprint resulting in unprecedented economies of scale. The converged footprint of LTE has made it an attractive technology baseline for several segments that had traditionally operated outside the commercial cellular domain. There is a growing demand for a more versatile M2M platform. The challenge for industrial deployment of IoT is the lack of convergence across the M2M architecture design that has not materialized yet. It is expected that LTE will remain as the baseline technology for wide-area broadband coverage also in the 5G area. The realization of 5G network is affecting many IoT protocols' initial design, especially at perception and network layers [64, 102]. Mobile operators now aim to create a blend of pre-existing technologies covering 2G, 3G, 4G, WiFi, and others to allow higher coverage and availability, as well as higher network density in terms of cells and devices with the key differentiator being greater connectivity as an enabler for M2M services [65]. 3GPP standard/5G-based backhaul has become a popular solution for connectivity problems in IoT systems. Munoz et al. indicates that the next generation of mobile networks (5G), will need not only to develop new radio interfaces or waveforms to cope with the expected traffic growth but also to integrate heterogeneous networks from End-to-End (E2E) with distributed cloud resources to deliver E2E IoT and mobile services [66]. Fantacci et al. has provided a backhaul solution through mobile networks for smart building applications [67]. The proposed network architecture will improve services for users and will also offer new opportunities for both service providers and network operators. As 5G has become available and is being adopted as the main backhaul infrastructure for IoT system, it will play a huge role in IoT perception and networking layers [68]. 5G has moved the focus to a user-centric service from a network-centric service unlike 4G and 3G. With massive multiple-input and multiple-output (MIMO) technologies deployed in 5G, network selection and rapid handovers are becoming essential in terms of supporting QoS and Quality-of-user Experience (QoE) aware services [69]. The handover between different network interfaces should be authenticated and the information exchange during the handover should be protected and private. Currently, SDN is considered as the mainstream for a higher efficiency through its centralized control capability in the 5G communication process [70]. With SDN, the control logic is removed from the underlying infrastructures to a management platform. Software and policies can be implemented on the central SDN controller to provide consistent and efficient management over the whole 5G network. One advanced and beneficial feature offered by SDN is that it can separate the control plane and data source by abstract, the control logic from the underlying switches and routers to the centralized SDN controller [71]. To address the Machine Type Communication (MTC) in IoT systems based on 5G network, several approaches are available [72, 73]:

1) A higher level of security for devices is achievable by utilizing new security mechanisms being embedded with Subscriber Identity Module (SIM).
2) It is recommended to implement and employ physical-layer security adopting RF fingerprinting.
3) Using asymmetric security schemes to transfer the burden of required computations to the network domain or gateways with high computing capabilities.

2.6.3 Fog and Edge Computing

Although powerful, the cloud model is not the best choice for environments where internet connectivity is limited or operations are time-critical. In scenarios such as patient care, milliseconds have fatal consequences. As in the vehicle-to-vehicle communications, the prevention of collisions and accidents relies on the low latency of the responses. Cloud computing is not consistently viable for many IoT applications, and so, it is replaced by fog computing. Fog computing, also known as fogging, is a decentralized computing infrastructure in which the data, compute, storage, and applications split in an efficient way between the data source and the cloud.

Fog computing extends cloud computing and services alike, to the edge of the network, by bringing the advantages and the power of the cloud to where the data initially arise. The main goal of fogging is to improve efficiency and also to reduce the quantity of data that moves to the cloud for processing, analysis, and storage. In fogging, data processing takes place in a router, gateway, or data hub on a smart device, which sends it further to sources for processing and return transmission, therefore reducing the bandwidth payload to the cloud.

The back-and-forth communication between IoT devices and the cloud can negatively affect the overall performance and security of the IoT asset. The distributed approach of fogging addresses the problem of the high amount of data coming from smart sensors and IoT devices, which would be costly and time-consuming to send to the cloud each time. Among other benefits, fog computing offers better security by protecting the fog nodes with the same policy, controls, and procedures used in other parts of the IT environment and by using the same physical safety and cyber security solutions [74]. Fog networking complements cloud computing and allows for short-term analytics at the edge while the cloud performs resource-intensive, longer-term analytics. Computation moves even closer to the edge and becomes deeply-rooted in the very same devices that created the data initially, and so, generates even greater possibilities for M2M intelligence and interactions.

The movement of computation from the fog to the actual device opens the path-to-edge computing. That is a distributed architecture in which the processing of client data takes place at the outer edge of the network, in the proximity of the originating source. Mobile computing, low cost of computer components and the absolute quantity of IoT devices drive the move toward edge computing. Time-sensitive data is processed at the point of origin by an intelligent and resource-capable device or sent to a broker server located in close geographical proximity to the client. Less time-sensitive data travels to the cloud for historical analysis, big data analytics, and long-term storage. One of the greatest benefits of edge computing is that it removes network bottlenecks by improving time to action and response time down to milliseconds, while also conserving network resources.

The edge computing concept is not without its flaws though. Edge computing raises a high amount of security, licensing and configuration challenges and concerns. The vulnerability to some attack vectors like malware infections and security exploits increases because of the nature of the distributed architecture. Smart clients can have hidden licensing costs, where the base version of an edge client might initially have a low price, additional functionalities could be licensed separately and will drive the price up. Also, decentralized and poor device management leads to configuration drift by the administrators. They can inadvertently create security holes by not consistently updating the firmware or by failing to change the default password on each edge device [75].

2.6.4 Quantum Security, AI, and Predictive Data Analytics

With the technological advancements of quantum computing, Artificial Intelligence (AI), and cognitive systems, and with the continuous development and mass adoption of IoT ecosystem, the current security practices and methodologies will become a part of the past. Quantum computing, not only can it break through any form of security that is known to humankind, but it can also offer the solution to finding the formula for tight security. IoT will vastly benefit from these technology advancements, especially from the quantum mechanics science on a microchip. Further research is recommended, once the technology matures and evolves, to discover how the security of the future impacts on things around and especially on the IoT ecosystem.

2.6.5 Network Slicing

Network slicing is the concept of slicing a physical network into several logical planes to facilitate the various IoT services to customize their differentiated on-demand services with the same physical network [91]. The main aim of this paradigm is to reinforce different service requirements such as latency, bandwidth, and reliability of heterogeneous IoT applications to utilize the resources such as storage, computing, and bandwidth of the IoT device platforms [92]. The complexity of the IoT service integration with core network resources could be alleviated using a standardized network slicing mechanism as proposed by the Next Generation Mobile Network (NGMN). A typical network slicing process could be described under three layers, namely service instance layer, network slice instance layer and resource layer which follow the principles automation, isolation, customization, elasticity, programmability, end-to-end, and hierarchical abstraction [94].

The evolvement of the network slicing concept has reached the depths of 5G Information Centric Networking (ICN) model, which consists of five functional planes (FPs), namely; FP1 – service business plane, FP2 – service orchestration and management plane, FP3 – IP/ICN orchestrator plane, FP4 – domain service orchestration and management plane and FP5 – infrastructure plane. FP1 interfaces with external 5G users in providing various service Application Programme Interfaces (APIs) which realize the objective and relevant services to accomplish that objective with inputs such as service type, demand patterns, Service Level Agreements (SLAs)/QoS/QoE requirements. The service requests forwarded by FP1 are communicated to the FP3 as service requirements by FP2. The FP3 interfaces with a domain controller to virtualize compute, storage, and network resources to meet the service requirements conveyed from FP2. FP4 supports the management of IP and ICN services belonging to different technological domains such as 4G/5G Radio Access Network (RAN), Multi-Protocol Label Switching (MPLS) and edge technologies, while FP5 enables the service rules in end-to-end manner.

The entities operating in network slicing infrastructure, such as network slice manager and host platforms are attributing the vulnerabilities exploitable by impersonation attacks, DoS, SCA attacks and the interoperability of different security protocols and policies [93]. An IoT user may access different slices depending on both the requirements and the intended outcomes. Thus, the access granting control for different slices is a critical juncture in the perspective of security. The plausibility for isolating the slices for constricting the deliberate hacking attempts at resources operating at each plane should be focused. Due to the facts that a network slice is a composite of the actual physical infrastructure and the processes should be dynamic, adaptive, and flexible for servicing the intended functions, the assurance of user confidentiality, privacy, integrity, and availability are challenging. However, authentication is the most effective mechanism to be used for enhancing the robustness of the network slices toward attacks. Among the 5G Security –as-a-Service (SaaS) concepts, micro-segmentation, deception of the attacker and AI deployments for monitoring, attack detection and remediation are emerging initiatives for securing network slices [95].

2.7 Conclusions

IoT technology is the most discussed paradigm in the research community these days. Its potential to connect all the devices in the world and to create a large information system that would offer services to improve the quality of human beings exponentially has made the concept much more popular. The integration of various technologies and devices with different architectures are creating interoperability issues with the components in the IoT architecture. These issues and the highly diversified type of services are creating security concerns which disperse into all three layers of IoT architecture: Perception, Network, and Application. Hence, the security measures to be taken should be developed while analyzing the threats and vulnerabilities at each layer.

Mitigating risks associated with security breaches are possible, if security receives consideration from early product planning and design, and if some basic prevention mechanisms are in place. Enactment and standardization will simplify the manufacturing and development processes, give the market an incentive for mass-adoption and also increase the security posture of IoT products and services. Security will have to be inbuilt so that IoT can withstand a chance against the threats that technological advancements will bring.

References

1 Alaba, F., Othman, M., Hashem, I., and Alotaibi, F. (2017). Internet of things security: a survey. *Journal of Network and Computer Applications* 88: 10–28.

2 Kim, H. and Lee, E.A. (2017). Authentication and authorization for the internet of things. *IT Professional* 19 (5): 27–33.

3 Jurcut, A., Coffey, T., Dojen, R., and Gyorodi, R. (2008). Analysis of a key-establishment security protocol. *Journal of Computer Science and Control Systems* 2008: 42–47.

4 Jurcut, A.D., Coffey, T., and Dojen, R. On the prevention and detection of replay attacks using a logic-based verification tool. In: *Computer Networks*, vol. 431 (eds. A. Kwiecień, P. Gaj and P. Stera), 128–137. Switzerland: Springer International Publishing.

5 Jurcut, A.D., Liyanage, M., Chen, J. et al. (2018). On the security verification of a short message service protocol. Presented at the. In: *2018 IEEE Wireless Communications and Networking Conference (WCNC)*. Barcelona, Spain: (April, 2018).

6 Deogirikar, J. and Vidhate, A. (2017). Security attacks in IoT: a survey. In: *2017 International Conference on I-SMAC (IoT in Social, Mobile, Analytics and Cloud)*. Coimbatore, India: (February 2017).

7 Pasca, V., Jurcut, A., Dojen, R., and Coffey, T. (2008). Determining a parallel session attack on a key distribution protocol using a model checker. In: *ACM Proceedings of the 6th International Conference on Advances in Mobile Computing and Multimedia (MoMM '08)* Linz, Austria (24–26 November). New York, USA: ACM.

8 Jurcut, A.D., Coffey, T., and Dojen, R. (2017). A novel security protocol attack detection logic with unique fault discovery capability for freshness attacks and interleaving session attacks. *IEEE Transactions on Dependable and Secure Computing* https://doi.org/10.1109/TDSC.2017.2725831.

9 Liyanage, M., Braeken, A., Jurcut, A.D. et al. (2017). Secure communication channel architecture for software defined Mobile networks. *Journal of Computer Networks* 114: 32–50.

10 Jurcut, A., Coffey, T., and Dojen, R. (2014). Design requirements to counter parallel session attacks in security protocols. Presented at the. In: *12th IEEE Annual Conference on Privacy, Security and Trust (PST'14)*. in Toronto Canada (July 2014).

11 Jurcut, A.D., Coffey, T., and Dojen, R. (2014). Design guidelines for security protocols to prevent replay & parallel session attacks. *Journal of Computers & Security* 45: 255–273.

12 Jurcut, A.D., Coffey, T., and Dojen, R. (2012). Symmetry in security protocol cryptographic messages – a serious weakness exploitable by parallel session attacks. Presented at the . In: *7th IEEE International Conference on Availability, Reliability and Security (ARES'12)*. Prague, Czech Republic: (August 2012).

13 Jing, Q., Vasilakos, A.V., Wan, J. et al. (2014). Security of the internet of things: perspectives and challenges. *Wireless Networks* 20 (8): 2481–2501.

14 Zhao, K. and Ge, L. (2013). A survey on the internet of things security. Presented at the. In: *2013 9th International Conference on Computational Intelligence and Security (CIS)*. in Leshan, China (December 2013).

15 Weis, S.A., Sarma, S.E., Rivest, R.L., and Engels, D.W. (2004). Security and privacy aspects of low-cost radio frequency identification systems. In: *Security in Pervasive*

Computing (eds. D. Hutter, G. Muller, W. Stephan and M. Ullmann), 201–212. Springer.

16 Kumar, T., Liyanage, M., Ahmad, I. et al. (2018). User privacy, identity and trust in 5G. In: *A Comprehensive Guide to 5G Security* (eds. M. Liyanage, I. Ahmad, A.B. Abro, et al.), 267. Wiley.

17 Kavun, E.B. and Yalcin, T. (2010). A lightweight implementation of keccak hash function for radio-frequency identification applications. In: *International Workshop on Radio Frequency Identification: Security and Privacy Issues*. New York, USA (23–24 June 2015): Springer.

18 Zhang, Y., Shen, Y., Wang, H. et al. (2015). On secure wireless communications for IoT under eavesdropper collusion. *IEEE Transactions on Automation Science and Engineering* 13 (3): 1281–1293.

19 Massis, B. (2016). The internet of things and its impact on the library. *New Library World* 117 (3/4): 289–292.

20 Liu, Y., Cheng, C., Gu, T. et al. (2016). A lightweight authenticated communications scheme for a smart grid. *IEEE Transactions on Smart Grid* 7 (3): 1304–1313.

21 Kumar, T., Porambage, P., and Ahmad, I. (2018). Securing gadget-free digital services. *Computer* 51 (11): 66–77.

22 Horrow, S. and Anjali, S. (2012). Identity management framework for cloud based internet of things. In: *Proceedings of the First International Conference on Security of Internet of Things, SecurIT' 12*. Kollam, India (17–19 August 2012): ACM.

23 Lin, X., Sun, X., Wang, X. et al. (2008). TSVC: timed efficient and secure vehicular communications with privacy preserving. *IEEE Transactions on Wireless Communications* 7 (12): 4987–4998.

24 Lin, X. and Li, X. (2013). Achieving efficient cooperative message authentication in vehicular ad hoc networks. *IEEE Transactions on Vehicular Technology* 62 (7): 3339–3348.

25 Zhou, J., Dong, X., Cao, Z. et al. (2015). 4S: a secure and privacy-preserving key management scheme for cloud-assisted wireless body area network in m-healthcare social networks. *Information Sciences* 314: 255–276.

26 Sen, J. (2011). Privacy preservation Technologies in Internet of things. In: *Proceedings of International Conference on Emerging Trends in Mathematics, Technology, and Management*, (18–20 November 2011).

27 Zhou, J., Dong, X., Cao, Z. et al. (2015). Secure and privacy preserving protocol for cloud-based vehicular DTNs. *IEEE Transactions on Information Forensics and Security* 10 (6): 1299–1314.

28 Roman, R., Alcaraz, C., Lopez, and Sklavos, N. (2011). Key management systems for sensor networks in the context of the internet of things. *Computers and Electrical Engineering* 37 (2): 147–159.

29 Zhou, J., Cao, Z., Dong, X. et al. (2015). TR-MABE: White-Box Traceable and Revocable Multi-Authority Attribute-Based Encryption and Its Applications to Multi-Level Privacy-Preserving e-Heathcare Cloud Computing Systems. *IEEE INFOCOM*.

30 Lu, R., Lin, X., Zhu, H. et al. (2010). Pi: a practical incentive protocol for delay tolerant networks. *IEEE Transactions on Wireless Communications* 9 (4): 1483–1492.

31 Paillier, P. (1999). Public key cryptosystems based on composite degree residuosity classes. In: *Eurocrypt '99 Proceedings of the 17th international conference on Theory*

and application of cryptographic techniques. Prague, Czech Republic (2–6 May 1999): ACM.

32 Groth, J. and Sahiai, A. (2008). Efficient noninteractive proof systems for bilinear groups. In: *Advances in Cryptology®EUROCRYPT*. Istanbul, Turkey (13–17 April 2008): Springer.

33 Akhunzada, A., Gani, A., Anuar, N.B. et al. (2016). Secure and dependable software defined networks. *Journal of Network and Computer Applications* 61: 199–221.

34 Miorandi, D., Sicari, S., De Pellegrini, F., and Chlamtac, C. (2012). Internet of things: vision, applications and research challenges. *Ad Hoc Networks* 10 (7): 1497–1516.

35 Porambage, P., Okwuibe, J., Liyanage, M. et al. (2018). Survey on multi-access edge computing for internet of things realization. *IEEE Communication Surveys and Tutorials* 20 (4): 2961–2991.

36 Garcia-Morchon, O., Hummen, R., Kumar, S. et al. (2012) Security considerations in the ip-based internet of things, draft-garciacore-security-04.

37 Singh, J., Pasquier, T., Bacon, J. et al. (2015) Twenty security considerations for cloud-supported internet of things.

38 Bekara, C. (2014). Security issues and challenges for the IoT-based smart grid. *Procedia Computer Science* 34: 532–537.

39 Kouicem, D., Bouabdallah, A., and Lakhlef, H. (2018). Internet of things security: a top-down survey. *Journal of Computer Networks* 141: 199–121.

40 Desai, D. and Upadhyay, H. (2014). Security and privacy consideration for internet of things in smart home environments. *International Journal of Engineering Research and Development* 10 (11): 73–83.

41 Kumar, J.S. and Patel, D.R. (2014). A survey on internet of things: security and privacy issues. *International Journal of Computer Applications* 90 (11).

42 Stamp, M. (2011). *Information Security*, 2e. Hoboken, N.J: Wiley.

43 Borgia, E. (2014). The internet of things vision: key features, applications and open issues. *Computer Communications* 54: 1–31.

44 Xu, L., Collier, R., and O'Hare, G.M.P. (2017). A survey of clustering techniques in wsns and consideration of the challenges of applying such to 5g iot scenarios. *IEEE Internet of Things Journal* 4: 1229–1249.

45 Wu, M., Lu, T.-J., Ling, F.-Y. et al. (2010). Research on the architecture of internet of things. In: *2010 3rd International Conference on Advanced Computer Theory and Engineering (ICACTE)*. Chengdu, China: (20–22 August 2010).

46 Feki, M.A., Kawsar, F., Boussard, M., and Trappeniers, L. (2013). The internet of things: the next technological revolution. *Computer* 46: 24–25.

47 Contiki. http://www.contiki-os.org (accessed 16 July 2019).

48 Brillo. https://developers.google.com/brillo (accessed 16 July 2019).

49 Tinyos. http://www.tinyos.net (accessed 16 July 2019).

50 Openwsn. http://openwsn.atlassian.net (accessed 16 July 2019).

51 Riot. http://www.riot-os.org (accessed 16 July 2019).

52 Porambage, P., Manzoor, A., Liyanage, M. et al. (2019). Managing Mobile Relays for Secure E2E Connectivity of Low-Power IoT Devices. *IEEE Consumer Communications & Networking Conference*, Las Vegas, USA (11–14 January 2010). IEEE.

53 Perera, C., Jayaraman, P.P., Zaslavsky, A. et al. (2014). Mosden: an internet of things middleware for resource constrained mobile devices. In: *47th Hawaii International Conference on System Sciences*. Hawaii, USA (6–9 January 2014): IEEE.

54 Zhou, H. (2012). *The Internet of Things in the Cloud: A Middleware Perspective*, 1e. Boca Raton, FL, USA: CRC Press, Inc.

55 Xu, L., Lillis, D., O'Hare, G.M., and Collier, R.W. (2014). A user configurable metric for clustering in wireless sensor networks. In: *SENSORNETS*. Lisbon, Portugal (7–9 January 2014): SciTePress.

56 Ngu, A.H., Gutierrez, M., Metsis, V. et al. (2017). Iot middleware: a survey on issues and enabling technologies. *IEEE Internet of Things Journal* 4: 1–20.

57 Srivastava, V. and Motani, M. (2005). Cross-layer design: a survey and the road ahead. *IEEE Communications Magazine* 43: 112–119.

58 Zhang, Q. and Zhang, Y.Q. (2008). Cross-layer design for qos support in multihop wireless networks. *Proceedings of the IEEE* 96: 64–76.

59 Zanella, A., Bui, N., Castellani, A. et al. (2014). Internet of things for smart cities. *IEEE Internet of Things Journal* 1: 22–32.

60 Roman, R., Zhou, J., and Lopez, J. (2013). On the features and challenges of security and privacy in distributed internet of things. *Computer Networks* 57 (10): 2266–2279.

61 Lin, I., I.C. and Liao, T.C. (2017). A survey of blockchain security issues and challenges. *International Journal of Network Security* 19: 653–659.

62 Underwood, S. (2016). Blockchain beyond bitcoin. *Communications of the ACM* 59: 15–17.

63 How blockchain can change the future of IoT (2016). https://venturebeat.com/2016/11/20/how-blockchain-can-change-the-future-of-iot (accessed 16 July 2019).

64 Chiang, M. and Zhang, T. (2016). Fog and iot: an overview of research opportunities. *IEEE Internet of Things Journal* 3: 854–864.

65 Liyanage, M., Ahmad, I., Abro, A.B. et al. (eds.) (2018). *A Comprehensive Guide to 5G Security*. Wiley.

66 Munoz, R., Mangues-Bafalluy, J., Vilalta, R. et al. (2016). The cttc 5g end-to-end experimental platform: integrating heterogeneous wireless/optical networks, distributed cloud, and iot devices. *IEEE Vehicular Technology Magazine* 11: 50–63.

67 Fantacci, R., Pecorella, T., Viti, T., and Carlini, C. (2014). A network architecture solution for efficient iot wsn backhauling: challenges and opportunities. *IEEE Wireless Communications* 21: 113–119.

68 Ahmad, I., Kumar, T., Liyanage, M. et al. (2018). Overview of 5G security challenges and solutions. *IEEE Communications Standards Magazine* 2 (1): 36–43.

69 Xu, L., Xie, J., Xu, X., and Wang, S. (2016). Enterprise lte and wifi interworking system and a proposed network selection solution. In: *2016 ACM/IEEE Symposium on Architectures for Networking and Communications Systems (ANCS)*. Santa Clara, USA (17–18 March 2016): ACM/IEEE.

70 Liyanage, M., Gurtov, A., and Ylianttila, M. (eds.) (2015). *Software Defined Mobile Networks (SDMN): Beyond LTE Network Architecture*. New York: Wiley.

71 Liyanage, M., Abro, A.B., Ylianttila, M., and Gurtov, A. (2016). Opportunities and challenges of software-defined mobile networks in network security. *IEEE Security and Privacy* 14 (4): 34–44.

72 Tehrani, M.N., Uysal, M., and Yanikomeroglu, H. (2014). Device-to-device communication in 5g cellular networks: challenges, solutions, and future directions. *IEEE Communications Magazine* 52: 86–92.

73 Shariatmadari, H., Ratasuk, R., and Iraji, S. (2015). Machine-type communications: current status and future perspectives toward 5g systems. *IEEE Communications Magazine* 53: 10–17.

74 Fog Computing and the Internet of Things: Extend the Cloud to where the Things Are. http://www.cisco.com/c/dam/en_us/solutions/trends/iot/docs/computing-overview.pdf (accessed 25 June 2019).

75 Rouse, M. (2016). Edge computing. http://searchdatacenter.techtarget.com/definition/edge-computing. (accessed 25 June 2019).

76 CEN-CENELEC-ETSI Smart Grid Coordination Group. (2014). SGCG/M490/G Smart Grid Set of Standards Version 3.1, Oct-2014. ftp://ftp.cencenelec.eu/EN/EuropeanStandardization/HotTopics/SmartGrids/SGCG_Standards_Report.pdf (accessed 25 June 2019).

77 Ahmad, I., Kumar, T., Liyanage, M. et al. (2018). Towards gadget-free internet services: a roadmap of the naked world. *Telematics and Informatics* 35 (1): 82–92.

78 Rajakaruna, A., Manzoor, A., Porambage, P. et al. (2018). Lightweight Dew Computing Paradigm to Manage Heterogeneous Wireless Sensor Networks with UAVs. *arXiv preprint arXiv:*1811.04283.

79 Rahimi, H., Zibaeenejad, A. and Safavi, A.A. (2018). A Novel IoT Architecture based on 5G-IoT and Next Generation Technologies. GlobeCom-IoT. https://arxiv.org/ftp/arxiv/papers/1807/1807.03065.pdf (accessed 25 June 2019).

80 Srilakshmi, A., Rakkini, J., Sekar, K.R., and Manikandan, R. (2018). A comparative study on internet of things (IoT) and its applications in smart agriculture. *Pharmacognosy Journal* 10 (2): 260–264.

81 Williams, R., McMahon, E., Samtani, S. et al. (2017). Identifying vulnerabilities of consumer internet of things (IoT) devices: a scalable approach. In: *2017 IEEE International Conference on Intelligence and Security Informatics (ISI)*. Beijing, China (22–24 July 2017): IEEE.

82 Frustaci, M., Pace, P., Aloi, G., and Fortino, G. (2018). Evaluating critical security issues of the IoT world: present and future challenges. *IEEE Internet of Things Journal* 5 (4): 2483–2495.

83 Fiore, U., Castiglione, A., De Santis, A., and Palmieri, F. (2017). Exploiting battery-drain vulnerabilities in mobile smart devices. *IEEE Transactions on Sustainable Computing* 2 (2): 90–99.

84 Xiao, Q., Gibbons, T., and Lebrun, H. (2009). RFID technology. *Security Vulnerabilities, and Countermeasures* https://doi.org/10.5772/6668 : https://www.researchgate.net/publication/221787702_RFID_Technology_Security_Vulnerabilities_and_Countermeasures (accessed 25 June 2019).

85 Khalajmehrabadi, A., Gatsis, N., Akopian, D., and Taha, A. (2018). Real-time rejection and mitigation of time synchronization attacks on the global positioning system. *IEEE Transactions on Industrial Electronics* 65 (8): 6425–6435.

86 Ning, H., Liu, H., and Yang, Y. (2015). Aggregated-proof based hierarchical authentication scheme for the internet of things. *IEEE Transactions on Parallel and Distributed Systems* 26 (3): 657–667.

87 Hao, P., Wang, X., and Shen, W. (2018). A collaborative PHY-aided technique for end-to-end IoT device authentication. *IEEE Access* 6: 42279–42293.

88 Shahzad, M. and Singh, M.P. (2017). Continuous authentication and authorization for the internet of things. *IEEE Internet Computing* 21 (2): 86–90.

89 Aman, M.N., Chua, K.C., and Sikdar, B. (2017). Mutual authentication in IoT systems using physical unclonable functions. *IEEE Internet of Things Journal* 4 (5): 1327–1340.

90 Chauhan, J., Seneviratne, S., Hu, Y. et al. (2018). Breathing-based authentication on resource-constrained IoT devices using recurrent neural networks. *Computer* 51 (5): 60–67.

91 Ni, J., Lin, X., and Shen, X.S. (2018). Efficient and secure service-oriented authentication supporting network slicing for 5G-enabled IoT. *IEEE Journal on Selected Areas in Communications* 36 (3): 644–657.

92 Ravindran, R., Chakraborti, A., and Amin, S. (2017). 5G-ICN: delivering ICN services over 5G using network slicing. *IEEE Communications Magazine* 55 (5): 101–107.

93 Harel, R., and Babbage, S. (2016). 5G Security Recommendations Package #2: Network Slicing, published by NGMN Alliance, Ver. 01. https://www.ngmn.org/fileadmin/user_upload/160429_NGMN_5G_Security_Network_Slicing_v1_0.pdf (accessed 25 June 2019).

94 Siriwardhana, Y., Porambage, P., Liyanage, M. et al. (2019). Micro-Operator driven Local 5G Network Architecture for Industrial Internet. *IEEE Wireless Communications and Networking Conference*, Marrakech, Morocco (15–18 April 2019). IEEE.

95 Dotaro, E. (2018). 5G Network Slicing and Security. IEEE SDN newsletter, https://sdn.ieee.org/newsletter/january-2018/5g-network-slicing-and-security (accessed 25 June 2019).

96 Yao, X., Chen, Z., and Tian, Y. (2015). A lightweight attribute-based encryption scheme for the internet of things. *Future Generation Computer Systems* 49: 104–112.

97 Kumar, T., Braeken, A., Liyanage, M., and Ylianttila, M. (2017). Identity privacy preserving biometric based authentication scheme for naked healthcare environment. In: *2017 IEEE International Conference on Communications (ICC).* Paris, France (21–25 May 2017): IEEE.

98 Braeken, A., Liyanage, M., and Jurcut, A.D. (2019). Anonymous lightweight proxy based key agreement for IoT (ALPKA). *Wireless Personal Communications* 106 (2): 1–20.

99 Manzoor, A., Liyanage, M., Braeken, A. et al. (2018). Blockchain based Proxy Re-Encryption Scheme for Secure IoT Data Sharing. *IEEE International Conference on Blockchain and Cryptocurrency (ICBC 2019)*, Seoul, South Korea (14–17 May 2019). IEEE.

100 Liyanage, M., Salo, J., Braeken, A. et al. (2018). 5G privacy: scenarios and solutions. In: *2018 IEEE 5G World Forum (5GWF).* Silicon Valley, USA (9–11 July 2018): IEEE.

101 Kumar, T., Braeken, A., Jurcut, A.D., Liyanage, M., and Ylianttila, M. (2019). AGE: Authentication in Gadget-Free Healthcare Environments. *Information Technology and Management Journal*, Springer US.

102 Xu, L., Jurcut, AD., and Ahmadi, H. (2019). Emerging Challenges and Requirements for Internet of Things in 5G. 5G-Enabled Internet of Things. CRC Press.

Part II

IoT Network and Communication Authentication

3

Symmetric Key-Based Authentication with an Application to Wireless Sensor Networks

An Braeken

Abstract

This chapter starts with an introduction on different methods to provide key establishment and authentication using symmetric key-based mechanisms limited to hashing, xoring and encryption/decryption operations. Based on an idea coming from the context of multi-server authentication, and already applied in several IoT contexts, we present a new key management protocol for wireless sensor networks with hierarchical architecture, using solely symmetric key-based operations. The protocol establishes confidentiality, integrity, and authentication. It supports various communication scenarios, has limited storage requirements, is highly energy efficient by minimizing the number of communication phases and cryptographic operations, and avoids message fragmentation as much as possible. With the pre-installation of an individual secret key with the base station and some additional key material different for cluster head and cluster node, all possible keys in the network can be efficiently on-the-fly computed and updated. We discuss the differences with the well-known LEAP key management system for wireless sensor networks.

3.1 Introduction

When a common secret is shared between two entities, symmetric key schemes in authenticated encryption (AE) mode, like for instance AES-GCM or AES-CCM, can be applied. However, in order to construct such an authenticated key, called the session key, either public key-based or symmetric key-based mechanisms can be used. The advantage of symmetric key-based mechanisms is that they require much less computational efforts and smaller messages, compared to the public key-based variants. The main challenge is how to efficiently construct this key in a scalable way. For instance, it would be very impractical if the device needs to store secret information corresponding to each different device with which it wants to communicate. Typically, a trusted third party (TTP) is involved in these protocols in order to manage the process.

There are different architectures to be considered, either a one-to-one communication or one-to-many communication setting. In a one-to-one communication, two devices construct a common secret key, with the aid of the TTP. Each device only needs to share

IoT Security: Advances in Authentication, First Edition.
Edited by Madhusanka Liyanage, An Braeken, Pardeep Kumar, and Mika Ylianttila.
© 2020 John Wiley & Sons Ltd. Published 2020 by John Wiley & Sons Ltd.

a secret key with the TTP. The Needham-Schroeder protocol [1] solves this problem. However, this protocol was found to be vulnerable for a replay attack [2], and was finally fixed in order to lead to the symmetric variant of the Kerberos protocol [3].

Recently, many advances have been made in protocols describing one-to-many communication. In particular, the two-factor and three-factor based multi-server login protocols have evolved a lot since 2016, by including, in addition, anonymity and untraceability features [4]. These protocols can very easily be adopted in an IoT setting. For instance, in [5], we have shown how the technique can be used to provide different stakeholders (residents, recurring guests, or temporary guests) end-to-end secure access to the IoT devices in a smart home, managed by the home owner in an anonymous way. Here, the IoT devices take the role of gateway, receiving the requests from stakeholders who are in possession of a user device and thus enable two-factor or three-factor authentication. In [6], we also address the smart home scenario, but describe a key construction between IoT devices and gateway guaranteeing anonymity of the IoT device, without direct involvement of users. Consequently, the role of the IoT device in both [5] and [6] is different.

In this chapter, we want to demonstrate how the same idea can be used to construct a key management protocol for wireless sensor networks (WSNs), being an important part of IoT. The popularity of WSNs has grown tremendously during the last decade, due to the multitude of application areas. Security in these networks has also been extensively researched and standardized. However, these standardizations often do not include mechanisms for key management. Nevertheless, key management is essential in this whole procedure in order to send secure and authenticated messages. Minimal security features to be established are confidentiality, integrity, and authentication. Efficient mechanisms are required, due to the limited bandwidth, processing power, storage capacities, and battery age. Efficiency implies scalability and adjustability [7]. Scalability in the sense that the key management protocol is able to include additional nodes in a secure manner during the network's lifetime. Adjustability implies a proper mechanism to deal with network condition changes.

The key management protocol proposed in this chapter, called symmetric authenticated key agreement (SAK), is for sensor networks with a hierarchical architecture, consisting of base station (BS), cluster heads (CHs), and cluster nodes (CNs). Only symmetric key-based operations are used to establish confidentiality, integrity, and authentication. Based on a preshared individual key between each sensor node and the BS and some additional key material different for CN and CH, secret keys are derived between CN and CH, two CNs of the same cluster, a group key among all the CHs and the BS, a group key among all CNs belonging to a certain CH, and a group key among a selected group of CNs inside a cluster.

The mechanisms to derive these keys are constructed in such a way that the number of communication phases among the involved entities and message fragmentation is limited as much as possible. Moreover, the storage requirements for the key derivation in each entity is very small, related to other previous work. If a node is captured, then awaiting the update of the BS for each individual node in the cluster, a change mode is proposed to derive new values for the keys among all trusted nodes in the same cluster with and without the CH.

The remainder of the chapter is organized as follows. Section 3.2 provides an overview of the related work in key management protocols for WSNs. Section 3.3 explains the network architecture with some key definitions and assumptions used. In Section 3.4, we explain the different phases of our scheme in normal mode in detail. We show how to include authentication in Section 3.5. Section 3.6 discusses the system in change mode, offering an answer to situations where the system is attacked or where a node changes its place, is added, or removed from the network. Sections 3.7 and 3.8 respectively describe the security analysis and the corresponding efficiency. Finally, section 3.9 concludes the paper.

3.2 Related Work

Key management in wireless sensor networks is already extensively described for both distributed and hierarchical architectures of the nodes [8, 9]. We focus on the last one, since it was shown to be more energy-efficient and performant [10, 11]. In general, there are three main approaches for key management, using symmetric key cryptography, public key cryptography, and hybrid. Especially, in a hierarchical architecture, it can make sense to use a hybrid approach where the most computational heavy operations are performed by the CH. Often, authentication and integrity are obtained by digital signatures. However, as there is a huge performance difference between symmetric key and public key cryptography, it is still interesting to look at symmetric key-based solutions, due to the limited energy, communication bandwidth, and computational capacities of the sensor nodes.

There have been several proposals published for symmetric key-based hierarchical wireless sensor networks. Within the domain of symmetric-key based, authenticated key management protocols, the Localized Encryption and Authentication Protocol (LEAP) [12] is the most complete one. It describes procedures to derive keys for the most common communication scenarios, being between two CNs, a group key for all nodes in a cluster, and a network key among all nodes. A detailed efficiency comparison between LEAP and our protocol is given in Section 3.8.

References [13, 14] present randomized approaches for the key management, which have no guarantee of successful key establishment but limit the impact of a compromised node. Both protocols ([13, 14]) have huge storage requirements in each of the nodes in the network.

Partial solutions for authenticated symmetric key-based solutions can be found in [15–18]. In [15], a method using one-way functions is explained to derive an authenticated pairwise key between CN and CH. Group keys are not derived in this paper. In [16], each node receives, dependent of its place in the sensor topology, an evaluation of a point on a pre-defined bivariate polynomial of the BS. For the computation of the group key, Lagrange interpolation is used, but a lot of unicast communications are required to send the group key to the individual members. Moreover, no method is explained on how to use this group key in an authenticated way. Finally, the key management in [17, 18] is based on a generator matrix, predefined at the BS. The computation of the group key requires the involvement of the BS and cannot be performed by the CH alone. Again no mechanism is described to guarantee authenticated usage of the group key. In addition, the protocols from [15–18] are restricted to star networks, and thus no mechanism is described to compute a shared key between two CNs of the same cluster.

For our protocol, we use a trick where CNs and their corresponding CH share different content, that when combined uniquely determines the sender of a message. As explained before, this idea is inspired from a protocol for a biometric-based multi-server login [4].

3.3 System Model and Assumptions

3.3.1 Design Goals

The design goals of SAK are similar to LEAP [12] and should satisfy the following criteria.

– SAK should support authentication, integrity, and confidentiality for the most common communication scenarios. These include the communication between two CNs of the same cluster, between a CN and a CH, among nodes in the same cluster, among all the nodes in the network.
– If one or a limited number of sensor nodes are compromised, the entire network should still be able to-survive further with a limited number of adaptations.
– Due to the resource-constrained nature of the sensor nodes, the protocol should be designed as efficiently as possible. This aims a to minimize the communication phases and the number of expensive cryptographic operations as much as possible.
– As there is a relatively high packet loss in sensor networks, message fragmentation should be avoided as much as possible. Using the adaptation layer 6LoWPAN for the compression of IP headers, a payload can contain approximately 70 bytes [19].

3.3.2 Setting

We consider a sensor network with a three-layered architecture, being the BS, the *CH* nodes C_j, and the CNs L_i^j belonging to a particular CH C_j. Let us assume that each sensor node (CHs and CNs) participating in the network, has a pre-installed secret shared key with the BS, an identifier, some auxiliary key material derived from the identifier and different in structure for CHs and CNs, and a common network key k_n. This information is stored in each sensor node, before the node is put into the field. The position of the nodes, i.e. combination of CN and CH, does not need to be known in advance, only the fact that a node will behave as CH or CN. Since CHs need to perform a considerably higher amount of processing, more powerful nodes are used for this task and distinction is easy.

Once the nodes are put in the field, the CH broadcasts its identity ID_j and a timestamp t_s. Next, the key management protocol will describe how each node L_i^j will be able to construct three on-the-fly-different types of keys.

– Group node key kn_j. This key is shared among all nodes and CH in the cluster with CH, C_j.
– Multicast key km_j: This key is shared among a selected group of nodes L_i^j by the CH C_j.
– Individual cluster key k_{ij}. This key is uniquely shared between L_i^j and the CH C_j.
– Pairwise key k_{i1i2}^j. This key is shared between the nodes L_{i1}^j and L_{i2}^j in the same cluster.

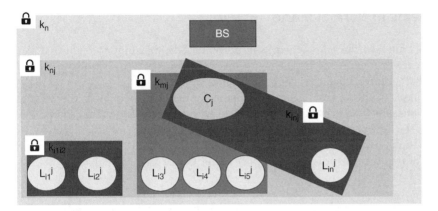

Figure 3.1 This figure shows an example of the different possible keys in the scheme with base station BS, cluster head C_j and corresponding cluster nodes L_i^j. The network key is denoted by k_n, the group node key by k_{nj}, the multicast key by k_{mj}, the pairwise key by $k_{i_1j_1}$ and the individual cluster key by k_{inj}.

When the BS sends a timestamp t_b to the CHs in the network, the CHs are also able to derive a group key shared among the CHs and the BS, called the group cluster key kg. In addition, the CH can also derive on-the-fly the group node key kn_j, the multicast key km_j, and the individual cluster key k_{ij}. Moreover, we also explain how we include authentication into the communication. Fig. 3.1 illustrates the involved entities for examples of different types of keys.

The communication capacity between the different entities in the nodes differs due to the difference in resources. The BS can in any case unicast or broadcast information directly to the CHs or the CNs. A CH communicates to the BS in a multi-hop way, while it is able to unicast or broadcast directly to all nodes in its cluster. A sensor node communicates to its CH in single hop or multiple-hop, dependant on its distance.

3.3.3 Notations

We represent a hash function by H. The encryption operation of message m under a key K to obtain the ciphertext c is denoted as $c = E_K(m)$, and the corresponding decryption operation as $m = D_K(c)$. Furthermore, the concatenation of values m_1 and m_2 is denoted by $m_1 \| m_2$ and the xor operation by $m_1 \oplus m_2$.

3.3.4 Attack Model

The attackers may come from inside or outside the network. They are able to eavesdrop on the traffic, inject new messages, replay and change old messages, or spoof other identities. They may try to take the role of CN or CH.

Attackers can also compromise and capture a node, such that all security material is revealed. Note that we do not discuss the mechanisms to detect malbehavior of a node, e.g. by storing trust tables in each node. Therefore, we refer to the literature on intrusion detection mechanisms [20] to detect abnormal behaviour of a compromised node. On the other hand, we describe how to act when such a situation occurs, corresponding to the description of the scheme in change mode, Section 3.6.

3.4 Scheme in Normal Mode

We first describe how the scheme works in normal mode, meaning no real threats or indications exist to assume a node is behaving maliciously. In the next section, we explain how to deal otherwise, called the scheme in change mode.

Several phases in the key management scheme are distinguished. First, there is the installation phase, where every node is installed with a unique individual shared key with the BS, the network key, and some auxiliary key material different for CN and CH. This information is independent of the final location in the network. It only requires the knowledge of role the sensor is going to play, either CN or CH. Note that the BS acts as TTP.

The following phases correspond with the construction of the different types of keys, using this key material. We distinguish the derivation of the group node key, the individual cluster key, the pairwise key, the multicast key, and the group cluster key. In Section 3.5, we also explain how authentication is obtained in the scheme. These different phases are now described into more detail.

3.4.1 Installation Phase

The installation differs for CH and CN. We will discuss both types. Let K_n and K_m be two master secret keys of the BS.

3.4.1.1 Installation of CH

The following material is installed at each CH, called C_j.

- The network key k_n.
- The identifier ID_j.
- An individual shared key with the BS, being k_j. In order to decrease the storage requirements of the BS, these keys can be derived from a master key K_m and the corresponding identity of the node, i.e. $k_j = H(K_m \| ID_j)$.
- The auxiliary key material consists of

$$H_1 = H(ID_j \| H(K_n))$$
$$H_2 = H(K_m) \tag{3.1}$$

3.4.1.2 Installation of CN

The following material is installed at each CN.

- The network key k_n.
- The individual shared key with the BS, $k_i^j = H(K_m \| ID_i^j)$.
- Define

$$A_i = H(ID_i^j \| H(K_m \| K_m)). \tag{3.2}$$

Then, the following auxiliary key material, is also stored at each CN.

$$B_i^j = H(K_m) \oplus A_i^j$$
$$C_i^j = k_i^j \oplus H(A_i^j)$$
$$D_i^j = H(K_n) \oplus k_i^j \tag{3.3}$$

Note that any direct reference to the original identity ID_i^j is removed from the node's memory. From now on, the node will communicate under the identity B_i^j. Consequently, the supplier is not able to follow its nodes in the network. In order to increase the difficulty of a node's capture attack, we will show later that it is sufficient only to protect k_i^j and k_n by means of software obfuscating techniques, for example [22, 23]. We consider these measures outside the scope of this paper.

Finally, we also want to draw attention to the division of the key material. Both CH and CN each possess one secret master key of the BS and a variable which is derived from the other secret master key.

3.4.2 Group Node Key

The CH first chooses a random value t_s and computes the group key

$$kn_j = H(ID_j \| H(K_n)) \oplus t_s = H_1 \oplus t_s \qquad (3.4)$$

The CH C_j broadcasts $E_{k_n}(ID_j \| t_s \| H(kn_j))$.

All the CNs in the neighbourhood of C_j can now easily derive this group key. First the message is decrypted using the network key. Then, $H(K_n)$ is extracted from D_i^j and used to construct the group key, kn_j (see Equation 3.4). The result is compared with the last part of the decrypted message in order to verify the integrity.

This process is typically initiated by the CH, but can in theory also be activated by another CN. We show in Section 3.5 how authentication of the CH can be included in this process and how the CH can use this key in an authenticated way.

3.4.3 Individual Cluster Key

This phase leads to the derivation of the individual shared key k_{ij} between L_i^j and C_j and requires only one communication phase between L_i^j and C_j. We may assume here that L_i^j is aware of the identity ID_j of C_j, since it has received the broadcast message $E_{k_n}(ID_j \| t_s)$ from the previous step. We now describe in more detail the different steps to be performed by the participants. There are two possibilities, either initiated by the CN or the CH.

3.4.3.1 CN to CH

The following steps are executed by L_i^j.

- First $H(K_n)$ is derived from D_i^j.
- In addition, L_i^j also computes from its stored values

$$H(A_i^j) = C_i^j \oplus k_i^j \qquad (3.5)$$

- Next, L_i^j chooses a random value k_{ij}, which will serve as an individual cluster key and computes:

$$M_1 = H(ID_j \| H(K_n)) \oplus H(H(A_i^j) \| k_{ij})$$
$$M_2 = k_{ij} \oplus H(A_i^j)$$

- The following message is then sent to C_j:

$$B_i^j\|M_1\|M_2 \tag{3.6}$$

The C_j receives this message and performs the following computations:

$$A_i^j = B_i^j \oplus H(K_m)$$
$$k_{ij} = M_2 \oplus H(A_i^j)$$
$$\text{Check:}\quad H(H(A_i^j)\|k_{ij}) = M_1 \oplus H(ID_j\|H(K_n))$$

This check corresponds with an identity check since only the node with the correct pre-installed values by the BS is able to derive the identity related information in the form of $H(A_i^j)$. In addition, the integrity on the derived key is also verified. Only if the right value for k_{ij} is found, this equality holds. If the check is positive, then C_j can conclude that the node with identifier B_i^j belongs to the network and its corresponding shared key equals to k_{ij}.

3.4.3.2 CH to CN

We now assume that the CH is aware of the presence of the CN B_i^j in its cluster. For instance, the CN could have already sent some information to the CH before, as shown in the previous step. The CH then performs the following steps to derive the individual cluster key.

- The CH first extracts A_i^j from the sensor node's public parameter B_i^j, by $A_i^j = B_i^j \oplus H(K_m)$.
- In order to add randomness to it, a random parameter r is chosen and the individual cluster key is defined by

$$k_{ij} = H(H(A_i^j)\|r\|H(ID_j\|H(K_n))). \tag{3.7}$$

Note that the last part is required to make sure only CH C_j is able to establish this message.
- The CH sends to the CN L_i^j

$$r\|ID_j\|H(r\|k_{ij}) \tag{3.8}$$

The CN L_i^j receives this message and computes $H(A_i^j)$ (Equation 3.5) and the value $H(ID_j\|H(K_n))$. Consequently, the shared key k_{ij} from Equation 3.7 can be determined. Finally, in order to check the integrity of the key, the value $H(r\|k_{ij})$ is computed and checked with the last part of the received message. If the integrity control is positive, then L_i^j can conclude that the CH C_j wants to share the key k_{ij}.

3.4.4 Pairwise Key Derivation

Here, node L_{i1}^j establishes a key with L_{i2}^j. There is a straightforward way to realize this if you let the CH generate the key and securely send it to the involved nodes in its cluster using k_{i1j} and k_{i2j}. However, we propose a protocol here that allows the construction of a pairwise key, only shared between the sensor nodes and not with the CH.

The idea is based on the fact that the nodes of the cluster and not the CH share the secret value $H(K_n)$. On the other hand, the CH shares with each individual node the value $H(A_i^j)$. Consequently, the node L_{i1}^j needs to request unique information from the CH (based on A_{i2}^j) on node L_{i2}^j, corresponding with phase 1. If L_{i1}^j is correctly authenticated, this info is sent from CH to L_{i1}^j in phase 2, which puts L_{i1}^j in the knowledge of uniqe information only shared by L_{i2}^j. Finally in phase 3, the key can be constructed using this information and the value $H(K_n)$ by L_{i1}^j in order to uniquely share a key with L_{i2}^j without involvement of the CH. Let us now describe these three phases into more detail.

– Phase 1: $L_{i1}^j \rightarrow C_j$ Here, L_{i1}^j requests some unique information, able to distinguish L_{i2}^j from the cluster group. Therefore, it first generates a random value r and sends a key request KR. The first part of the message is to derive the individual cluster key k_{i1j} as described in Equation 3.6.

$$B_{i1}\|M_1\|M_2\|E_{k_{i1j}}(KR\|B_{i2}\|r)$$

– Phase 2: $C_j \rightarrow L_{i1}^j$ C_j first derives the key k_{i1j} as described in the previous paragraph. Then, if C_j can correctly decrypt the last part of the message, the following computations are performed:
  • A new value for $k_{i1j} = H(H(A_{i1})\|r\|H(ID_j\|H(K_n)))$.
  • A value to distinguish B_{i2}, asked by B_{i1}: $H(H(A_{i2})\|B_{i1}\|r)$. Only the CH is able to compute this value since the other CNs have no knowledge on $H(K_m)$.
  The value $r\|ID_j\|E_{k_{i1j}}(r\|H(H(A_{i2})\|B_{i1}\|r))$ is sent to L_{i1}^j.
– $L_{i1}^j \rightarrow L_{i2}^j$ First, the key k_{i1j} should be derived, as explained in the previous paragraph, in order to decrypt the message. Note that instead of sending $H(r\|k_{ij})$ as in Equation 3.8, only a ciphertext is sent. If after decryption, the first part of the plain text contains the random value r, the integrity of the key is verified.
  Next, a random value k_{i1i2} is chosen, which will serve as pairwise key. An additional random value r_2 is chosen. Now, the following computations are performed.

$$M_0 = r\|E_{H(H(A_{i2})\|B_{i1}\|r)}(r_2\|r)$$
$$M_2 = k_{i1i2} \oplus H(B_{i2}\|r_2) \oplus H(K_n)$$
$$V_1 = H(k_{i1i2} \oplus B_{i1})$$

The values $B_{i1}\|M_0\|M_2\|V_1$ are sent to L_{i2}^j.
– Phase 3: Key derivation by L_{i2}^j In the last step, the receiver node L_{i2}^j, starts with decrypting the last part of M_0, by computing the key $H(H(A_{i2})\|B_{i1}\|r)$. If the last part of the decrypted message is equal to the random value r, the node L_{i2}^j concludes that the key request is coming from a validated node of the cluster since only the CH is able to compute the value of $H(A_{i2})$. The originator of the message is verified since it is included in the key computation. After successful decryption, the node L_{i2}^j can continue its computations with the random value r_2 from M_0. From M_2, the key k_{i1i2} can be easily computed. Finally, the integrity of the key is checked with comparing the outcome of the message $H(k_{i1i2} \oplus B_i^j)$ and the received V_1.

3.4.5 Multicast Key

In case of doubt on the trust of the CNs in its network, it might be interesting, for example, for the CH to select the group of receivers. Another possibility might be that the message is only meant for nodes with a certain functionality or a certain property. We describe the operations performed by the CH and the operations by the CNs.

3.4.5.1 Initiation by CH

Suppose there are n nodes in the cluster, with identities $B_{i1}, \ldots, B_{in}$. The CH, C_j, chooses a random key km_j. The following steps are now performed:

- First, all the shared keys $k_{ij} = H(H(A_i^j)\|H(km_j)\|H(ID_j\|H(K_n)))$ for $i \in \{i1, \ldots, in\}$ of the nodes in the cluster are computed (cf. Equation 3.7 with $r = H(km_j)$)
- Next, we determine the polynomial of degree n with Lagrange interpolation through the n points $(x_i^j, y_i^j) = (H(k_{ij}\|0), H(k_{ij}\|1))$ with $i \in \{i1, \ldots, in\}$, together with the point $(0, km_j)$. We derive n other points $P_{1j}, \ldots, P_{nj}$ on this polynomial.
- The following message is then broadcast to the nodes L_i^j in the cluster.

$$P_{1j}\| \ldots \|P_{nj}\|H(km_j) \tag{3.9}$$

3.4.5.2 Derivation by CNs

Each L_i^j from the cluster with CH C_j needs to perform the following operations to derive the multicast key.

- First, the key $k_{ij} = H(H(A_i^j)\|H(km_j)\|H(ID_j\|H(K_n)))$ is computed based on the value $H(km_j)$.
- The group cluster key km_j can now be computed by finding the intersection with the Y-axis of the polynomial with degree n passing through the transmitted set of points, together with its own derived point $(x_i^j, y_i^j) = (H(k_{ij}\|0), H(k_{ij}\|1))$.
- Finally, if the hash of this value $H(km_j)$ corresponds with the last part of the message 3.9, the node L_i^j has confirmation on the derived key. If not, a message needs to be sent to the CH for verification.

3.4.6 Group Cluster Key

The BS is able to establish the group cluster key kg by just broadcasting a key request for a group cluster key, together with a random value t_b to all the CNs. The corresponding key is then

$$kg = H_2 \oplus t_b = H(K_m) \oplus t_b$$

We show how authentication can be guaranteed in here and how to use this key in an authenticated way in Section 3.5.

3.5 Authentication

Since all entities in the network know the network key and all CNs in the same cluster know the group node key, it is a challenge to obtain authentication in these communications. A classical method, as proposed in LEAP, is the use of one-way key chains [21] in each entity. This method has the disadvantage that it is not resistant against insider attacks and possesses large storage requirements. Here, we propose a different method and make distinction between the authenticated usage of kn_j and k_n by a CN, a CH, and the BS.

3.5.1 Authentication by CN

Due to the particular construction of the security material in each CN, every CN can authenticate itself with the CH (and the BS) by computing $H(H(A_i^j)\|H(m))$, with m the message to be sent including a counter/random/timestamp in order to ensure randomness. The message, $B_i^j\|H(H(A_i^j)\|H(m))$ will unambiguously authenticate the CN with the CH and/or BS. Only in case of a problem, the situation is communicated in the group.

3.5.2 Authenticated Broadcast by the CH

Suppose the CH C_j wants to send the message m, authenticated and encrypted, in its cluster, where the list of participants $(B_{i1}^j, \ldots, B_{in}^j)$ is known. These participants include the nodes that have sent messages to the CH during the last timeframe. There are two straightforward ways to deal with this.

- The CH can use the same method as described in the derivation of the multicast key in the previous section.
- The CH can use the group key kn_j and again exploit the particular construction of the security materials of the CNs and CH (cf. equations 3.1 and 3.3), by adding to the transmitted message m, the following values:

$$H(H(A_{i1})\|H(m)\|H_1)\| \ldots \|H(H(A_{in})\|H(m)\|H_1).$$

Note that the message m should include a counter to avoid replay. Each part in this message corresponds with a different CN $L_{i1}^j, \ldots, L_{in}^j$ in the cluster of C_j.

It is clear that the second method is more efficient. However, both methods have the disadvantage that the message linearly increases with the number of participating CNs. There are two possible ways to overcome this issue:

- Restricting the number of the CNs in each cluster. For instance, if we assume the hash values of size 160-bit (20 bytes), a message of size 10 bytes, and a payload of 70 bytes as with 6LoWPAN, there can be at most 3 CNs in the cluster in order to not fragment the message ($3 \times 20 + 10 \leq 70$).
- Distribution of the authentication check. The CH divides the group of CNs into groups $G_1, \ldots, G_g$ of size 3. To each message, the group number is added in

consecutive order. If this order is not respected or the authentication check of a node L_i^j belonging to that group mentioned in the message is unsuccessful, doubts are sent to the CH and BS in unicast mode. Consequently, each message is checked by a part of the nodes belonging to the cluster and it is assumed that no more than 3 nodes are malicious.

3.5.3 Authenticated Broadcast by the BS

This authentication should pass the two layers in hierarchical order. First, the BS should be authenticated by the CHs and next the CHs should be authenticated by the CNs in its cluster. The second authentication is discussed above and thus we focus now on the first one. For the authentication of the BS to the CHs, the same technique as described above can be used, adding to the message the values

$$H(H(ID_{j1}\|H(K_m))\|H(m))\| \dots$$
$$\|H(H(ID_{jn}\|H(K_m))\|H(m)),$$

where each part corresponds with each CH $C_{j1}, \dots, C_{jn}$ in the network. However, as the CHs are less constrained than the CNs, packet fragmentation should not form too large an issue.

3.6 Scheme in Change Mode

We distinguish three different situations. In the first and second situation, we assume that a CN and CH respectively are captured or compromised. The third situation deals with an honest node that changes from cluster, is removed or added to the network.

3.6.1 Capture of CN

First, it should be mentioned that the derivation by a CN of an individual cluster key, pairwise key, and group node key is only possible if the key k_i^j is known. Consequently, the focus of protection (e.g. using techniques of code obfuscation for example) should be put on this value.

Let us now suppose that the CN is captured or compromised. This has the following consequences.

- No further pairwise keys and individual cluster keys are allowed to be derived with the captured or compromised node.
- The group node key and the network key are known. Consequently, no information in the global network can be secured with the network key and in the local network with the group node key.
- Also $H(K_n)$ can be derived. This value is used to establish group node keys in all other clusters. Consequently, from the moment $H(K_n)$ is leaked, a node that wants to securely share the group node key with a new cluster needs to use its individual cluster key. Note that $H(K_n)$ is also used to derive the pairwise key between two CNs without involvement of the CH.

To summarize, inside the infected cluster only existing pairwise keys and individual cluster keys, different from the captured node, remain secure. Using the multicast key, the infected node can be excluded. In the other not infected clusters, only existing group node keys, pairwise keys and individual cluster keys remain secure. A group key cannot be updated using a broadcast message of the CH.

Consequently, the first step to execute by the BS is to share the information on the compromised node by broadcast communication in an authenticated way. Next, in order to make the system fully functional again, the BS should update the value K_n and k_n. In practise, it means that first all CNs should receive an updated version of both k_n and $H(K_n)$, allowing them to update the values C_i^j and D_i^j. Also the CHs should receive an updated value of k_n and $H_1 = H(ID_j \| H(K_n))$. Since the shared key between BS and each CN and CH is not leaked, the updated info can be communicated using unicast communication to each entity. Fortunately, this situation will be very rare.

3.6.2 Capture of CH

We can assume that this situation is very unlikely as the CH is supposed to be a more powerful device, possessing more possibilities to guarantee secure storage of the secret parameters. However, if the CH is captured, $H(K_m)$ is leaked and thus an attacker can create new valid combinations for A_i^j and B_i^j, thus creating fake nodes which cannot be verified by legitimate CHs. However, they can be detected by the BS in case the BS is storing the registered IDs of the CNs since the newly maliciously constructed A_i parameters do not involve the registered IDs. Note that the other CHs cannot create valid individual cluster key or pairwise key requests, since they do not know $H(K_n)$, required to build the message M_1 from Equation 3.6. Moreover, the other nodes in the network can still continue to communicate securely by using the key $H(kn_j \| H(K_n))$.

Consequently, the variable K_m requires an update. The BS sends a message to all the CHs by unicast communication using the current shared key k_j between CH and BS, containing an updated version of $H(K_m)$ and $k_j = H(K_m \| ID_j)$. Also for each node, the new versions of $A_i^j, B_i^j, C_i^j, D_i^j, k_i^j$ should be calculated by the BS and the update values for $B_i^j, C_i^j, D_i^j, k_i^j$ should be sent to the CN using unicast communication with the old k_i^j.

Finally, it must be said that when a malicious CN and CH collaborate, the two main important secrets $H(K_n)$ and $H(K_m)$ are revealed and thus require a complete update. However, the knowledge of these both secrets, still do not reveal the individual shared keys k_i^j for CNs and k_j for CHs with the BS.

3.6.3 Changes for Honest Nodes

Here, we discuss the situations where an honest node replaces one cluster with another and where a node is added to or removed from the network.

3.6.3.1 Key Update for Honest Node Replacement

An honest sensor node L_i^j leaves the cluster C_j in order to join another cluster C_j'. The node broadcasts a HELLO message for a group node key request. Based on that, the procedure to determine the group node key starts, as explained in the previous section. The identity ID_j and corresponding random value t_s of the previous C_j are now replaced with the newly received values of C'_j. Once the node receives this new information, it can participate to all other phases in the cluster with CH, C'_j, in order to derive the other different keys.

3.6.3.2   Node Removal and Addition

A node can be securely removed from the network if the secret keys k^j_i and k_n are erased from the memory. As mentioned before, the other stored values are worthless without the value k^j_i. No other changes need to be done at the other nodes in the cluster or the CH.

In order to include a node into the network, it should simply install the same material as discussed in the installation phase of the CN (cf. (3.3)). Next, it can join any CH in its neighbourhood by broadcasting a HELLO message for a group node key request as discussed in the situation of node replacement.

3.7   Security Analysis

First of all, the two security related design goals, as mentioned in Section 3.3, are realized. Firstly, we have explained above how confidentiality, integrity, and also authentication is supported for communications between two CNs of the same cluster, between a CN and a CH, among nodes in the same cluster, among all the nodes in the network. Secondly, the scheme in change mode explains the procedures to be followed when a limited number of nodes in the network are compromised and thus shows the survivability feature of our system. Let us now discuss the resistance of the system against the most important attacks.

3.7.1   Resistance Against Impersonation Attack

In this type of attack, the adversary (inside or outside the network) tries to take over the role of CN or CH.

- CH role: It would mean that the adversary is capable of generating a group node key, an individual cluster key, or information for a pairwise key establishment. Although, since each of these operations rely on the knowledge of both $H(ID_j\|H(K_n))$ and $H(K_m)$, only the legitimate CH is able to do this.
  Even inside, CNs are not able to derive $H(K_m)$, since A^j_i is not installed at node L^j_i. Also CHs inside the network are not aware of $H(ID_j\|H(K_n))$.
- CN role: A node, not installed with valid information from the BS, is not able to send legitimate info to the CH since it cannot derive $H(A^j_i)$ and $H(ID_j\|H(K_n))$, required to identity itself.
  Again, an inside CN knows $H(ID_j\|H(K_n))$, but not $H(A^j_i)$ from that node and thus cannot take over the role of that sensor node identified with B^j_i. Also an inside CH cannot take over the role of a sensor node for a communication to the BS, since it does not know and is not able to derive k^j_i from the messages send in the network.

However, we should note that in the very rare situation where a malicious CN and malicious CH collaborate, this attack would become possible.

3.7.2 Resistance Against Node Capture

If a node is captured and the stored keys k_i^j, k_n (protected by code obfuscation or other techniques) cannot be retrieved, the whole system is still secure. In case the key k_i^j can be retrieved somehow, the procedure from the scheme in change mode, as explained in the previous section, should be followed.

If the CH is captured, although even more difficult since it is considered to have more resources and thus more capabilities to protect itself with, for instance, tamper proof hardware, the CNs inside the cluster can still communicate securely using the key $H(kn_j \| H(K_n))$, since the CH is not aware of $H(K_n)$. The procedure for key material update, as explained in the scheme in change mode, should be also followed.

3.7.3 Resistance Against Replay Attacks

Since the computation of an individual cluster key or the request for info at the CH in a pairwise key phase is performed in a one-phase communication using random values, a replay attack will be noticed immediately. Note that the random values could be also replaced by counters, timestamps, or hashes of the transmitted message if these include a counter.

3.8 Efficiency

As explained in Section 3.2, LEAP can be considered as the only complete key management system based on symmetric key operations. Let us compare the efficiency of the proposed key management scheme with LEAP for the features: number of communication phases to install the different keys, storage requirements, and packet fragmentation.

3.8.1 Number of Communication Phases

Table 3.1 enumerates the different keys used in the schemes and discusses the most important characteristics of the establishment process, in particular, the number of communication phases and the timing for establishment. As can be concluded from Table 3.1, we have one additional key, the multicast key km_j, compared with LEAP. For most of the keys, the derivation in LEAP requires more communication phases. Moreover, SAK is also more flexible since the keys are ad hoc constructed and not limited to the construction at installation time like LEAP. Only the pairwise keys among nodes in the same cluster require one additional communication phase. This follows from the fact that our protocol has the additional feature that the CH is not aware of the established key. In case the CH would generate the key and forward it to both nodes, the key could also be constructed in two phases.

Note that the impact of node capture and the corresponding update procedure is more or less similar. This follows on from the fact that all communication in SAK is also authenticated, which makes it possible to localize the source of the problem as in LEAP.

Table 3.1 Comparison of number of communication phases between SAK and LEAP

Key	Type	SAK	LEAP
k_n	Network key	Installed	Installed
k_i^j	Key with BS	Installed	Installed
km_j	Multicast key	1 phase and ad hoc	Nonexistent
kn_j	Group key	1 phase and ad hoc	Constructed by CH and in unicast comm. sent to other nodes of cluster
k_{ij}	Key CN-CH	1 phase and ad hoc	2 phases, only in limited time frame at installation time
k_{i1i2}	Pairwise key	3 phases and ad hoc at installation time	2 phases, only in limited time frame

3.8.2 Storage Requirements

In SAK, each CN has to store three auxiliary parameters (B_i^j, C_i^j, D_i^j) and 2 critical parameters (k_i^j, k_n). By critical parameters, we mean parameters to be stored by preference in tamper proof hardware or protected by means of code obfuscation techniques. Initial values for them are pre-installed. This key material allows the node to efficiently construct any type of keys, on-the-fly and ad hoc.

In LEAP, the individual and network key are also pre-installed. The other keys need to be computed in a strict timeframe and are also stored at the node, preferably in tamper-proof hardware. Moreover, in order to guarantee authentication, every node needs to store a one-way hash chain with length corresponding to the number of messages to be send in authenticated way.

3.8.3 Packet Fragmentation

When the 6LoWPAN protocol is used, packets with a payload length of 70 bytes or less are never fragmented. However, when a packet is larger, fragmentation occurs. The number of required fragments N_f is given by

$$N_f = 1 + (s - 64)/72, \tag{3.10}$$

with s the size of the payload expressed in bytes. This follows on from the fact that the first fragment is limited to 64 bytes and all subsequent fragments are a maximum of 72 bytes.

For the construction of group key kn_j and the individual cluster key k_{ij}, there is no fragmentation. For the pairwise individual key, the packets sent by the CNs are fragmented in 2, the one from the CH to the CN is not fragmented. Finally, the packets transmitted in the multicast key and for authenticating the group key, fragmentation is also required.

3.9 Conclusions

This chapter describes the SAK protocol, which is a complete protocol to establish the key management for all possible types of communications in an hierarchical three-layered wireless sensor network. Confidentiality, integrity, and authentication are obtained using symmetric key cryptography for all entities in the network.

We compare our protocol with the LEAP protocol and show that the number of communication phases for most of the keys in SAK is lower, the generation of the keys is more flexible and ad hoc, the storage requirements are less severe and smaller, and the packet fragmentation is also reasonable.

Acknowledgement

This work is supported by the TETRA grant of the Flanders agency for Innovation by Science and Technology, and the Short Term Scientific Mission performed under COST Action IC1303.

References

1 Needham, R. and Schroeder, M. (1978). Using encryption for authentication in large networks of computers. *Communications of the ACM* 21 (12): 993–999.

2 Lowe, G. (1995). An attack on the Needham-Schroeder public key authentication protocol. *Information Processing Letters* 56 (3): 131–136.

3 Tbatou, Z., Asimi A., Asimi, Y. et al. (2015). V5: Vulnerabilities and perspectives. *Third World Conference on Complex Systems (WCCS)*, Marrakech, Morocco (23–25 November 2015). IEEE.

4 Braeken, A. (2015). Efficient Anonym Smart Card Based Authentication Scheme for Multi-Server Architecture. *International Journal of Smart Home* 9 (9): 9.

5 Braeken, A., Porambage, P., Stojmenovic, M. et al. (2016). eDAAAS: Efficient distributed anonymous authentication and access in smart homes. *International Journal of Distributed Sensor Networks* 12 (12).

6 Kumar, P., Braeken, A., Gurtov, A. et al. (2017). Anonymous secure framework in connected smart home environments. *IEEE Transactions on Information Forensics and Security* 12 (4): 968–979.

7 Xiao, Y., Rayi, V.K., Sun, B. et al. (2007). A survey of key management schemes in wireless sensor networks, Science direct. *Computer Communications* 30 (11–12): 2314–2341.

8 Bala, S., Sharma G. and Verna, A.K. (2013). Classification of Symmetric Key Management Schemes for Wireless Sensor Networks. *International Journal of Security and Its Applications* 7 (2): 117–138.

9 Chen, C.-Y. and Chao, H.-C. (2014). A Survey of Key Distribution in Wireless Sensor Networks. *Security and Communication Networks* 7 (12): 2495–2508.

10 Das, S.R., Perkins, C.E. and Royer, E.M. (2000). Performance comparison of two on-demand routing protocols for ad hoc networks. *Proceedings of the 19th Annual Joint Conference of the IEEE Computer and Communications Societies (INFOCOM 2000)* Tel Aviv, Israel (26–30 March 2000) IEEE.

11 Gupta, P. and Kumar, P.R. (2000). The capacity of wireless networks. *IEEE Transactions on Information Theory* 46 (2): 388–404.

12 Zhu, S., Setia, S. and Jajodia, S. (2003). LEAP: Efficient Security Mechanisms for Large-Scale Distributed Sensor Networks. *Proceedings of the Tenth ACM conference on Computer and Communications Security*, Washington, DC, USA (25–29 October 2004). ACM.

13 Durresi, A., Bulusu, V., Paruchuri, V. et al. (2006). WSN09-4: key distribution in mobile heterogeneous sensor networks. *Proceedings of the IEEE Global Telecommunications Conference (GLOBECOM 2006)*, San Francisco, USA (27 November–1 December 2006). IEEE.

14 Hussain, S., Kausar, F. and Masood, A. (2007). An efficient key distribution scheme for heterogeneous sensor networks. *Proceedings of the International Wireless Communications and Mobile Computing Conference (IWCMC 2007)*, Hawaii, USA (12–16 August 2007). ACM.

15 Ou, G., Huang, J. and Li, J. (2011). *A Key-Chain Based Key Management Scheme for Heterogeneous Sensor Network*, pp. 358–361. https://www.researchgate.net/publication/224212208_A_key-chain_based_key_management_scheme_for_heterogeneous_sensor_network/citation/download (accessed 25 June 2019).

16 Zhang, Y., Shen, Y. and Lee, S.K. (2010). A Cluster-Based Group Key Management Scheme for Wireless Sensor Networks. *12th International Asia-Pacific Web Conference*, Busan, Korea (6–8 April 2010).

17 Li, L. and Wang, X. (2010). *A high security dynamic secret key management scheme for Wireless Sensor Networks. Third International Symposium on Intelligent Information Technology and Security Informatics*, Jian, China (2–4 April 2010). IEEE.

18 Shnaikat, K.N. and Qudah, A.A.A. (2014). Key management techniques in wireless sensor networks. *International Journal of Network Security and Its Applications (IJNSA)* 6 (6),6.

19 Smeets, R., Aerts, K., Mentens, N. et al. (2014). A cryptographic key management architecture for dynamic 6LowPan networks. *Proceedings of the 9th ICAI*, Eger, Hungary (1 February 2014).

20 Butun, I., Morgera, S.D., and Sankar, R. (2014). A Survey of Intrusion Detection Systems in Wireless Sensor Networks. *IEEE Communications Surveys and Tutorials* 16 (1): 266–282.

21 Lamport, L. (1981). Password authentication with insecure communication. *Communications on the ACM* 24 (11): 770–772.

22 Alarifi, A. and Du, W. (2006). Diversify sensor nodes to improve resilience against node compromise. *Proceedings of the fourth ACM workshop on Security of ad hoc and sensor networks*, Alexandria, USA (30 October 2006). ACM.

23 Collberg, C., Thomborson, C. and Low, D. (1997). A taxonomy of obfuscating transformations. Computer Science Technical Reports 148.

4

Public Key Based Protocols – EC Crypto

Pawani Porambage, An Braeken, and Corinna Schmitt

Abstract

Elliptic curve cryptography (ECC) is extensively applied in various security protocols for authentication and key management. ECC is a public key or asymmetric key cryptographic approach which is based on elliptic curve theory. ECC was introduced to minimize computational costs while providing equal and faster layers of security than other familiar operations (such as modular exponentiation) and with smaller keys. The technique of ECC has numerous applications in authentication protocols concerning RFIDs, digital signatures, wireless sensor networks, smart cards, and other authentication techniques. In this chapter, we describe the utilization of ECC for designing security protocols in terms of authentication, key establishment, signcryption, and secure group communication.

4.1 Introduction to ECC

Information security in any system is determined based on the basic CIA principles, which include confidentiality, integrity, and availability. In order to achieve these properties, a network needs three key functionalities such as data encryption, user authentication, and secure channel establishment [16]. These functionalities are interrelated and they use security keys with different perspectives. Typically, in encryption and decryption functions, the cryptographic operations are performed in two ways. In the first category, which is the symmetric key cryptography, the sender and the receiver share a common secret key (e.g. DES, AES). The second type is called as the asymmetric or Public Key Cryptography (PKC) where each user has a pair of keys (e.g. ECC, RSA). This pair consists of a public key, which can be widely disseminated and known by others, and a private key, which is known only by the owner. Both types can be used for user and device authentication. Moreover, secure channels are obtained based on successful user (or device) authentication, which is followed by establishing secret keys between the communication entities.

Unlike symmetric key algorithms, PKC does not need a secure channel for initial key exchange between two parties. However, since the security strength of PKC systems are mostly reliant on the complexity of mathematical problems, they are significantly more resource consuming than the symmetric key algorithms. Numerous PKC solutions are

IoT Security: Advances in Authentication, First Edition.
Edited by Madhusanka Liyanage, An Braeken, Pardeep Kumar, and Mika Ylianttila.

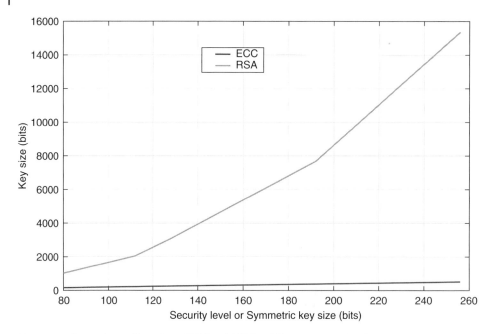

Figure 4.1 Comparison of key sizes of ECC and RSA for different security levels [11].

involved in key management schemes, applications, Internet standards and protocols. Among them, Elliptic Curve Cryptography (ECC) is an efficient PKC scheme with the most suitable adaptations for low-performing resource-constrained networking devices [6]. Compared to other expensive PKC schemes, like the well-known Rivest Shamir Adleman (RSA) public-key algorithm, ECC has faster computational time, smaller keys, and uses less memory and bandwidth. This is explained in the graph in Figure 4.1, where ECC can obtain the similar security level as RSA using much smaller keys.

4.1.1 Notations

While designing ECC algorithms, the Elliptic Curves (ECs) are defined over a finite field by an equation using two variables with coefficients, which are the elements of the finite field [11]. Consequently, all the variables, coefficients and curve points fall below the same finite abelian group, G. The resultant points of the curve operations are also restricted in the same abelian group. A special point 0, known as the zero element or point of infinity, is considered the identity element of the group. ECC is formulated with EC point addition, point scalar multiplication and, additive and multiplicative inverses on ECs over prime integer fields or binary polynomial fields. Modulo arithmetic is the foundation for all the EC point operations. The implementation of ECC on WSNs is performed over prime integer fields, since binary polynomial field operations are too costly for low-power sensors.

We consider the ECs defined over prime fields Z_p, where p is a large prime number. The variables and coefficients will have the values between 0 and $p - 1$ and calculations are performed in *modulo p*. Let $a, b \in Z_p$ and $4a^3 + 27b^2 \neq 0$. Then the EC is defined as;

$$y^2 mod\ p = (x^3 + ax + b)mod\ p \tag{4.1}$$

Once p, a, and b are selected, a group of EC points $E_p(a, b)$ are defined so they satisfy Equation 4.1. Then a base point generator $G = (x_1, y_1)$ is chosen so that the order of G is a very large value n and $n * G = 0$. The key building block of ECC is the scalar point multiplication which is $Q = k * P$, where k is a positive integer and P and Q are points in the EC. The value $k * P$ is computed by adding point P for $k - 1$ times and the resulting point Q is obtained. However, the recovery of k, knowing the points P and Q is a hard or computationally infeasible problem which is known as the Elliptic Curve Discrete Logarithmic Problem (ECDLP). In real-time applications k is made large in order to overcome guessing and brute force attacks.

Nevertheless, ECC is still not a widely deployed cryptographic scheme and its theoretical foundation is not very popular in standardized protocols. On the other hand, many ECs are already patented which makes it far more challenging to define new ones. Weak random number generators may also lead to successful attacks and built-in traps can be hidden behind bad-curve designs.

4.1.2 ECC for Authentication and Key Management

With the rapid expansion of the connected smart objects, authentication has become crucial to secure IoT and prevent malicious attacks [7]. In the key establishment protocols, this property defines whether peers are authenticated during the negotiation process or not. Authentication is the process of identifying an object or person as a legitimate entity to use a particular product or service. It is a prerequisite for authorization or the access control, which determines whether an entity can access resources or participate in a given communication. With the heterogeneous devices and distributed nature, the authentication protocols in IoT should not only be resistant to malicious attacks, but they should also be lightweight to enable deployment in less performing IoT devices [8]. Authentication is an essential feature in key establishment protocols which can be classified as symmetric vs. asymmetric techniques.

The shared secret-based authentication is a classic symmetric scheme where two parties are statically configured with a common shared secret mapped to their respective identities. Under asymmetric techniques, there are four variants such as static public key authentication, certificate-based authentication, cryptographically generated identifiers, and identity-based authentication. In every case, a node proves its identity by providing a proof of knowledge of the corresponding private key. In the first two categories, the authentication is implicitly ensured by the ownership of corresponding public-private keys or certificates. In the third category, the authentication identifiers are generated using the public key of the node. In the last asymmetric technique, opposite to the previous category, a node's public key is derived from its identity.

The rest of the chapter is built as follows. Section 4.2 describes ECC-based implicit certificates and their utilization for authentication and key establishment in the resource constrained IoT devices. Section 4.3 and Section 4.4 respectively describe ECC-based signcryption and ECC-based group communication with the examples of relevant protocols. Finally, Section 4.5 provides the implementation aspects of ECC and Section 4.6 discusses the main limitations of ECC.

Table 4.1 Size comparison between ECC and RSA public key and certificates.

Security level	Public key size (bits)		Certificate size (bits)		
	ECC	RSA	ECQV	ECDSA	RSA
80	160	1024	193	577	2048
112	224	2048	225	673	4096
128	256	3072	257	769	6144
192	384	7680	385	1153	15360
256	512	15360	522	1564	30720

4.2 ECC Based Implicit Certificates

Digital certificates advocate the establishment of identity in secure communications. Similar to the conventional or explicit certificates such as X.509, implicit certificates are made up of three parts [1]: identification data, a public key and a digital signature, which binds the public key to the user's identification data and verifies that the binding is accepted by a trusted third-party. In an explicit certificate, the public key and digital signature are two distinct elements. In contrast, the public key and digital signature are included in implicit certificates and allow the recipient to extract and verify the public key of the other party from the signature segment. This will significantly reduce the required bandwidth since there is no need to transmit both the certificate and the verification key.

The most important advantages of using implicit certificates over the conventional certificates are the smaller size and faster processing. Table 4.1 specifies the comparable key sizes for symmetric and asymmetric cryptosystems based on equivalent security strengths (i.e., symmetric key size). Elliptic Curve Digital Signature Algorithm (ECDSA) is a variant of the Digital Signature Algorithm (DSA) that operates in elliptic curve groups. Elliptic Curve Qu-Vanstone (ECQV) [3] is another type of implicit certificate scheme with smaller certificate sizes, lower computational power and very fast processing time for generating certified public keys. Accordingly, the sizes of ECQV and ECDSA-signed certificates are substantially smaller than RSA due to the reduced public key size of ECC.

4.2.1 Authentication and Key Management Using ECC Implicit Certificates

ECC-based implicit certificates have been used to design the lightweight and secure key establishment and authentication protocols for resource-constrained sensor networks. As a starting point, in [13], the authors have proposed a two-phase implicit certificate-based key establishment protocol for resource-constrained sensors deployed in generic WSNs. In the first phase, sensor nodes receive implicit certificates from the cluster head which acts as a certificate authority (CA). The certificate generation process is inspired by the design principles of the ECQV implicit certificate scheme. The second phase contains the key establishment component where sensors use implicit certificates to establish pair-wise keys with neighbouring sensor nodes. The concepts behind this

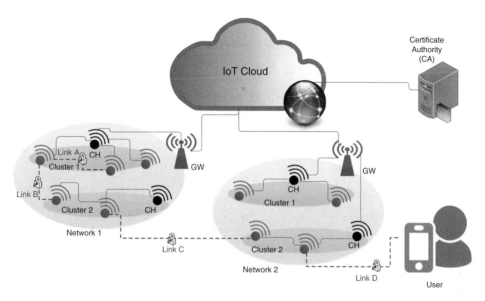

Figure 4.2 Network architecture for device/user authentication and key establishment.

work are extended in [15] by using implicit certificates for authenticating devices and users under the umbrella of IoT.

Using the schemes presented in [13] and [15], a pervasive authentication and key establishment scheme is designed for IoT networks in [14]. The scheme is called as *PAuthKey* and uses ECQV implicit certificates. Figure 4.2 illustrates the considered network architecture to design the protocols while considering four types of communication links: Link A - between two sensor nodes in the same cluster; Link B - between two sensor nodes in distinctive clusters and in the same network; Link C - between two sensor nodes located in distinctive clusters and different networks; Link D - between a user and a sensor node.

4.2.1.1 Phase 1: Registration and Certificate Acquisition

During this phase, all sensors receive the security credentials from the respective cluster heads (CH) which is acting as a local CA. The message flow of the certificate acquisition is described below (Figure 4.3). The starting `Requester Hello` message includes node identity (U) and cipher suites which are supported by the requester and installed on the sensor nodes in off-line mode (e.g., *CERT_ECC_160_WITH_ AES_128_SHA1* requests for certificates in 160-bit EC curves, 128-bit AES for bulk encryption, and SHA1 for hashing). At a successful requester identity verification, CA agrees to one cipher suite from the received options, and sends `CA Hello` message with its public key Q_{CA} as an unprotected message to approve the initiation of the handshake.

Upon receiving `CA Hello` message, the requester generates a random number $r_U \in [1, ..., n-1]$ and computes EC point $R_U = r_U \times G$, where the EC domain parameters support the negotiated cipher suites. The node produces a random cryptographic nonce N_U, calculates Message Authentication Code (MAC) value (*i.e.*, $MAC [R_U, U, N_U]$)

Certificate Requestor (U)   **Certificate Authority (CA)**

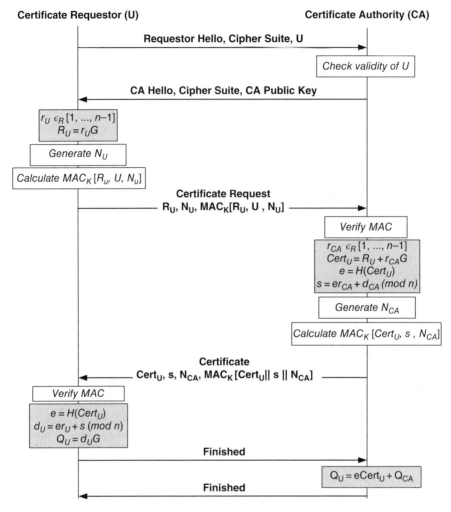

Figure 4.3 Message flow of registration phase [14].

using the common authentication key K, and sends those two along with the Certificate Request message.

Then CA verifies the MAC value and nonce N_U to identify the integrity and the message freshness. If both are successfully verified, CA generates a random number $r_{CA} \in$
$[1, ..., n-1]$, computes the certificate $Cert_U = R_U + r_{CA} \times G$ and private key reconstruc-
tion value $s = er_{CA} + d_{CA} \pmod{n}$, where d_{CA} is CA's private key and $e = H(Cert_U)$.
Certificate message includes the certificate $Cert_U$, s, a random nonce N_{CA}, and the
MAC on $[Cert_U, s, N_{CA}]$.

If MAC and N_{CA} are correct, U calculates $e = H(Cert_U)$ using the same hash function.
Then the node U can compute its own private key $d_U = er_U + s \pmod{n}$ and public key
$Q_U = d_U \times G$.

Node U's Finished message contains an encrypted message digest of previous
handshake messages using the requester public key Q_U. CA is also capable of computing

U's public key Q_U; $Q_U = eCert_U + Q_{CA}$. The following derivation proves that both equations give exactly the same Q_U as computed by node U.

$$Q_U = d_U G$$
$$= (er_U + s(mod\ n))\ G$$
$$= (er_U + er_{CA} + d_{CA}(mod\ n))\ G$$
$$= e(r_U + r_{CA}(mod\ n))\ G + d_{CA}G \tag{4.2}$$
$$= e(r_U G + r_{CA} G) + Q_{CA}$$
$$= e(R_U + r_{CA} G) + Q_{CA}$$
$$= eCert_U + Q_{CA}$$

CA uses public key Q_U for encrypting previous messages and answers with the `Finished` message to complete the handshake of the pre-authentication phase. Finally, the sensor nodes possess the security credentials to start secure communication with the internal and the external network entities (*i.e.*, end-users and sensor nodes).

4.2.1.2 Phase 2: Authentication and Key Establishment

Phase 2 is described in terms of scenario 1 (Figure 4.4), which is the communication between two sensor nodes in the same cluster. It is assumed that all nodes in the same cluster are aware of the identities of the other nodes.

First, the client node U sends the `Client Hello` message accompanied with its identity to the server node V. The server replies with the `Server Hello` message along with the cipher suites it supports and CH's identity, which is the CA for both nodes. If the client does not have the security credentials from the given cipher suite, it has to retrieve them from the CA. Otherwise, the client can continue the handshake by sending its certificate $Cert_U$. Similar to the registration phase handshake, random cryptographic nonce N_U and MAC values are used to preserve the freshness and integrity of the message.

Upon receiving the client's certificate, the server first verifies the MAC value and then computes the client's public key Q_U using its certificate; $e = H(Cert_U)$ and $Q_U = eCert_U + Q_{CA}$. This is proven according to Eq. 4.2. Then V generates a random nonce N_V and sends it along with $Cert_V$ and $MAC_K[Cert_V, N_V, V]$. In the meantime, V computes the pairwise key K_{UV} from its private key d_V and U's public key Q_U, where $K_{UV} = d_V Q_U$. Similar to V, upon receiving the message, U verifies the MAC and if the verification is successful it computes Q_V and $K_{UV} = d_U Q_V$.

Finally, the exchange of the `Finished` messages conclude the handshake, which are encrypted by the common key K_{UV}. At the end of six message transfers, the two edge nodes can authenticate each other, and establish a common secret key and a secure communication link that can be used for securing further data acquisitions between the client and the server. In the same manner, the rest of the scenarios will follow similar approaches with slight modifications as described in [14].

4.3 ECC-Based Signcryption

A signcryption scheme simultaneously generates encryption of a message and a corresponding signature in one single phase. As a consequence, cryptographic properties

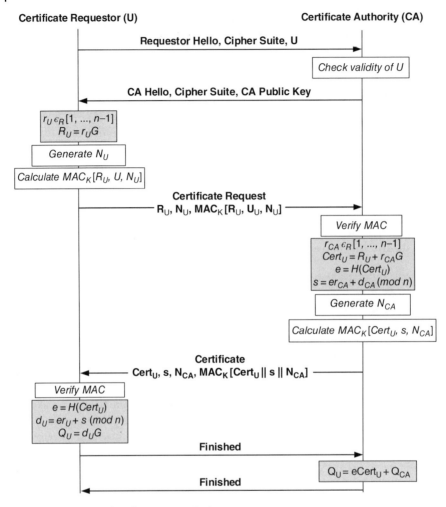

Figure 4.4 Message flow for scenario 1 [14].

like confidentiality, integrity, authentication and non-repudiation are more efficiently obtained compared to the traditional approach of first encrypt and then sign the message. Recently, signcryption schemes with multiple receivers, able to establish anonymity of sender and receivers, have gained a lot of popularity. This follows on from the fact that nowadays, people pay more and more attention to personal privacy. Therefore, it is important to derive suitable solutions that allow hidden involvement of the user in certain communications. Several application domains exist for the multi-receiver signcryption schemes with anonymity of the receivers, like e-voting systems, network conference meetings, broadcast DVD, pay-TV, etc. For example, a service provider can broadcast a ciphertext for a TV program or DVD. Each authorised device who has subscribed can then use his own private key to decrypt the ciphertext. In this situation, it is clear that the anonymity of the receiver is an essential requirement.

We now shortly describe the security features of the signcryption scheme ASEC, which is proposed in [5]. Next, the different steps of the scheme are described. We refer to [5] for a more detailed analysis on the security and performance.

4.3.1 Security Features

It is explained in [5] that the ASEC scheme provides the following security features.

- The main goal of ASEC is to provide anonymity to sender and receivers, based on basic EC operations. The anonymity for the receivers in the scheme is established with respect to external and internal users of the system. This means that not only attackers, but also authenticated and intended receivers cannot reveal the other intended receivers. Since the identity of the sender is encrypted, only the intended receivers can recover the sender's identity after correct decryption of the ciphertext. This leads to the anonymity of the sender.
- The authentication of the entities and integrity of the message is guaranteed. It is impossible for an attacker to change the identity of sender/receivers or to change the content of the message without being noticed by the legitimate senders.
- The scheme is also unforgeable, which means that only the intended sender can produce a valid message-signature pair.
- Thanks to the particular construction of the private and public keys of the entities, a public verifiable scheme is created. This means that any third party without knowledge of the private key of sender or receiver(s) can validate the signcrypted message.
- ASEC also satisfies the forward secrecy feature. Even if the private key of the sender is revealed, an attacker will not be able to derive the key and to decrypt the message. The key is to build using a random value, which is later deleted from the memory of the sender after completion of the protocol and can never be reconstructed thanks to the ECDLP.
- The scheme is also resistant against the decryption fairness problem [10], meaning that either all authorized receivers can correctly decrypt the message or not, since the content required for decryption is similar for all receivers.
- Finally, the master public key can be reused after applying instances of the scheme.

4.3.2 Scheme

The signcryption scheme consists of three different phases. The first phase is the key generation of the private and public keys of the participants. The second part handles the signcryption, executed by the sender. In the third phase, receivers perform the operations individually, which are referred to as the unsigncryption process.

4.3.2.1 Key Generation

The key generation is performed by a trusted PKG and involves the user identities to construct the private key of the users. Let us denote the master private key of the scheme by d_k and the master public key by $Q_k = d_k G$. This pair is used to derive the key pairs of the entities in the scheme. The parameter r is a random value in F_q and is required to revoke the user's credentials and issue new credentials without changing the private master key or the identity of the user.

- The private key of the sender is determined as $d_S = H(H(ID_S\|r)d_kG)$, and the corresponding public key as $Q_S = d_SQ_k = d_Sd_kG$.
- The private key of the receiver U_j is determined as $d_j = H(H(ID_j\|r)d_kG)$, and the corresponding public key as $Q_j = d_jQ_k = d_jd_kG$. Let us consider that $U = \{U_1, \dots, U_n\}$ contains the intended receivers of the message.

The private keys are securely delivered to the different participants in off-line mode. The master public key is stored in each node. For each user U_i of the system, the values $(ID_i, Q_i, H(ID_i\|Q_i))$ are made publicly available and are independently checked on changes by a third party.

4.3.2.2 Signcryption

The sender S selects the group of receivers U and looks up the corresponding identities ID_j and public keys Q_j for $j \in \{1, \dots, n\}$. A check on the hash value $H(ID_j\|Q_j)$ is made to guarantee the integrity and authenticity of the published data. Then the following steps are performed.

1. A random $r_1 \in F_p^*$ is selected. The corresponding point $k_s = r_1Q_k$ is computed.
2. A random $r_2 \in F_p^*$ is selected. The corresponding point $R_2 = r_2Q_k$ is computed.
3. $\forall j \in \{1, \dots, n\}$, the EC point $S_j = r_1H(ID_j)Q_j = (x_j, y_j)$ is computed.
4. S chooses a random key K.
5. S constructs the polynomial of degree n that passes through the points $\{(0, K), (H(x_1), H(y_1)), \dots, (H(x_n), H(y_n))\}$ by means of Lagrange interpolation.
6. S derives n other points $P = (P_1, \dots, P_n)$ on this polynomial.
7. The encryption $C = E_K(M\|ID_S\|R_2)$ is performed.
8. The hash $h = H(k_s\|R_2\|P\|C)$ is computed.
9. The value $s = r_1 - hr_2$ is computed.
10. The value $s_2 = r_1 - hd_s$ is computed.
11. This value s_2 will be encrypted, $C_2 = E_K(h\|s_2)$
12. The message $P\|R_2\|C\|C_2\|h\|s$ is broadcasted.

4.3.2.3 Unsigncryption

Let U_j be a receiver of the message. U_j performs the following steps.

1. The EC point $T = sQ_k + hR_2$ is calculated.
2. Let $k_s = T$.
3. The hash operation $h = H(k_s\|R_2\|P\|C)$ is executed. If this value corresponds with h of the transmitted message, the validity of the message is approved and the process can be continued.
4. The EC point $S_j = d_jH(ID_j)T = (x_j, y_j)$ is calculated.
5. U_j constructs the polynomial of degree n that passes through the points $\{(H(x_j), H(y_j)), P_1, \dots, P_n\}$ by means of Lagrange interpolation.
6. The intersection of this polynomial with the Y-axis $(0, K)$ defines the key K.
7. The decryption $M\|ID_S\|R_2 = D_K(C)$ is performed.
8. If the last field corresponds with the transmitted value of R_2, it means that the receiver belongs to the intended group of receivers.
9. The decryption $h\|s_2 = D_K(C_2)$ is executed.

Table 4.2 Different steps in ASEC

Signcryption	Unsigncryption
1. Choose r_1, $k_S = r_1 Q_k$	1. $T = sQ_k + hR_2$
2. Choose r_2, $R_2 = r_2 Q_k$	2. $k_S = T$
3. $\forall j$, $S_j = r_1 H(ID_j)Q_j = (x_j, y_j)$	3. $H(k_S\|R_2\|P\|C) = h$?
4. Choose K	4. $S_j = d_j H(ID_j)T = (x_j, y_j)$
5. Construct pol through $(0, K)$, S_j's	5. Construct pol with S_j, P
6. Derive n other points P	6. Find K as $(0, K)$
7. $C = E_K(M\|ID_S\|R_2)$	7. $M\|ID_S\|R_2 = D_K(C)$
8. $h = H(k_S\|R_2\|P\|C)$	8. Check R_2
9. $s = r_1 - hr_2$	9. $h\|s_2 = D_K(C_2)$
10. $s_2 = r_1 - hd_S$	9. Check $k_s = s_2 Q_k + hQ_S$
11. $C_2 = E_K(h\|s_2)$	
12. Send $P\|R_2\|C\|C_2\|h\|s$	

10. First, the same h as the received one should be obtained. Next, the identity of the sender is now checked by first requesting the public key Q_S belonging to ID_S. Then the equality $s_2 Q_k + hQ_S = k_s$ should be checked. If so, ID_S possesses the public key Q_S and its identity is verified.

The different steps of the signcryption and unsigncryption mode of ASEC are summarized in Table 4.2.

4.4 ECC-Based Group Communication

Rather than device-to-device communications, group communications in the form of broadcasting and multicasting incur efficient message deliveries among resource-constrained sensor nodes in IoT-enabled WSNs. Secure and efficient key management is significant to protect the authenticity, integrity, and confidentiality of multicast messages. In [12], two group key establishment protocols are developed for secure multicast communications among resource-constrained devices in IoT. We limit the explanation in this chapter to the second and most efficient protocol. We start with a short discussion on background and some assumptions considered in the system and then explain the different steps in the scheme. We refer to [12] for a more detailed analysis on the security and performance.

4.4.1 Background and Assumptions

In the scheme, it is assumed that the underlying communication technology and sensor nodes support multicast group formation and message transactions. It is also considered that all network entities possess common security associations (i.e., cipher suites) and perform identical cryptographic operations (e.g., hashing ($h()$), encoding, decoding).

Common EC parameters are embedded in all the network entities that participate in the communication scenario. The initiator (I) is considered a main powered resource-rich entity (e.g., gateway node) and has higher processing power and memory capacity than the rest of the nodes in the multicast group. The initiator is also aware of the constitution of the group (i.e., knowing the identities of the legitimate nodes). The initiator is supposed to know the public keys of all the nodes and vice versa. The sleeping patterns of the nodes and path losses in the communication links are not being considered since they are outwith the scope of the key objective of this paper. Therefore, it is assumed that the members of the multicast group will eventually receive the initiator requests and the rest of the messages without failures.

The adversary is able to eavesdrop the controlling messages exchanged between the different entities in the scheme. He may fraudulently act as a legitimate intermediate device during the key establishment between the initiator and the and the other nodes, and launch MITM attacks. Alternatively, an adversary who is external or internal to the network may re-transmit the previous key establishment messages to generate replay attacks and interrupt the normal operations of the nodes. If the adversary captures a node, he may uncover the secret group keys stored in that node. The initiator is considered to be honest and not to be captured as this device is more powerful.

4.4.2 Scheme

The message flow of multicast key establishment of the scheme is shown in Figure 4.5. We note that the scheme exploits the concepts of ECIES to establish a shared secret key among the multicast group.

Step 1: First, the size (n) and the composition of the multicast group $U = \{U_1, U_2, \ldots, U_{(n-1)}\}$ are determined by the initiator. Then a random value r is generated, where $R = rG$. EC points S_js are computed using r and the public keys Q_j of the group members: $S_j = d_iQ_j + R$, where $j = 1$ to $n-1$. Next, the EC point $S_j = (x_j, y_j)$ is encoded into the point (u_j, v_j) as follows: $u_j = h(x_j)$; $v_j = h(y_j)$. For each node U_j, the value $\bar{u}_j = \{\oplus_{i\neq j}u_i\} \oplus v_j$ is computed. The group key is then defined by $k = h(\oplus_i u_i)$. Denote $P = (\bar{u}_1 \| \cdots \| \bar{u}_{n-1})$ and let $Auth = h(k\|R\|P)$. The new multicast message for group U is generated and transmitted by the initiator with the calculated values $(Auth, C, R, U, P)$. Additionally, the digital signature is appended to preserve message authentication and integrity. Note that the same R value can also be reused as the parameter R in the signature scheme (see e.g. Schnorr Signature scheme).

Step 2: When the sensor node U_j receives the broadcast message, initially, it checks whether it is included in the multicast group U. Then the digital signature and the counter C are checked. If both are correctly verified, S_j is computed using the received random value R and the node's private key d_j: $S_j = d_jQ_i + R$. The EC point S_j is converted to the point (u_j, v_j) using the same encoding as in step 1. After that U_j can compute $k = h(\bar{u}_j \oplus u_j \oplus v_j)$ and verify the integrity by checking if $h(k\|R\|P)$ corresponds with the received $Auth$ value.

Step 3: Each sensor node should send an acknowledgement message $Ack_j = h(k, Q_j)$ to finish the handshake. Later, by verifying the acknowledgement message, the initiator can ensure the authenticity of that particular group member and the accurate derivation of group key k.

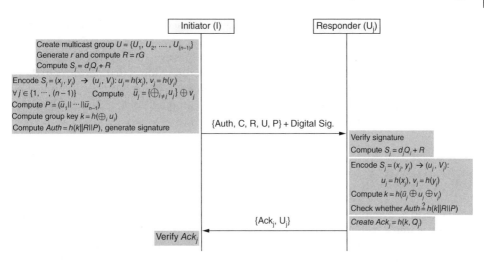

Figure 4.5 Message flow of scheme.

After three steps the shared secret key is known by the initiator and the other members in the multicast group.

4.5   Implementation Aspects

For the implementation of the security protocol, one needs to choose a particular elliptic curve. Several standards exist where curves are presented, e.g. ANSI X9.62 (1999), IEEE P1363 (2000), SEC 2 (2000), NIST FIPS 186-2 (2000), ANSI X9.63 (2001), Brainpool (2005), NSA Suite B (2005), ANSSI FRP256V1 (2011). The curves in each of these standards are selected in such a way that the ECDLP is difficult. However, as pointed out in [2, 4], protection against the ECDLP does not guarantee that the resulting scheme is secure. Also, the implementation of the curve plays a very important role. It might be possible that the implementation produces incorrect results for some rare curve points, leaks secret data when the input isn't a curve point or leaks secret data through branch timing and/or cache timing. Resistance against this type of attack is called ECC security. Most of these attacks can be ruled out by a good choice of parameters and definition of the curve. It has been demonstrated in [2] that all of the curves proposed in the standards are vulnerable with regards to ECC security. A list of secure curves is also provided.

In particular, the elliptic curve, Curve25519, offering ECC security with 128 bits security level is worth mentioning. It is one of the fastest ECC curves and is not covered by any known patents. The reference implementation is public domain software. Besides its speed, it has some other very interesting features.

- The public keys are very short as they can be represented by only 32 bytes. In most of the schemes, the keys are represented by 64 bytes, although able to be compressed by half as suggested by Miller [9]. However, the time for compression is also quite noticeable and usually not reported.
- There is no need to add additional protection against any timing attacks. The standard implementation is already avoiding all input-dependent branches, all

input-dependent array indices, and other instructions with input-dependent timings.
- There is no need for time-consuming key validation as every 32-byte string can act as a valid public key. If this is not the case and no key validation is first performed, there are attacks which are able to break the key agreement scheme.
- The software code is very small. In particular, the compiled code, including all necessary tables, is around 16 kilobytes.

4.6 Discussion

Although ECC schemes guarantee a high level of security, they could still contain an easily-exploitable vulnerability if they are applied without an additional level of protection. Therefore, an additional layer of protection needs to be added to assure complete security towards clogging attacks. Moreover, there are other limitations of ECC schemes due to the low degree of maturity of the techniques and less adaptability. These will also create difficulties and challenges in practical deployments of ECC solutions. Security of the ECC scheme is entirely dependant on the curve. As a result, poor curve designs might hugely affect the security in entire systems. Despite the above discussed limitations, as discussed in the previous sections, ECC is used in many security protocols for numerous resource-constrained networking devices that require lightweight solutions. Although, in this book chapter, we discussed some selected areas of ECC use cases, there can be many other areas where it is applicable.

References

1 Explaining Implicit Certificates. Certicom cooperation. https://www.certicom.com/content/certicom/en/code-and-cipher/explaining-implicit-certificate.html (accessed 25 June 2019).

2 SafeCurves: choosing safe curves for elliptic-curve cryptography. https://safecurves.cr.yp.to (accessed 25 June 2019).

3 SEC4: Elliptic Curve Qu-Vanstone Implicit Certificate Scheme (ECQV). (2013). Version 0.97. http://secg.org/sec4-1.0.pdf (accessed 25 June 2019).

4 Bernstein, D.J. (2018). Curve25519: New Diffie-Hellman speed records. International Workshop on Public Key Cryptography. https://rd.springer.com/chapter/10.1007/11745853_14 (accessed 25 June 2019).

5 Braeken, A. and Porambage, P. (2015). Asec: Anonym signcryption scheme based on ecoperations. *International Journal of Computer Applications* 5 (7):90–96.

6 Hankerson, D., Menezes, A.J. and Vanstone, S. (2003). *Guide to Elliptic Curve Cryptography*. Berlin, Heidelberg: Springer-Verlag.

7 Kothmayr, T., Schmitt, C., Hu, W. et al. (2013). Dtls based security and two-way authentication for the internet of things. *Ad Hoc Networks* 11 (8): 2710–2723.

8 Liu, J., Xiao, Y. and Chen, C.P. (2012). Authentication and access control in the internet of things. *Distributed Computing Systems Workshops (ICDCSW), 2012 32nd International Conference*, Macau, China (18–21 June 2012). IEEE.

9 Miller, V.S. (1985). Use of elliptic curves in cryptography. *Conference on the theory and application of cryptographic techniques*, Linz, Austria (April 1985). Springer.

10 Pang, L. and Li, H. (2013). nmibas: a novel multi-receiver id-based anonymous signcryption with decryption fairness. *Computing and Informatics* 32 (3): 441–460.

11 Porambage, P. (2018). Lightweight Authentication and Key management of Wireless Sensor Networks for Internet of Things. http://jultika.oulu.fi/files/isbn9789526219950 .pdf (accessed 17 July 2019).

12 Porambage, P., Braeken, A., Schmitt, C. et al. (2015). Group key establishment for secure multicasting in iot-enabled wireless sensor networks. *2015 IEEE 40th Conference on Local Computer Networks (LCN)*, Washington, USA (26–29 October 2015). IEEE.

13 Porambage, P., Kumar, P., Schmitt, C. et al. (2013). Certificate based pairwise key establishment protocol for wireless sensor networks. *2013 IEEE 16th International Conference on Computational Science and Engineering (CSE)*, Sydney, Australia (3–5 Decmber 2013). IEEE.

14 Porambage, P., Schmitt, C., Kumar, P. et al. (2014). Pauthkey: A pervasive authentication protocol and key establishment scheme for wireless sensor networks in distributed iot applications. *International Journal of Distributed Sensor Networks* 10 (7): 357430.

15 Porambage, P., Shmitt, C., Kumar, P. et al. (2014). Two-phase authentication protocol for wireless sensor networks in distributed iot applications. *Wireless Communications and Networking Conference (WCNC)*, Istanbul, Turkey (6–9 April 2014). IEEE.

16 Stallings, W. (2010). *Cryptography and Network Security: Principles and Practice*, 5e. Upper Saddle River, NJ: Prentice Hall Press.

5

Lattice-Based Cryptography and Internet of Things
Veronika Kuchta and Gaurav Sharma

5.1 Introduction

Post-quantum cryptography is an essential research topic which became more popular since the start of research on quantum computing. Quantum computers are highly powerful machines which take advantage of subatomic particles which exist in more than one state at any time. Such machines are able to process information in an incomparably faster time than the fastest computers. IBM and Google are the leading companies in this race for the first quantum computer that will then be made publicly available and extremely useful. The main feature of such a powerful computer is that it will be able to perform calculations which are almost impossible to be simulated by a conventional computer. A computer with this feature will ewasily be able to break all of the current cryptographic constructions which have proven to be secure under number-theoretical assumptions. A possible solution to this problem can be offered by the following research fields which are assumed to be resistant against quantum attacks:

- **Hash-based Cryptography.** A typical example for this field is given by the Merkle's hash-tree public-key signature scheme, which was introduced in 1979. While on the one side hash-based cryptographic schemes offer an efficient solution to certain cryptographic problems, the main disadvantage of those schemes is the large size of signatures.
- **Code-based Cryptography.** McEliece's hidden Goppa-code public-key encryption represents a typical scheme of this research field. The scheme was introduced in 1979. The main idea of such cryptosystems, which are based on error-correcting codes, is to construct a secure and efficient one-way function.
- **Lattice-based Cryptography.** Another research topic of post-quantum cryptography is given by the lattice-based cryptography which also has the feature of conjectured quantum attacks resistance. Furthermore, lattice-based cryptoschemes profit mostly from algorithmically simple and highly parallelizable constructions.
- **Multi-variate quadratic equations Cryptography.** Introduced in the 1980s, this type of scheme represents asymmetric cryptography where the public keys are defined as a set of multi-variate polynomials and the security of such schemes is based on a problem of solving multi-variate quadratic equations over finite fields.

IoT Security: Advances in Authentication, First Edition.
Edited by Madhusanka Liyanage, An Braeken, Pardeep Kumar, and Mika Ylianttila.
© 2020 John Wiley & Sons Ltd. Published 2020 by John Wiley & Sons Ltd.

- **Isogeny-based Cryptography.** This represents the newest direction in post-quantum cryptography, where the cryptoschemes are based on supersingular elliptic curve isogenies. The security of those systems is based on the difficulty of constructing an isogeny between two supersingular curves having the same number of points as input.
- **Secret-Key cryptography.** The Rijndael cipher, developed in 1998, also known as the Advanced Encryption Standard (AES) is, nowadays, still the leading example for this type of cryptography.

We will focus here on one specific topic and its relation to the Internet of Things (IoT). We should consider the fact that we are at the very beginning of the application process of quantum-resistant algorithms to IoT. Therefore, in this chapter we want to make the reader familiar with notions and the main constructions of lattice-based cryptography and share our ideas on how to use these schemes in IoT.

5.1.1 Organization

In Section 2, we will have a closer look into the main definitions of lattice-based cryptography and introduce the reader to the solutions of certain essential problems in this topic. In section 3, we discuss the main cryptographic primitives constructed from lattices. In section 4 we show the relation between lattice-based cryptoschemes and IoT and provide some examples for these relations.

5.2 Lattice-Based Cryptography

As we know, cryptography requires average-case intractability, meaning that there are problems for which random instances are hard to solve. This intractability differs from the worst-case notion of hardness, where a problem is considered to be hard if there exists some intractable instances. This latter notion of hardness is considered to be NP-complete. A trivial conclusion follows that if there are problems which are hard in worst case, they often appear to be easier on the average.

A crucial contribution to this problem, according to lattice, has been made by Ajtai [3] who proved that certain problems are hard on the average if there are some lattice-related problems which are hard in the worst case. Using this result, cryptographers can construct schemes which are note feasible to break, unless all instances of some certain lattice problems are easy to solve.

5.2.1 Notations

Let $\mathbb{Z}_q = \mathbb{Z}/q\mathbb{Z}$ denote the quotient ring modulo q, where $q \in \mathbb{Z}^+$ is a positive integer. Elements in $\mathbb{Z}_q$ are given as $x + q\mathbb{Z}$, with $x \in \mathbb{Z}$. It holds that $\mathbb{Z}_q$ is an additive group supporting scalar multiplication by integers, i.e. $s \cdot x \in \mathbb{Z}_q$ for an integer $s \in \mathbb{Z}$ and $x \in \mathbb{Z}_q$.

We use bold capital letters to denote matrices, such as $\mathbf{A}, \mathbf{B}$ and bold lower-case letters to denote column vectors, such as $\mathbf{x}, \mathbf{y}$. To indicate horizontal concatenation of vectors and matrices we use the following notation: $[\mathbf{A}|\mathbf{Ax}]$.

5.2.2 Preliminaries

5.2.2.1 Lattices

Let $B = \{\mathbf{b}_1, \ldots, \mathbf{b}_n\} \subset \mathbb{R}^n$ be the basis of a lattice L which consists of n linearly indepen-dent vectors. The n-dimensional lattice L is then defined as $L = \sum_{i=1}^{n} \mathbb{Z}b_i$.

The i-th minimum of lattice Λ, denoted by $\lambda_i(L)$ is the smallest radius r such that L contains i linearly independent vectors of norms $\leq r$. (The norm of vector b_i is defined as

$$\|\mathbf{b}_i\| = \sqrt{\sum_{j=1}^{n} c_{i,j}^2},$$ where $c_{i,j}, j \in \{1, \ldots, n\}$ are coefficients of vector $\mathbf{b}_i$. We denote by $\lambda_1^{\infty}(L)$

the minimum distance measured in the infinity norm, which is defined as $\|\mathbf{b}_i\|_{\infty} := \max(|c_{i,1}|, \ldots, |c_{i,n}|)$. Additionally, we recall $\|B\| = \max\|b_i\|$ and its fundamental paral-lelepiped is given by $P(B) = \left\{ \sum_{i=1}^{n} a_i b_i \mid \mathbf{a} \in [0, 1)^n \right\}$. Given a basis B for a lattice L and a vector $\mathbf{a} \in \mathbb{R}^n$ we define $\mathbf{a} \bmod L$ as the unique vector in $P(B)$ such that $\mathbf{a} - (\mathbf{a} \bmod L) \in L$. If L is a lattice, its dual lattice is defined as

$$L^* = \{\hat{\mathbf{b}} \in \mathbb{R}^n \mid \forall \mathbf{b} \in L, \langle \hat{\mathbf{b}}, \mathbf{b} \rangle \in \mathbb{Z}\}.$$

5.2.2.2 Integer Lattices

The following specific lattices contain $q\mathbb{Z}^m$ as a sub-lattice for a prime q. For $A \in \mathbb{Z}_q^{n \times m}$ and $s \in \mathbb{Z}_q^n$, define:

$$\Lambda_q(A) := \{e \in \mathbb{Z}^m | \exists s \in \mathbb{Z}_q^n, \quad \text{where} \quad A^T s = e \bmod \quad q\},$$
$$\Lambda_q^{\perp}(A) := \{e \in \mathbb{Z}^m | Ae = 0 \bmod q\},$$

Many lattice-based works rely on Gaussian-like distributions called Discrete Gaussians. In the following paragraph we recall the main notations of this distribution.

5.2.2.3 Discrete Gaussians

Let L be a subset of $\mathbb{Z}^m$. For a vector $c \in \mathbb{R}^m$ and a positive $\sigma \in \mathbb{R}$, define

$$\rho_{\sigma,c}(x) = \exp\left(-\pi \frac{\|x - c\|^2}{\sigma^2}\right) \quad \text{and} \quad \rho_{\sigma,c}(L) = \sum_{x \in L} \rho_{\sigma,c}(x).$$

The discrete Gaussian distribution over L with center c and parameter σ is given by $\mathscr{D}_{L,\sigma,c}(y) = \frac{\rho_{\sigma,c}(y)}{\rho_{\sigma,c}(L)}, \quad \forall y \in L$. The distribution $\mathscr{D}_{L,\sigma,c}$ is usually defined over the lattice $L = \Lambda_q^{\perp}(A)$ for $A \in \mathbb{Z}_q^{n \times m}$. Ajtai [3] showed how to sample an uniform matrix $A \in \mathbb{Z}_q^{n \times m}$ with a basis B_A of $\Lambda_q^{\perp}(A)$ with low Gram-Schmidt norm. Gentry et al. [18] defined an algorithm for sampling from the introduced discrete Gaussian $\mathscr{D}_{\Lambda,\sigma,c}$ for a given basis B of the n dimensional lattice Λ with a Gaussian parameter $\sigma \geq \|B\| \cdot \omega(\sqrt{\log m})$, where $\widetilde{B}$ is the orthonormal basis of B, defined as follows:

5.2.2.4 The Gram-Schmidt Norm of a Basis [1]

Let B be a set of n vectors $B = \{b_1, \ldots, b_n\} \in \mathbb{R}^n$ with the following functionalities:

– $\|B\|$ denotes the norm of the longest vector in B, i.e. $\|B\| := \max_i \|b_i\|$ for $1 \leq i \leq m$.

- $\widetilde{B} := \{\widetilde{b}_1, \ldots, \widetilde{b}_m\} \subset \mathbb{R}^m$ is the Gram-Schmidt orthogonalization of the vectors $b_1, \ldots, b_m$.

The Gram-Schmidt norm is denoted by $\|\widetilde{B}\|$.

5.2.3 Computational Problems

In the next definitions we recall the most popular computational problems on lattices which can also be found in the report of Peikert [37]. The most well-studied problem is the Shortest Vector Problem (SVP):

Definition 5.1 *(Shortest Vector Problem (SVP))* Given an arbitrary basis B of some lattice $L = L(B)$, find a shortest non-zero vector $\mathbf{v} \in L$ for which holds $\|\mathbf{v}\| = \lambda_1(L)$.

Apart from the main definition of the SVP problem, there are many approximation problems which are parameterized by an approximation factor $\gamma \geq 1$ which is represented as a function in the lattice parameter n, i.e. $\gamma = \gamma(n)$. The corresponding approximation problem for the SVP problem is defined as follows:

Definition 5.2 *(Approximate Shortest Vector Problem (SVP$_\gamma$))* Given a basis B of an n-dimensional lattice $L = L(B)$, find a nonzero vector $\mathbf{v} \in L$, such that holds $\|\mathbf{v}\| = \gamma(n) \cdot \lambda_1(L)$.

It is known that many cryptoschemes can be proved secure under the hardness of certain lattice problems in the worst case. But there is no known proof for the search version of SVP$_\gamma$. But there are many proofs which are based on the following decision version of approximate-SVP problem:

Definition 5.3 *(Decisional Approximate-SVP (GapSVP$_\gamma$))* Given a basis B of an n-dimensional lattice $L = L(B)$ where either $\lambda_1(L) \leq 1$ or $\lambda_1(L) > \gamma(n)$, decide which is the case.

Another approximate version of SVP is recalled in the following definition:

Definition 5.4 *(Approximate Shortest Independent Vectors Problem (SIVP$_\gamma$))* Given a basis B of a full-rank n-dimensional lattice $L = L(B)$, output a set $\mathbf{S} = \{\mathbf{s}_i\} \subset L$ of n linearly independent vectors, where $\|\mathbf{s}_i\| \leq \gamma(n)\lambda_1(L)$.

Special case (presented in [13]) of short integer solution (SIS) problem introduced by Ajtai [3]. Another particularly important computational problem for cryptographic constructions is the *Bounded-Distance Decoding* BDD$_\gamma$ problem.

Definition 5.5 *(Bounded Distance Decoding Problem (BDD))* Given a basis B of an n-dimensional lattice $L = L(B)$ and a target point $\mathbf{t} \in \mathbb{R}^n$ with the guarantee that $dist(\mathbf{t}, L) < d = \lambda_1(L)/(2\gamma(n))$, find the unique lattice vector $\mathbf{v} \in L$, such that holds $\|\mathbf{t} - \mathbf{v}\| < d$.

The main difference between BDD_γ and SVP_γ is the uniqueness of the solution to the earlier problem, while the target of the latter can be an arbitrary point.

Most modern lattice-based cryptographic schemes rely on the following average-case problems, Short Integer Solution (SIS) and Learning with Errors (LWE) problems, their analogues defined over rings. They involve analytic techniques such as Gaussian probability distribution.

Definition 5.6 *(Learning with Errors Problem (LWE))* For an integer q and error distribution χ, the goal of $LWE_{q,\chi}$ in n dimensions the problem is to find $\mathbf{s} \in \mathbb{Z}_q^n$ with overwhelming probability, given access to any arbitrary *poly*(n) number of samples from $A_{\mathbf{s},\chi}$ for some random $\mathbf{s}$.

In matrix form, this problem looks as follows: collecting the vectors $\mathbf{a}_i \in \mathbb{Z}_q^n$ into a matrix $\mathbf{A} \in \mathbb{Z}_q^{n \times m}$ and the error terms $e_i \in \mathbb{Z}$ and values $b_i \in \mathbb{Z}_q$ as the entries of the m-dimensional vector $\mathbf{b} \in \mathbb{Z}_q^m$ we obtain the input $\mathbf{A}$, $\mathbf{b} = \mathbf{A}^t s + e \bmod q$.

Definition 5.7 *(One-Dimensional Short Integer Solution (SIS))* Given a uniformly vector $v \in_r \mathbb{Z}_q^m$, find a vector $z \in \mathbb{Z}^m$, such that $\|z\| \leq t$ and also $\langle v, z \rangle \in \lceil -t, t \rceil + q\mathbb{Z}$. The formal notation of the problem is $1D - SIS_{q,m,t}$.

Definition 5.8 *(Ring-SIS$_{q,n,m,\beta}$)* Let $\mathcal{R}$ be some ring and $\mathcal{K}$ some distribution over $\mathcal{R}_q^{n \times m}$. Given a random matrix $A \in \mathcal{R}_q^{n \times m}$ sampled from $\mathcal{K}$, find a non-zero vector $\mathbf{v} \in \mathcal{R}_q^m$ such that $A\mathbf{v} = 0$ and $\|\mathbf{v}\|_2 \leq \beta$.

5.2.4 State-of-the-Art

In this chapter we want to briefly survey some of the previous significant works in lattice cryptography. The particularly ground-breaking work of Ajtai [3] provided the first worst-case to average-case reductions for lattice problems. In that work, Ajtai introduced the average-case short integer solution (SIS) and showed that solving it is at least as hard as approximating various lattice problems in the worst case. In a later work [4], Ajtai and Dwork presented a lattice-based public-key encryption scheme which became the basic template for all lattice-based encryption schemes.

Almost at the same time, a concurrent work had been published by Hoffstein, Pipher, Silverman [25] introducing the NTRU public-key encryption scheme. It was the first construction using polynomial rings. The advantages of that construction are the practical efficiency and particularly compact keys. The NTRU system is parameterized by a certain polynomial ring $R = \mathbb{Z}[X]/(f(X))$, where $f(X) = X^n - 1$ for a prime n, or $f(x) = X^n + 1$ for n which is a power of two, with a sufficiently large modulus q, which defines the quotient ring $R_q = R/qR$.

Around the same time, Goldreich, Goldwasser and Halevi [20] published a paper on public-key encryption scheme and a digital-signature scheme, both based on lattices. The main idea behind their constructions was that a public key is a "bad" basis of some basis consisting of long and non-orthogonal lattice vectors, while the secret key is a "good" basis of the same lattice consisting of short lattice vectors.

Later on, Oded Regev [40] provided improvements to the results of Ajtai and Dwork's work by introducing Gaussian measures and harmonic analysis over lattices. The main

consequences of these new techniques were simpler algorithms and tighter approximation factors for the underlying worst-case lattice problems. In another important work,
Regev [41] introduced the average-case learning with error (LWE) problems, for which
Oded Regev was awarded the Goedel Prize in 2018. In the same paper, the author introduced a new cryptosystem which can be proved secure under the new LWE assumption.
This construction had the favourable feature of more efficient public keys, secret keys
and ciphertexts, where the efficiency of the former was improved from $\mathcal{O}(n^4)$ to $\mathcal{O}(n^2)$
and the ciphertext efficiency improved from $\mathcal{O}(n^2)$ to $\mathcal{O}(n)$.

5.3 Lattice-Based Primitives

Some core constructions which were built upon (ring)-SIS/LWE assumptions are listed
below:

5.3.1 One-Way and Collision-Resistant Hash Functions

An example of this is the SWIFFT construction by Lyubashevsky et al. [34]. It is an
instantiation of the ring-SIS-based hash function and is highly efficient in practice, as it
uses the fast Fourier transform in $\mathbb{Z}_q$. Without going too deeply into details, we sketch
the construction as follows: Let $\mathbb{W} \in \mathbb{Z}_q^{n \times n}$ be an invertible quadratic matrix over $\mathbb{Z}_q$.
The SWIFFT hash function maps a key consisting of t/n vectors $(\mathbf{v}_1, \mathbf{v}_{t/n})$ and a vector
$\mathbf{s} \in \{0, \dots, d-1\}^t$ to the product $\mathbf{W} \cdot f_A(\mathbf{s}) \bmod q$, where $\mathbf{A}$ is not an uniformly chosen matrix, but a block-matrix generated using structured block as shown in [34]. The
collision resistance property follows from the usage of worst case lattice problems on
ideal lattices. A collision-resistant hash function is a highly significant cryptographic
tool, which plays the role of an important building block in many cryptoschemes. From
a hash function, we move to more specific cryptographic constructions, discussed in the
following sections.

5.3.2 Passively Secure Encryption

This topic contains all encryption schemes which are IND-CPA secure. As the name
already reveals, passively secure schemes involve a passive eavesdropper, who learns
no more information about the messages, except seeing the public key and the ciphertext. The first such scheme has been introduced by Regev [41]. Gentry, Peikert and
Vaikuntanathan [18] defined the dual version of Regev's encryption. The main feature of
their construction is that public keys are uniformly random with many possible secret
keys. But since Regev's encryption scheme [41] served as a basic building block for
many successful schemes, we believe that it is useful to have this scheme recalled here.
This scheme is based on LWE problem, where the size of the public keys is $\widetilde{\mathcal{O}}(n^2)$ bits
and that of the ciphertext is $\widetilde{\mathcal{O}}(n)$ bits, where the ciphertext is an encryption of a single bit and n denotes the dimension of the underlying LWE problem. Let m be the
number of samples and χ denote an error distribution over $\mathbb{Z}$. The three algorithms
(KeyGen, Encrypt, Decrypt) of Regev's LWE cryptoscheme are given as follows:

– Since the construction is underlying an LWE problem, the secret key is a uniformly
  and random LWE secret $\mathbb{s} \in \mathbb{Z}_q^n$. The corresponding public key is a vector of m

samples $(\mathbf{a}_i, b_i = \langle \mathbf{s}, \mathbf{a}_i \rangle + e_i) \in \mathbb{Z}_q^{n+1}$ which are drawn from the LWE distribution $D_{\mathbf{s}, \chi}$. The m samples are collected into a matrix

$$\hat{\mathbf{A}} = \begin{bmatrix} \mathbf{A} \\ \mathbf{b}^t \end{bmatrix} \in \mathbb{Z}_q^{(n+1) \times m}$$

It holds $\mathbf{b}^t = \mathbf{s}^t \mathbf{A} + \mathbf{e}^t$, with $\mathbf{e} = (e_1, \dots, e_m) \in \mathbb{Z}_q^m$. The secret and public keys satisfy the relation $(\mathbf{s}, 1)^t \cdot \hat{\mathbf{A}} = \mathbf{e}^t \approx 0 \bmod q$.

- Encryption algorithm takes as input, a message bit $\mu \in \{0, 1\}$ and a public key $\hat{\mathbf{A}}$, chooses a uniformly random $\mathbf{x} \in \{0, 1\}^m$ and generates the ciphertext $\mathbf{c} = \hat{\mathbf{A}} \cdot \mathbf{x} + (0, \mu \cdot \lceil q/2 \rceil) \in \mathbb{Z}_q^{n+1}$.
- Decryption algorithm uses the secret key $\mathbf{s}$ and computes $(-\mathbf{s}, 1)^t \cdot \mathbf{c} = (-\mathbf{s}, 1)^t \cdot \hat{\mathbf{A}} \cdot \mathbf{x} + \mu \cdot \lfloor q/2 \rfloor = \mathbf{e}^t \cdot \mathbf{x} + \mu \cdot \lfloor q/2 \rfloor \approx \mu \cdot \lfloor q/2 \rfloor \bmod q$. The decryption is correct as long as the maximal size of $\langle \mathbf{e}, \mathbf{x} \rangle$ is q/4.

As mentioned before, the security of this scheme relies on the hardness of LWE assumption with parameters n, q, χ, m.

Later on, Gentry et al. [18] constructed an LWE-based encryption scheme which is dual to Regev's encryption [41]. The dualism is represented as follows: in [41] the public keys are generated corresponding to a non-uniform LWE distribution with an unique secret key, while there are several options of ciphertext randomness which yield the same ciphertext. In the dual LWE [18], the public keys are uniformly and random, having many possible secret keys, while the encryption randomness is unique and outputs a certain ciphertext.

An enhancement of LWE cryptoscheme with smaller public and secret keys was provided by Lindner and Peikert [30], which uses a code-based cryptosystem of Alekhnovich [5] and adapts it to LWE.

5.3.3 Actively Secure Encryption

This is represented by encryption schemes which are indistinguishable against chosen-ciphertext attacks, in other words, they are IND-CCA secure. Fujisaki and Okamoto showed a technique on how to convert any IND-CPA cryptoscheme into an IND-CCA secure public key encryption scheme. A lattice-based instantiation of the Fujisaki-Okamoto technique was introduced by Peikert [36]. The pioneer work of actively secure encryption over lattices (LWE assumption) defined in a standard model, was defined by Peikert and Waters [38]. Their construction was based on a concept called lossy trapdoor function family, where the public key pk of a function $f_{pk} : X \to Y$ can either be generated along with a trapdoor sk, and f_{pk} is a function that can be inverted using sk, or it can be generated without any trapdoor. Finally, the public keys generated in these two different ways are indistinguishable.

5.3.4 Trapdoor Functions

These are functions which are easy to evaluate but hard to invert. They can be generated with some trapdoor information. Gentry et al. [18] were the first to show that certain types of trapdoor functions can be constructed from lattice problems. These trapdoor functions can be used in many applications such as digital signature schemes, identity-based and attribute-based encryption. The basic idea of such constructions in

[34] is to generate a collection of pre-image sampleable functions which are collision resistant and one-way trapdoor functions. To do so, let $\mathbf{B}$ be the public basis of lattice Λ. Then, in order to evaluate a function $f = f_{\mathbf{B}}$ from the collection of those preimage sampleable functions in a random lattice point $\mathbf{v} \in \Lambda$, we need to disturb it by a short error vector $\mathbf{e}$, such that $f(\mathbf{v}) = \mathbf{y} = \mathbf{v} + \mathbf{e}$. Inverting $f(\mathbf{v})$ means decoding the function to some lattice point $\mathbf{v}'$ which is not necessarily equal to $\mathbf{v}$, because of the disturbance there is more than one preimage of f.

5.3.5 Gadget Trapdoor

The core idea of gadget trapdoors is that they are represented by special matrices, called "gadget matrices" whose use makes solving LWE and SIS easy. This special primitive which is a favour of cryptographic trapdoors is significant for many constructions that have used gadget matrices as a building block. The easiest way to represent a gadget trapdoor $\mathbf{g}$ is a vector of powers of two, i.e. $\mathbf{g} = (1, 2, 4, \ldots, 2^{l-1}) \in \mathbf{Z}_q^l$, where $l = \log_2 q$. For a given $u \in \mathbb{Z}_q$, it is easy to find a solution to the equation

$$\langle \mathbf{g}, \mathbf{x} \rangle = u \bmod q$$

by decomposing u into its binary representation in $\{0, \ldots, q - 1\}$. This SIS problem can be adapted to an LWE problem as follows: given any vector $\mathbf{b}^t \approx s \cdot \mathbf{g}^t \bmod q$ with errors in the interval $[-q/4, q/4)$, discover the most-significant bit of $s \in \mathbb{Z}_q$.

The one-dimensional SIS and LWE problem as described above can be extended to n-dimensional SIS and LWE using the block-diagonal gadget matrix defined as

$$\mathbf{G} := \mathbf{I}_n \otimes \mathbf{g}^t = diag(\mathbf{g}^t, \ldots, \mathbf{g}^y) \in \mathbb{Z}_q^{n \times nl}.$$

The LWE problem is given by an approximate vector $\mathbf{b}^t \approx \mathbf{s}^t \cdot \mathbf{G} = (s_1 \mathbf{g}, \ldots, s_n \mathbf{g})^t$, from which we have to recover vector $\mathbf{s} = (s_1, \ldots, s_n)$. The corresponding SIS problem is to find a short solution to the equation $\mathbf{G} \cdot \mathbf{x} = \mathbf{u}$, given a vector $\mathbf{u} \in \mathbb{Z}_q^n$, meaning that we have to find short solutions for $\mathbf{g}^t \cdot \mathbf{x}_i = u_i \in \mathbf{Z}_q$ for each $i \in [n]$. This procedure to find a SIS solution can be expressed using a decomposition function

$$\mathbf{G}^{-1} : \mathbb{Z}_q^n \to \mathbb{Z}^{nl}; \qquad \mathbf{u} \mapsto \text{short solution} \quad \mathbf{x}, \text{ s.t.:} \quad \mathbf{Gx} = \mathbf{u} \bmod q.$$

This gadget trapdoor is also useful to generate trapdoors for a parity-check matrix $\mathbf{A} \in \mathbb{Z}_q^{n \times m}$, where such a matrix is essential for several lattice-based cryptoschemes and its trapdoor is a sufficiently short integer matrix. Let $\mathbf{R} \in \mathbb{Z}_q^{m \times nl}$ be such a trapdoor of $\mathbf{A}$, then the following equation holds: $\mathbf{AR} = \mathbf{HG} \bmod q$, where $\mathbf{H} \in \mathbf{Z}_q^{n \times n}$ is an invertible matrix, normally called the tag of the trapdoor $\mathbf{R}$. When the quality of a trapdoor is increasing, it's maximum norm is decreasing.

5.3.6 Digital Signatures without Trapdoors

The first paper in this topic was published by Lyubashevsky and Micciancio [33]. The authors developed a one-time signature which is convertible to many-time signature schemes using tree-hashing techniques. The security of their scheme is based on hardness of Ring-SIS assumption as defined in Def. 8. Because of the key size, signature size, message length and running time, the key generation, signature generation and verification procedure are all quasi-linear in the security parameter, i.e. $\tilde{\mathcal{O}}(\lambda)$. Due to the polylogarithmic factors used in the [33] scheme, that construction appears not to be practical in use. Therefore, the search for more efficient signatures lead to further contributions. Lyubashevsky [32] presented the first approach of a three-move identification scheme which can be converted into a non-interactive signature scheme applying Fiat-Shamir characteristics, where the resulting scheme is defined in the Random Oracle Model (ROM). In the following paragraph, we sketch the main idea of the lattice-based identification protocol from [32] as it represents a basic tool for several cryptographic constructions.

The prover's secret key is given by a short integer matrix $S \in \mathbb{Z}^{m \times l}$ and the corresponding public key is $B = A \cdot S \in \mathbb{Z}^{n \times l}$ for a given public parameter $A \in \mathbb{Z}^{n \times m}$. The interactive protocol consists of the following steps:

1. The prover chooses a "somewhat short" vector $v \in \mathbb{Z}^m$ and sends $y = A \cdot v$ to the verifier.
2. The verifier chooses a random challenge $c \in \mathbb{Z}^l$ being a short vector and sends it to the prover.
3. The prover computes $u = S \cdot c + v$
4. The prover decides whether to accept or to reject u. This procedure is called rejection sampling and is better explained in the next paragraph.
5. If the prover accepts u and sends it to the verifier, the latter checks $A \cdot u = B \cdot c + y$ and that y is sufficiently short.

5.3.6.1 Rejection Sampling

The goal of rejection sampling is to achieve independence from the distribution of u from the secret S. To do so, rejection sampling algorithm shifts the distribution of u to be a discrete Gaussian distribution with the center at zero instead of $S \cdot c$. To avoid the problem when u gets rejected too often, we assume that y has a discrete Gaussian distribution with parameters which are proportional to $\|S \cdot u\|$ such that the two distributions, one centered at zero and another at $S \cdot u$ have sufficient overlap. In a later work, Ducas et al. [15] provided further refinement of that idea.

Applying Fiat-Shamir heuristic to the interactive protocol yields a signature scheme with signature length of 60 kilobits making it practical for implementation.

5.3.7 Pseudorandom Functions (PRF)

This primitive was first introduced by Goldreich, Goldwasser, Micalli [21]. A pseudorandom function (PRF) with a secret seed s is a map from a certain domain to a certain range with the essential property that given a randomly chosen function from the family of such PRFs, it is infeasible to distinguish it from a uniformly random function given access to an oracle. The first PRF from lattice problems was constructed by Banerjee, Peikert, Rosen [8]. The authors first introduced the derandomized version of LWE, called learning with rounding (LWR). The main difference between this problem and LWE is that the error is deterministic. We shortly recall the LWR problem as follows: Let $s \in \mathbb{Z}_q^n$ be a secret and the LWR problem is to distinguish noisy and random inner products with s from a random value, i.e. for a given $a_i \in \mathbb{Z}_q^n$ and the rounding function $\lfloor \cdot \rceil_p : \mathbb{Z}_q \to \mathbb{Z}_p$ defined as $\lfloor x + q\mathbb{Z} \rceil_p = (x \cdot (q/p) + p\mathbb{Z}$, the samples are given as follows:

$$(a_i, b_i = \lfloor \langle a_i, s \rangle \rceil)_p \in \mathbb{Z}_q^n \times \mathbb{Z}_p$$

We recall the very first lattice-based PRF from [8] based on LWE problem. This construction is particularly randomness-efficient and practical. The function is given by a rounded subset product. Let $\{\mathbf{S}_i\}_{i \in [l]}$ be a set of secret keys given as short Gaussian-distributed matrices, an uniformly random vector $\mathbf{a} \in \mathbb{Z}_q$. Then, the lattice-based PRF is given by the following function:

$$F_{\mathbf{a},\{\mathbf{S}_i\}}(x) = \left\lfloor \mathbf{a}^{tr} \cdot \prod_{i=1}^{l} \mathbf{S}_i^{x_i} \right\rfloor_p$$

An important feature of key homomorphism for lattice-based PRFs has been introduced by Boneh et al. [10]. The authors provided the first standard-model construction of key-homomorphic PRF secure under LWE assumption. This additional property plays a significant role in distributing the function of key generation center, as it satisfies the following function $F_{k_1}(x) + F_{k_2}(x) = F_{k_1+k_2}(x)$, for different secret keys k_1, k_2.

Furthermore, some other more advanced constructions have been developed in the last decade. These are:

5.3.8 Homomorphic Encryption

The first concept of fully homomorphic encryption (FHE) has been proposed by Rivest, Adleman and Dertouzos in 1978 [42]. The main idea of homomorphic encryption is to allow computation on encrypted data. In other words: given some data m, one can compute a ciphertext which encrypts an evaluated data $f(m)$ for any desired function f. The first FHE from lattices was introduced by Gentry [17], which is based on average-case assumption about ideal lattices. Brakerski and Vaikuntanathan [12] introduced a new version of FHE from lattices, which is based on the standard LWE assumption. We shortly recall their scheme as it was used in many applications and served as a building block for several constructions.

The scheme in [12] supports addition and multiplication of multiple ciphertexts, while each ciphertext encrypts a single bit. Exploiting the homomorphic property of a scheme, we can evaluate any Boolean circuit. The secret key in this scheme is an LWE secret $\mathbf{s} \in \mathbb{Z}^n$. The encryption of a bit $\mu \in \{0, 1\}$ is given by an LWE sample for an odd modulus q. The error term is an encoding of the message as its least-significant bit. For the ciphertext $\mathbf{c} \in \mathbb{Z}_q^n$ the following relations hold: $\langle \mathbf{s}, \mathbf{c} \rangle = \mathbf{s}^t \cdot \mathbf{c} = e \bmod q$, with $e \in \mathbb{Z}$ being a small error, and $e \in \mu + 2\mathbb{Z}$. To decrypt $\mathbf{c}$, we simply compute $\langle \mathbf{s}, \mathbf{c} \rangle$ and lift the result to the unique representative of $e \in \mathbb{Z} \cup [-q/2, q/2)$. We output $\mu = e \bmod 2$.

As mentioned earlier, the scheme in [12] supports addition and multiplication, where the former operation is a straightforward computation in these schemes, while the latter represents a bigger challenge, to achieve the feature of multiplication of different ciphertexts. For reasons of simplicity, we observe two different ciphertexts which we want to evaluate, exploiting homomorphic property of the scheme. When we add two ciphertexts $\mathbf{c}_1, \mathbf{c}_2$, we get an encryption of the sum of the corresponding plaintexts μ_1, μ_2, as we can see:

$$\langle \mathbf{s}, \mathbf{c}_1 + \mathbf{c}_2 \rangle = \langle \mathbf{s}, \mathbf{c}_1 \rangle + \langle \mathbf{s}, \mathbf{c}_2 \rangle = e_1 + e_2 \bmod q$$

And as shown above for this error term holds $e_1 + e_2 \in (\mu_1 + \mu_2) + 2\mathbb{Z}$. The problem here is that the number of different ciphertexts cannot be unbounded as the error magnitude will exceed the limit of $q/2$.

To multiply two ciphertexts, we use the mathematical construct called tensor product $\mathbf{c}_1 \otimes \mathbf{c}_2 \in \mathbb{Z}_q^{n^2}$, which is an encryption of $\mu_1 \cdot \mu_2$, under the secret key $\mathbf{s} \otimes \mathbf{s}$. Here also, the number of multiplication factors is bounded from the beginning. One of the main drawbacks of homomorphic encryption is that it always increases the error rate of a ciphertext. In the same paper [12], the authors introduced a technique to improve this issue, which is called 'key switching technique'. This technique allows the conversion of a ciphertext that encrypts some message μ into another ciphertext, that still encrypts the same message, but under some different secret key $\mathbf{s}'$ and uses the previously introduced gadget trapdoor. For further details we refer to the original paper.

Another idea for the solution, called bootstrapping, was introduced by Gentry [17]. The idea involves a technique which reduces the error rate of a ciphertext and allows unbounded homomorphic computation. The technique is to homomorphically evaluate the decryption function on a low-error encryption of the secret key, which represents a part of the public key.

An alternative scheme of homomorphic encryption was introduced by Gentry et al. [19] in 2013 which has some attractive properties. In order to perform homomorphic evaluations of the ciphertext, no key-switching technique is required. The scheme [19] can also be adapted to an identity-based-encryption or an attribute-based encryption. An extension of these schemes was proposed by [27], defining identity-based encryption and attribute-based encryption schemes in multi-identity and multi-authority settings, respectively.

5.3.9 Identity-Based Encryption (IBE)

An Identity-Based Encryption was first introduced by Shamir [44] and has the useful feature where any string can serve as a public key. The secret key is generated by an authority, called a key generation center (KGC), which has control over some master secret key. On input, this master secret key and some public string, the KGC derives a secret key for a user who can be identified by the mentioned public string. Every user can encrypt a certain message for a receiver, using the corresponding public string indicating the receiver's identity. The first lattice-based IBE scheme was proposed by Gentry et al. [18] and is defined in the random oracle model. Since real random functions are not practical in use, it is still useful to assume that a hash function can behave like a truly random function and so provide a security proof. Even though such models would not be secure, when the pretended random function is replaced by a real hash function, it is still better to provide a construction in the random oracle model, as it will enable the understanding of some obvious attacks. The construction in [18] is, so far, most efficient IBE scheme which is secure against quantum attacks. The main idea behind the [18] construction is that the secret keys of each user are represented by a signature of the user's identity generated by the KGC. The corresponding public key is simply the hash value of the user's identity string. To gain a better understanding of the basic IBE scheme, we provide a short overview of it. An IBE scheme encompasses the following steps:

– The master public key is represented by a uniformly random parity-check matrix $\mathbf{A} \in \mathbb{Z}_q^{n \times m}$, while the corresponding master secret key is a trapdoor for this matrix $\mathbf{A}$. There is a public hash function which takes as input an identity string id and maps it into a syndrome $\mathbf{y}_{id} \in \mathbb{Z}_q^n$, which is a n-dimensional vector with coefficients reduced modulo q. This syndrome is used in combination with the public key $\mathbf{A}$ to represent a user-specific public key $pk_{id} := \mathbf{A}_{id} = [\mathbf{A}|\mathbf{y}_{id}]$.

- Next, to extract a secret key for the above-mentioned identity string *id*, we use a trapdoor to run the Gaussian sampling algorithm as defined in [18]. The algorithm samples from a discrete Gaussian over an n-dimensional lattice Λ. Discrete Gaussians over lattices, are useful mathematical tools to study the complexity of lattice problems [35]. To sample from a discrete Gaussian the algorithm takes the parameters c, σ, as input representing the standard deviation and the center of a distribution $D_{L,\sigma,c}$. It also takes as input a trapdoor basis $\mathbf{B}$ of the lattice $L_{\mathbf{y}_{id}}^{\perp}(\mathbf{A})$ and outputs a lattice vector as long as the length of all the Gram-Schmidt (see Definition of Gram-Schmidt Norm in Section 2.2) vectors of basis $\mathbf{B}$ are exceeded by the standard deviation parameter. The Gaussian-distributed solution of key extraction is given by the vector $\mathbf{x}_{id}$ such that the following equation holds: $\mathbf{A} \cdot \mathbf{x}_{id} = \mathbf{y}_{id}$. The secret key $\mathbf{x}_{id}$ solves also the equation $\mathbf{A}_{id} \cdot (-\mathbf{x}_{id}, 1) = 0$.
- To encrypt a message bit $\mu = \{0, 1\}$ using identity *id* as a public key, the encryptor uses the technique of dual LWE scheme, the given user-specific public key $\mathbf{A}_{id}$ and computes the ciphertext as $\mathbf{c}_{id}^{t} \approx \mathbf{s}^{t} \cdot \mathbf{A} + (0, \mu + \lfloor q/2 \rfloor)^{t}$. The corresponding decryption of this ciphertext is computed using the user's secret key $\mathbf{x}_{id}$ as follows: $\mathbf{c}_{id}^{t} \cdot (-\mathbf{x}_{id}, 1) \approx \mu \cdot \lfloor q/2 \rfloor$.

An IBE scheme which is defined in the standard model, was later defined by Cash et al. [14]. In the same work, the authors provided a construction of an hierarchical identity-based encryption (HIBE) which allows any user to use their secret key in a secure way to delegate it to any subordinate user in hierarchy.

5.3.10 Attribute-Based Encryption

The first construction of attribute-based encryption (ABE) was introduced by Sahai and Waters [43]. It represents a generalization of identity-based encryption. There are two flavours of ABE, ciphertext-policy ABE (CP-ABE) and key-policy ABE (KP-ABE). In the first case (CP-ABE), the user's secret key is generated based on a certain set of attributes which satisfy a predicate formula. The ciphertext is computed by embedding that predicate formula, which is often called attribute policy. The user who is in possession of the attribute-based secret key is able to decrypt the received ciphertext if and only if their secret key's attribute set satisfies the predicate formula. In the second case (KP-ABE), the secret key is generated using an attribute predicate, while the attribute set is embedded into the ciphertext. The first lattice-based construction of ABE has been provided by Agrawal et al. [2] which is inherited from the lattice-based HIBE construction of [1]. An ABE for arbitrary predicates, expressed as circuits has been introduced by Gorbunov et al. [22]. The first lattice-based ABE scheme was proposed by Agrawal et al. [2] and represents an adaptation of the lattice-based HIBE scheme [1]. Here we recall this scheme.

First of all, we state that a finite field $\mathbb{F} = \mathbb{F}_{q^n}$ is isomorphic to a matrix sub-ring $\mathcal{H} \subseteq \mathbb{Z}_q^{n \times n}$, for a prime q and $\mathbb{F}_q = \mathbb{Z}_q$. Then for an attribute vector $\mathbf{h} = (\mathbf{H}_1, \ldots, \mathbf{H}_l) \in \mathcal{H}^l$ of length l the following holds:

- The master public key is given by the matrix $\overline{\mathbf{A}} \in \mathbb{Z}_q^{n \times \hat{m}}$ which is generated using a trapdoor as shown in Section 3.5, where $\hat{m} = m - nl$. Further components of the public key is a syndrome $\mathbf{u} \in \mathbb{Z}_q^b$ a uniformly random matrix $\mathbf{A} \in \mathbb{Z}_q^{n \times wl}$, which is a concatenation of l matrices, i.e.

$$\mathbf{A} = [\mathbf{A}_1 | \mathbf{A}_2 | \dots | \mathbf{A}_k],$$

where each $\mathbf{A}_i \in \mathbb{Z}_q^{n \times wl}$. The corresponding master secret key is the trapdoor that was used to generate $\overline{\mathbf{A}}$.

– As the scheme in [2] is a KP-ABE construction, the secret key is generated based on a predicate vector $\mathbf{p} = (\mathbf{P}_1, \dots, \mathbf{P}_l) \in \mathcal{H}^l$ using the trapdoor for $\overline{\mathbf{A}}$. A short integer matrix is defined using the Gadget trapdoor from Section 3.5:

$$\mathbf{S}_{\mathbf{p}}^t = [\mathbf{G}^{-1}(\mathbf{P}_1 \mathbf{G}), \dots, \mathbf{G}^{-1}(\mathbf{P}_l \mathbf{G})].$$

Using this matrix we define $\mathbf{B}_{\mathbf{p}} = \mathbf{A} \cdot \mathbf{S}_{\mathbf{p}}$ and sample a Gaussian-distributed solution using a sampling algorithm described in the previous Section 3.9., i.e. $[\overline{\mathbf{A}} | \mathbf{B}_{\mathbf{p}}] \cdot \mathbf{x}_{\mathbf{p}} = \mathbf{u}$.

– The encryption is generated using a set of attributes $\mathbf{h} \in \mathcal{H}^l$ as defined above. Using the Gadget trapdoor $\mathbf{G}$ from Section 3.5, an attribute-based trapdoor is defined as follows: $\mathbf{G}_{\mathbf{h}} = [\mathbf{H}_1 \mathbf{G} | \dots | \mathbf{H}_l \mathbf{G}]$. Using this trapdoor $\mathbf{G}_{\mathbf{h}}$, define $\mathbf{A}_{\mathbf{h}} = \mathbf{A} + \mathbf{G}_{\mathbf{h}}$. The attribute-based public key is defined as a concatenation $[\overline{\mathbf{A}} | \mathbf{A}_{\mathbf{h}} | \mathbf{u}]$. Using this key, a message $\mu \in \{0, 1\}$ is encrypted applying the dual LWE encryption with the corresponding LWE secret $\mathbf{s}$. The ciphertext is given as follows: $[\overline{c} | c_{\mathbf{h}} | c] = \mathbf{s}^t \cdot [\overline{\mathbf{A}} | \mathbf{A}_{\mathbf{h}} | \mathbf{u}] + (0, \mu \cdot \lfloor q/2 \rfloor)^t$.

– Decryption algorithm of this scheme by firstly multiplying the second component of the ciphertext by $\mathbf{S}_{\mathbf{p}}$ and secondly computing

$$[\overline{c} | c_{\mathbf{h}} \cdot \mathbf{S}_{\mathbf{p}}] \cdot \mathbf{x}_{\mathbf{p}} \approx [\overline{\mathbf{A}} | \mathbf{B}_{\mathbf{p}}] \cdot \mathbf{x}_{\mathbf{p}} \approx c - \mu \lfloor q/2 \rfloor$$

An ABE scheme arbitrary for any predicates represented as a priory bounded depth circuits based on a Lattice assumption was proposed by Gorbunov et al. [22]. The secret key in this scheme grows proportionally to the size of the circuit. An improvement of this scheme was later proposed by Boneh et al. [9].

5.4 Lattice-Based Cryptography for IoT

Nowadays, we live in a world where more and more smart devices such as mobile phones, TV, smart household appliances, and cars are connected to the Internet. Most of such devices are equipped with wireless sensors which establish an information flow between the Internet and the device. For instance, many smart medical devices collect health information from customers and forward it to an Internet-based database. Therefore, it is obvious that security and privacy of user's sensitive data, are the most significant challenges in such IoT systems. We know that next to symmetric cryptography, public key cryptography is one of the fundamental tools in IoT security. While the former provides security against quantum attacks, the latter is assumed to be vulnerable to these attacks. Lattice-based cryptography as one promising topic of quantum-based cryptography, offers solutions to IoT systems which one day may be exposed to quantum attacks.

IoT applications envision lattice-based cryptography as a promising candidate for strong security primitives. The future smart IoT devices will be equally vulnerable to quantum threats and it would be extremely difficult to update their security settings from classical to quantum-secure cryptography. Moreover, the computational and storage efficiency of lattice-based primitives also motivates the adoption of it for lightweight devices. Existing implementations of R-LWE on an 8-bit microcontroller can finish faster than RSA-1024 [31].

In [28], the authors reviewed the important role of attribute-based and identity-based authentication systems in IoT. They discussed the main opportunities and challenges which are faced in IoT by using attribute-based authentication schemes. The authors also mention the significant meaning of attribute-based signature (ABS) schemes in IoT, which represent an alternative to ABE schemes, where the signer generates a signature of a certain message using an attribute-based secret key. The verification succeeds if and only if the signer can prove to the verifier, that they have the right key based on a set of attributes, which satisfy a certain predicate. The lattice-based research on this area of cryptoscheme is not big; only few papers have been proposed in the recent years [16, 26, 45]. Therefore, because of the significant role of ABE and ABS schemes in IoT and the future perspective of quantum computers, the research topic of lattice-based ABE and ABS is particularly interesting and a direct application of these schemes to IoT is useful.

In [39], an R-LWE-based signature scheme 'BLISS' is implemented to ascertain the feasibility of BLISS for lightweight devices. The authors provided a scalable implementation of BLISS on a Xilinx Spartan-6 FPGA with optional 128-bit, 160-bit, or 192-bit security. Other variants of BLISS are also available in literature in order to speed up the key generation process.

Homomorphic encryption is another cryptographic primitive which attracts more and more IoT developers because of its useful features which allow the evaluation of encrypted data. Often sensitive data in the IoT has to be processed in the cloud. In [29], the authors showed how confidentiality can be guaranteed in the cloud by using homomorphic encryption. They also proposed an acceleration mechanism to speed up the homomorphic evaluation.

Pseudorandom functions play a significant role in IoT too, as shown recently in [24]. As mentioned by the authors, remote software update procedures are becoming more useful in automative industry. Therefore, security of those procedures are considered to play a core role in that sector. Message authentication codes and pseudorandom functions (PRFs) are particularly helpful in the provision of security of the remote software updates. Thus, with the future move to quantum computers, quantum-resistant constructions will be required, where lattice-based schemes, in this particular case, the PRFs could offer an efficient and practical solution.

The post-quantum key exchange is the most challenging phenomenon to address. The initial effort was made by Bos et al. [11] to transform classical TLS to the R-LWE version of TLS considering post-quantum requirements. With a small performance penalty their R-LWE key exchange was successfully embedded in TLS and implemented on web servers. Alkim et al. [7] presented an efficient implementation of R-LWE-based key exchange protocol, namely New Hope [6]. Initially, their proposal was to implement it on large Intel processors however, their implementation is finally optimized for ARM Cortex-M family of 32-bit microcontrollers. Moreover, within Cortex-M family, they choose Cortex-M0 as low-end and Cortex-M4 as high-end targets. The authors achieved 128-bit security in a post-quantum setting while implementing these embedded microcontrollers. Following the lattice-based construction of NTRUEncrypt which is accepted as IEEE standard P1363.1, Guillen et al. [23] implemented NTRUEncrypt for Cortex-M0 based microcontrollers. This ensures again the feasibility of lattice-based encryption on IoT devices. Another implementation of R-LWE-based encryption scheme is presented by Liu et al. [31] which is computationally equivalent to NTRUEncrypt. The authors achieved 46-bit security with encryption and decryption timings as 24.9 ms and 6.7 ms respectively, on an 8-bit ATxmega128 microcontroller.

5.5 Conclusion

Our main target in this contribution was to provide a general overview of lattice-based primitives and to summarize how these constructions can be applied to IoT. We first introduced the reader to the complex notions of lattice-based cryptography, yet tried to avoid details which are too specific for the concept of this chapter. Then, using these notions, we provided a summary of lattice-based constructions. Finally, in the last section, we reviewed the state-of-the-art applications of cryptographic primitives to the IoT systems, where most of the existing applications are based on classical number-theoretic assumptions, we want to catch the reader's attention to the importance of a switch to quantum-resistant constructions, where lattice-based constructions provide a powerful solution.

References

1 Agrawal, S., Boneh, D. and Boyen, X. (2010). Efficient lattice (H)IBE in the standard model. In: *Advances in Cryptology –EUROCRYPT 2010. Lecture Notes in Computer Science*, Vol. 6110 (ed. H. Gilbert), 553–572. Berlin, Heidelberg: Springer.

2 Agrawal, S., Freeman, D.M. and Vaikuntanathan, V. (2011). Functional encryption for inner product predicates from learning with errors. *Advances in Cryptology -CRYPTO 2011, Proceedings*. Santa Barbara, USA (14–18 August 2011). Springer.

3 Ajtai, M. (1996). Generating hard instances of lattice problems (extended abstract). *STOC '96 Proceedings of the twenty-eighth annual ACM symposium on Theory of Computing*, Philadelphia, USA (22–24 May). ACM.

4 Ajtai, M. and Dwork, C. (1997). A public-key cryptosystem with worst-case/average-case equivalence. *STOC '97 Proceedings of the twenty-ninth annual ACM symposium on Theory of computing*, El Paso, USA, (04–06 May 2019). ACM.

5 Alekhnovich, M. (2003). More on average case vs approximation complexity. 44th *Symposium on Foundations of Computer Science (FOCS 2003)*, Cambridge, USA (11–14 October 2003). IEEE Computer Society.

6 Alkim, E., Ducas, L., Pöppelmann, T. and Schwabe, P. (2016). Post-quantum key exchange a new hope. USENIX Security Symposium. https://www.usenix.org/system/files/conference/usenixsecurity16/sec16_paper_alkim.pdf (accessed 25 June 2019).

7 Alkim, E., Jakubeit, P., and Schwabe, P. (2016). New hope on arm cortex-m. *International Conference on Security, Privacy, and Applied Cryptography Engineering*, Hyderabad, India (14–18 December 2016). Springer.

8 Banerjee, A., Peikert, C. and Rosen, A. (2012). Pseudorandom functions and lattices. *Advances in Cryptology. EUROCRYPT 2012. Proceedings*, Cambridge, UK (15–19 April 2012). Springer.

9 Boneh, D., Gentry, C., Gorbunov, S. et al. (2014). Fully key-homomorphic encryption, arithmetic circuit ABE and compact garbled circuits. *Advances in Cryptology - EUROCRYPT 2014 - Proceedings, volume 8441 of Lecture Notes in Computer Science*, Copenhagen, Denmark (11–15 May 2014). Springer.

10 Boneh, D., Lewi, K., Montgomery, H.W. and Raghunathan, A. (2013). Key homomorphic prfs and their applications. *Advances in Cryptology - CRYPTO 2013. Proceedings, Part I, volume 8042 of Lecture Notes in Computer Science*, Santa Barbara, USA (18–22 August 2013). Springer.

11 Bos, J.W., Costello, C., Naehrig, M. and Stebila, D. (2015). Post-quantum key exchange for the tls protocol from the ring learning with errors problem. *2015 IEEE Symposium on Security and Privacy (SP)*, San Jose, USA (17–21 May 2015). IEEE.

12 Brakerski, Z. and Vaikuntanathan, V. (2011). E_cient fully homomorphic encryption from (standard) LWE. *IEEE 52nd Annual Symposium on Foundations of Computer Science*, Washington, USA (22–25 October 2011). IEEE Computer Society.

13 Brakerski, Z. and Vaikuntanathan, V. (2015). Constrained key-homomorphic prfs from standard lattice assumptions - or: How to secretly embed a circuit in your PRF. *Theory of Cryptography - Proceedings, Part II, volume 9015 of Lecture Notes in Computer Science*, Warsaw, Poland (23–25 March 2015). Springer.

14 Cash, D., Hofheinz, D., Kiltz, E. and Peikert, C. (2010). Bonsai trees, or how to delegate a lattice basis. *EUROCRYPT 2010, volume 6110 of LNCS*, French Riviera, France (30 May–3 June 2010). Springer.

15 Ducas, L., Durmus, A., Lepoint, T. and Lyubashevsky, V. (2013). Lattice signatures and bimodal gaussians. *Advances in Cryptology - CRYPTO 2013 - 33rd Annual Cryptology Conference*, Santa Barbara, USA (18–22 August 2013). Springer.

16 El Kaafarani, A. and Katsumata, S. (2018). Attribute-based signatures for unbounded circuits in the ROM and e_cient instantiations from lattices. *Public-Key Cryptography - PKC 2018, Proceedings, Part II*, Rio de Janeiro, Brazil (25–29 March 2018). Springer.

17 Gentry, C. (2009). Fully homomorphic encryption using ideal lattices. *Proceedings of the 41st Annual ACM Symposium on Theory of Computing, STOC*, Bethesda, USA (31 May–2 June 2009). ACM.

18 Gentry, C., Peikert, C. and Vaikuntanathan, V. (2008). Trapdoors for hard lattices and new cryptographic constructions. *STOC 08' Proceedings of the fortieth annual ACM symposium on Theory of computing*, Victoria, Canada (17–20 May 2008). ACM.

19 Gentry, C., Sahai, A., and Waters, B. (2013). Homomorphic encryption from learning with errors: Conceptually simpler, asymptotically faster, attribute based. *Advances in Cryptology –CRYPTO 2013*, Santa Barbara, USA (18–22 August 2013). Springer.

20 Goldreich, O., Goldwasser, S, and Halevi, S. (1997). Public-key cryptosystems from lattice reduction problems. *Advances in Cryptology - CRYPTO '97, 17th Annual International Cryptology Conference*, Santa Barbara, USA (17–21 August 1997). Springer.

21 Goldreich, O., Goldwasser, S. and Micali, S. (1986). How to construct random functions. *Journal of the ACM* 33 (4): 792–807.

22 Gorbunov, S., Vaikuntanathan, V. and Wee, H. (2013). Attribute-based encryption for circuits. *Symposium on Theory of Computing Conference, STOC 2013*, Palo Alto, USA (1–4 June 2013). ACM.

23 Guillen, O.M., Poppelmann, T., Mera, J.M.B. et al. (2017). Towards post-quantum security for iot endpoints with ntru. *Proceedings of the Conference on Design, Automation & Test in Europe*, Lausanne, Switzerland (27–31 March 2017). European Design and Automation Association.

24 Hirose, S., Kuwakado, H. and Yoshida, H. (2018). A pseudorandom-function mode based on lesamnta-lw and the MDP domain extension and its applications. *IEICE Transactions* 101-A (1):110–118.

25 Hoffstein, J., Pipher, J. and Silverman, J.H. (1998). NTRU: A ring-based public key cryptosystem. *Algorithmic Number Theory, Third International Symposium, ANTS-III, Proceedings*, Portland, USA (21–25 June 1998). Springer.

26 Kuchta, V., Sahu, R.A., Sharma, G. and Markowitch, O. (2017) On new zero-knowledge arguments for attribute-based group signatures from lattices. *Information Security and Cryptology –ICISC 2017 –20th International Conference*, Seoul, South Korea (29 November–1 December 2017). Springer.

27 Kuchta, V., Sharma, G., Sahu, R.A. and Markowitch, O. (2017). Multi-party (leveled) homomorphic encryption on identity-based and attribute-based settings. *Information Security and Cryptology - ICISC 2017*, Seoul, South Korea (29 November–1 December 2017). Springer.

28 Lam, K. and Chi, C. (2016). Identity in the internet-of-things (iot): New challenges and opportunities. *Information and Communications Security ICICS 2016 –Proceedings*, Singapore, Singapore (29 November–2 December 2016). Springer.

29 Leveugle, R., Mkhinini, A. and Maistri, P. (2018). Hardware support for security in the internet of things: From lightweight countermeasures to accelerated homomorphic encryption. *Information* 9 (5): 114.

30 Lindner, R. and Peikert, C. (2011). Better key sizes (and attacks) for lwe-based encryption. *Topics in Cryptology –CT-RSA 2011 - The Cryptographers Track at the RSA Conference 2011*, San Francisco, USA (14–18 February 2011). Springer.

31 Liu, Z., Pöoppelmann, T., Oder, T. et al. (2017). High-performance ideal lattice-based cryptography on 8-bit avr microcontrollers. *ACM Transactions on Embedded Computing Systems (TECS)* 16 (4): 117.

32 Lyubashevsky, V. (2008). Lattice-based identification schemes secure under active attacks. *Public Key Cryptography –PKC 2008, Proceedings, volume 4939 of Lecture Notes in Computer Science*, Barcelona, Spain (9–12 March 2008). Springer.

33 Lyubashevsky, V. and Micciancio, D. (2008). Asymptotically efficient lattice-based digital signatures. *Theory of Cryptography, Fifth Theory of Cryptography Conference*, New York, USA (19–21 March 2008). Springer.

34 Lyubashevsky, V., Micciancio, D., Peikert, C. and Rosen, A. (2008). SWIFFT: A modest proposal for FFT hashing. *Fast Software Encryption, 15th International Workshop, FSE*, Lausanne, Switzerland (10–13 February 2008). Springer.

35 Peikert, C. (2007). Limits on the hardness of lattice problems in l-p norms. *22nd Annual IEEE Conference on Computational Complexity CCC 2007*, San Diego, USA (13–16 June 2007). IEEE Computer Society.

36 Peikert, C. (2014). Lattice cryptography for the internet. *Post-Quantum Cryptography – 6th International Workshop, PQCrypto 2014*, Waterloo, Canada, (1–3 October 2014). Springer.

37 C. Peikert. (2016). A decade of lattice cryptography. *Foundations and Trends in Theoretical Computer Science* 10 (4): 283–424.

38 Peikert, C. and Waters, B. (2008). Lossy trapdoor functions and their applications. *Proceedings of the 40th Annual ACM Symposium on Theory of Computing*, Victoria, Canada (17–20 May 2008). ACM.

39 Pöoppelmann, T., Ducas, L. and Guneysu, T. (2014). Enhanced lattice-based signatures on reconfigurable hardware. *International Workshop on Cryptographic Hardware and Embedded Systems*, Busan, South Korea (23–26 September 2014). Springer.

40 Regev, O. (2003). New lattice based cryptographic constructions. *STOC '03 Proceedings of the thirty-fifth annual ACM symposium on Theory of computing*, San Diego, USA (09–11 June 2003). ACM.

41 Regev, O. (2005). On lattices, learning with errors, random linear codes and cryptography. *STOC '05 Proceedings of the thirty-seventh annual ACM symposium on Theory of computing*, Baltimore, USA (22–24 May 2005). ACM.

42 Rivest, R., Adleman, L. and Dertouzos, M. (1978). On data banks and privacy homomorphisms. *Foundations of secure computation* 4 (11): 169–180.

43 Sahai, A. and Waters, B. (2005). Fuzzy identity-based encryption. *EUROCRYPT 2005, Proceedings*, Aarhus, Denmark (22–26 May 2005). Springer.

44 Shamir, A. (1984). Identity-based cryptosystems and signature schemes. *Advances in Cryptology, Proceedings of CRYPTO '84*, Santa Barbara, USA (19–22 August 1984). Springer.

45 Wang, Q., Chen, S. and Ge, A. (2015). A new lattice-based threshold attribute-based signature scheme. *Information Security Practice and Experience – 11th International Conference, ISPEC 2015*, Beijing, China (5–8 May 2015). Springer.

Part III

IoT User Level Authentication

6

Efficient and Anonymous Mutual Authentication Protocol in Multi-Access Edge Computing (MEC) Environments

Pardeep Kumar and Madhusanka Liyanage

Abstract

Multi-access edge computing (MEC) is an evolving paradigm of the Internet of things (IoT) applications. The MEC is a complement to traditional cloud computing where services are extended closer to the network and so to the end users. As mobile users can use MEC services in an inter-domain, security is one of the challenging questions, how to protect IoT applications in MEC environments from abuses? In addition, considering real-world MEC supported IoT applications (e.g. airport) where a user is always on the move from one network to another network. This scenario also poses many security challenges. To mitigate this, an authentication mechanism can play an important role to defend MEC from unauthorized access. Thus, an authentication mechanism is needed that can support mobility for MEC users. Moreover, establishing a session key is also highly desirable between the MEC users and foreign-edge servers to enable secure communication in MEC environments. In addition, how to maintain users' anonymity is another important security requirement, as MEC users do not want to disclose their private information. To solve these issues, this chapter proposes a new efficient and anonymous mobility supported mutual authentication scheme in MEC environments. The scheme utilizes the password and smart-card as two-factor authentications and facilitates many services to the users such as user anonymity, mutual authentication, and secure session key establishment in mobility supported environments. In addition, it allows users to choose/update their password regularly, whenever needed. Security and performance evaluation show the practicality of the proposed scheme.

Keywords *multi-access edge computing; internet of things; authentication; secure mobility*

6.1 Introduction

Multi-access edge computing (MEC) is one of the new paradigms in the internet of things (IoT) applications [1]. Following [2], many billions of devices are going to be deployed and connected via the Internet to support many of the real-world IoT and cyber physical applications. In IoT applications, several low-cost sensors or smart objects aggregate data from their respective environments and send it to cloud servers, where the data will be processed and analyzed for further analysis of

IoT Security: Advances in Authentication, First Edition.
Edited by Madhusanka Liyanage, An Braeken, Pardeep Kumar, and Mika Ylianttila.

the applications. In this scenario, indeed cloud computing technology has shown its efficiency as it has enormous data processing, high-computational power, and data storage and so on. However, in many IoT applications (e.g. IoT-healthcare, airport, etc.), the security of such traditional cloud computing can become debatable when the processed data grows significantly as it will suffer more latency. That can result in a lack of efficiency, which is challenging in such traditional cloud environments. In other words, as more and more IoT devices are involved in people's lives, the traditional cloud architecture can barely meet their requirements of mobility support, location awareness, and low latency. Thanks to the new paradigm, i.e. MEC, has been brought in recently to solve the problem of traditional cloud computing. In MEC, traditional cloud computing responsibilities are extended closer to the edge of the network and so to the end users [2]. Therefore, with the new paradigm, an edge node is responsible for performing all the operations (e.g. computation, communication, and storage) closer to the end user by pooling networks' local resources. As shown in Figure 6.1, traditional cloud services are extended closer to the MEC layer, where edge nodes can host the end users and the resource-constrained IoT devices (e.g. wireless sensor network). Such multi-layered network models allow distributed computing to run as close as possible to the sensor network and to the end users. This multilayered network model can support a wide range of distributed applications, e.g. IoT-healthcare [3], IoT-airports [4], smart vehicular network [5], etc.

With the new MEC paradigm, the prospect of new challenges appears. Security is one the main concerns including privacy. For instance, consider an end user, who owns a local private cloud and would like to turn the local private cloud into the *edge node* as a *server* and lease spare resources from the local private cloud. This feature raises trust issues in a multilayered MEC network and makes it is one of the most important challenges of such a network model. In addition, as the end user moves its topology

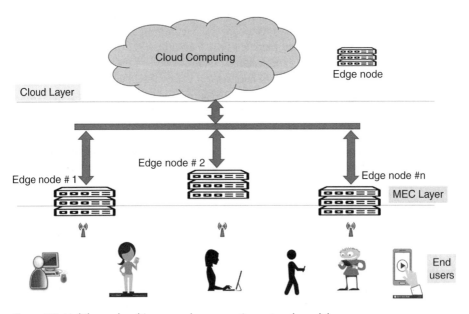

Figure 6.1 Multilayered multi-access edge computing network model.

and path changes regularly, as shown in Figure 6.1. In such settings, an attacker can counterfeit a rogue edge node in the network and pretend to be a legitimate edge node. Therefore, authentication is vital when a user is on the move in an MEC network.

In a system or network, an adequate authentication is the first measure that can verify users' identities. To achieve this, several schemes have been proposed, e.g. Imine et al. [6], Amor et al. [7], and the schemes proposed in [8, 9]. Most of these approaches, incur a high amount of computation overhead to perform device authentication. Moreover, most of these schemes do not support the mobility, except the scheme proposed in [8]. In addition, several authentication mechanisms have been proposed for the traditional multiserver environments in Hassan et al. [10], Han and Zhu [11], Tseng et al. [12], and Irshad et al. [13]. However, the scheme proposed in [11, 12] is vulnerable to many attacks as pointed out in [10, 14].

In this chapter, we design a new, efficient and anonymous mobility supported mutual authentication scheme for use in MEC environments. We present a secure mobility use-case, where various MEC nodes are deployed and a user is on move (e.g. from one network to another) within an airport setting [8]. The proposed scheme utilizes the password and smartcard (SC)-based approach. Our approach not only performs adequate authentication, but it also establishes a secure session-key between two end parties between all the communicating entities. Moreover, the proposed provides paramount and robust security services to the patient (i.e. a mobile node). Our preliminary analysis shows that our scheme is feasible for MEC environments where mobility is in high demand, for example, an airport, in-hospital, etc.

The rest of the chapter is organized as follows. Section 6.2 discusses the related work, and Section 6.3 describes the network model adversary model. The proposed scheme is discussed and presented in Section 6.4. Security analysis and performance evaluation, and conclusion are presented in Sections 6.5 and 6.6, respectively.

6.2 Related Work

This section discusses the most relevant state-of-the-art schemes in MEC and Fog computing proposed recently. In [6], Imine et al. proposed an efficient mutual authentication scheme for fog computing architecture. The scheme ensures mutual authentication and remedies to fog servers' misbehaviors, if any. In Imine et al.'s scheme, a fog server needs to hold only a couple of pieces of information to verify the authenticity of every user in the system. The main idea of this scheme is to utilize the blockchain technology and secret-sharing technique. The blockchain is maintained by fog/edge nodes and it allows end users to authenticate any fog node in the architecture. In addition, it allows fog nodes to establish mutual authentication with each other. Indeed, the blockchain has its own advantages but blockchain is a resource-consuming technology as it needs a large storage and very high-computational power.

Amor et al. proposed a privacy-preserving authentication scheme in an edge-fog environment [7]. The proposal basically relies on the introduction of a mutual authentication between fog users at the edge of the network and fog servers at the fog layer. The scheme utilizes pseudonym-based cryptography (PBC), elliptic curve discrete logarithm problem (ECDLP) and bilinear pairing to establish the session key. However, Amor et al.'s scheme incurred a high computation cost, as it has utilized the public key encryption and decryption operations, which are expensive for low-powered IoT devices.

In [8], the author proposed a secure and efficient mutual authentication scheme for the edge-fog-cloud network architecture that mutually authenticates fog users at the edge of the network. The scheme proposed in [8] also supports mobility to the fog users. The main idea of the scheme is to utilize a long-lived master secret key, allowing a fog user to communicate and authenticate with any of the fog servers in the network when a user is on the move. However, the master key is stored in the form of plain text in the SC, therefore, brute-force guessing, and SC stolen attacks may jeopardize the whole network.

In [10], a new unconditional anonymity identity-based user-authenticated key agreement scheme for an IoT multi-server environment is introduced by Hassan et al. The authors utilized a ring signature mechanism to allow users to anonymously authenticate themselves in the multiserver environments. However, the computation costs are significantly high. Han and Zhu [11] proposed a new identity-based mutual authentication protocol without bilinear pairings to improve the performance.

Tseng et al. [12] proposed a user authentication and key agreement protocol based on identity-based cryptosystem. They claimed that their protocol resists the ephemeral secret leakage (ESL) attacks in mobile multiserver environments. Furthermore, their protocol requires the lowest communication overhead. However, in [10], Hassan et al. pointed out that the Tseng et al. scheme does not provide anonymity.

Irshad et al. proposed an enhanced and provably secure chaotic map-based authenticated key agreement in multiserver architecture [13]. The authors claimed that their approach provides many security properties, e.g. mutual authentication and provides security against many attacks, e.g. replay attack, impersonation attack, known-key attack, and offline guessing attacks. However, in [14], Wu et al.'s pointed out that the scheme in [13] is susceptible to many attacks, e.g. offline password-guessing.

However, from the above-mentioned schemes, it can be noted that most of schemes are either vulnerable to security attacks or require high-computational complexities.

6.3 Network Model and Adversary Model

6.3.1 Network Model

Consider the environment of an airport setting, where several edge nodes are deployed within the airport boundaries to extend traditional cloud services to the users, to IoT devices or to other staff members. These edge nodes are connected securely with the main cloud or the server. A typical network model is shown Figure 6.1, where a user wants to access airport services in MEC environments. As shown in Figure 6.1, the user registers himself at a registration center (RC) and then he will be able to access the MEC services. In addition, the proposed model also supports secure mobility if a user leaves the home edge node (HEN) then he/she will be served by the other foreign edge node (FEN). In such a scenario, a new edge node needs to verify whether a joining node (i.e. user) is a legitimate user or at least, not a malicious external user. The joining node and new edge node, both need to establish a secure session key for securing the subsequent communication.

6.3.2 Adversary Model

An adversary can eavesdrop on the traffic, inject new messages, replay and change messages, or spoof other identities to disrupt the airport's services. Moreover, an attacker can have an illegal gain of the user's information via eavesdropping. However, the unauthorized user's goals might be to obtain illegitimate data access or control to the smart devices (i.e. user, HEN, and so on), for their own interest.

6.4 Proposed Scheme

In the proposed scheme, mainly three entities are involved, i.e. the user (U), the HEN, and the FEN. Before starting the scheme, we have made the following assumptions: (i) all the entities have identical cryptosystems (an encryption (E)/decryption (D), e.g. advanced encryption standard (AES) and a hash function (h), e.g. SHA-2/3). (ii) The HEN and FEN are the trusted entities, and neither disclose the identity of the users. (iii) The clocks of HEN, FEN, and the user are synchronized. The used notations and descriptions are shown in Table 6.1.

6.4.1 System Setup for the Edge Nodes Registration at the Registration Center

This phase invokes offline, the RC generates and assigns a unique identity for each edge node, i.e. ID_{Ei}. (here, i is 1,2,..n; and n is number of the edge nodes). Then, it generates a pool of keys for the edge nodes, i.e. KPE, of KE_i keys and the corresponding key index $Kidx_E$. Note that there should not be a common key (KE_i) in the pool KPE, as follows, $KE_i \cap KE_j = \emptyset$ and $i \neq j$. The RC loads in each an edge node unique key KE_i and its key index $Kidx_{Ei}$. As edge nodes are resource rich and temper-proof nodes, each edge node is preloaded with the master key $(Z = h(MEC_K))$ and the indexes of the keys, which are

Table 6.1 Symbols and descriptions.

Notations	Descriptions
$id_{Ui}, id_{HEN},$ and id_{FEN}	Identities of a user (U), home edge node (HEN), and foreign edge node (FEN)
PW_{Ui}	Password of user U.
$KPE, Kidx_{Ei}$	A pool of keys for the edge nodes, i.e. KPE, along with key indexes
MEC_K	Master key for the edge nodes
$E_K[M]$	M is encrypted (E) with symmetric Key K
$D_K[M]$	M is decrypted (E) with symmetric Key K
$h()$	One-way hash function
$HMAC\{M\}$	Hashed-Message authentication code on message M
$\|,$	Concatenation operation
$\oplus$	Ex-or operation

assigned to the deployed neighboring edge nodes. Note that the key index will be used to determine the address of a HEN when a user is visiting to the FEN (i.e. other edge node).

6.4.2 User Registration Phase

Step 1. User (U_i) selects id_{Ui}, random number m and password (PW_{Ui}), and computes $Q = h(m \oplus PW_{Ui})$. Then he submits registration request (id_{Ui} and Q) to the edge node where the user wants to register himself. This node from now onwards is called the HEN.

Step 2. Upon receiving the registration request from U_i, the HEN computes $B_i = h(id_{Ui}||Id_{HENi}||\$)$. Here $\$$ is a long integer, which is generated by the HEN. Now, it computes $C_i = E_{KE}(id_{Ui}||\$)$, and $D_i = h(id_{Ui}||h(m \oplus PW_{Ui}))$. Finally, the user receives an SC from the HEN. The SC contains $\{B_i, C_i, D_i, Kidx_{Ei}\}$.

Step 3. Upon receiving the SC, the user enters m to the SC, by doing so, the user does not have to remember the random number. Now the SC contains $\{B_i, C_i, D_i, m, h(.), Kidx_{Ei}\}$.

6.4.3 Login and User Authentication Phase

This phase is invoked when a user wants to access the services in the FEN. To do so, first the user needs to perform local verification, as follows.

Step 1. U_i inserts the SC to the card reader and enters id_{Ui} and PW_i. Now SC performs following procedures. It computes $D_i^* = h(id_{Ui}||h(m \oplus PW_{Ui}))$ and verifies $D_i^* = D_i$. Checks whether the condition is true or not. If yes, then goes to next step.

Step 2. Generates a random number r, computes $M = h(B_i \oplus t1)$ and $\alpha = E_M[h(id_{Ui}), id_{FEN}, r||t1]$. Now it sends $\{C_i, \alpha, t1, Kidx_{Ei}\}$ to the FEN. Here $t1$ is a timestamp of the user's system.

Step 3. Upon receiving the request message from U_i, the FEN, first checks the validity of timestamp $(t2 - t1) \leq \Delta T$, if it is true then goes to next step. Now, it computes $\mu = HMAC_Z[C_i, \alpha, t1,t2,Kidx_{Ei}]$ and sends $\{\mu$ and $C_i, \alpha,t1,t2,Kidx_{Ei}\}$ to the HEN. Here $t2$ is a timestamp of the FEN's system.

Step 4. After receiving the request message from FEN, the HEN, first checks the validity of timestamp $(t3 - t2) \leq \Delta T$, if it is true then goes to next step. It next computes $\mu^* = HMAC_Z[C_i, \alpha, t1,t2,Kidx_{Ei}]$ and verifies $\mu^* = \mu$. If the condition is true, then message is not altered.

Step 5. It decrypts C_i and obtains id_{Ui}^*, and $\$^*$. It computes $B_i^* = h(id_{Ui}^*||Id_{HENi}||\$)$. Subsequently, it computes M^*, decrypts α and obtains $h(id_{Ui}')$, id_{FEN}^*, r, $t1$. Now the HEN verifies whether the conditions $(h(id_{Ui}^*) = h(id_{Ui}'))$ and $id_{FEN}^* = id_{FEN})$ are true. If conditions are verified, then it assures that the user and FEN are the legal entities. Now the HEN notifies to FEN that the user is legal user, as follows. It computes a $\beta = E_Z[h(id_{Ui})||id_{HEN}||r||t3]$ and sends $\{\beta, t3\}$ to the FEN.

Step 6. After receiving the request message from HEN, the FEN, first checks the validity of timestamp $(t4 - t3) \leq \Delta T$, if it is true then goes to next step. It utilizes Z to decrypts β and obtains $h(id_{Ui})^*$, id_{HEN}^*, r^*, $t3^*$. Checks whether $t3^* = t3$, if not then aborts the system. It then computes a session key $SKey = h(h(id_{Ui})||id_{HEN}|| id_{FEN} ||r||t1||t4)$

for secure communication between the user and FEN. Generates $\pi = E_{SKey}[r\|t4]$ and sends $\{\pi, t4\}$ to the user.

Step 7. Upon receiving the message from FEN, the user first checks the validity of timestamp $(t5 - t4) \leq \Delta T$, if it is true then goes to next step. Now, it computes $S_{Key}{}^* = (h(h(id_{Ui}\|id_{HEN}\|id_{FEN}\|r\|t1\|t4))$ and decrypts π and obtains r^* and $t4^*$. The user verifies, $r^* = r$ and $t4^* = t4$. If yes, then the user believes that FEN is a legitimate node, otherwise not.

6.4.4 Password Update Phase

The password update phase is invoked whenever a user wants to update old password (PW_{Ui}), as follows

Step 1. User U_i inserts SC into the card-reader and enters identity (id_{Ui}) and password (PW_{Ui}).

Step 2. The SC validates the U_i's entered id_{Ui} and PW_{Ui} with the stored values by computing the following: $D_i{}^* = h(id_{Ui}\|h(m \oplus PW_{Ui}))$.

Step 3. It verifies whether $D_i{}^* = D_i$ is true. If not, then rejects the password update request. Otherwise, it goes to next step.

Step 4. Upon receiving the U_i's new password $(PW_{new}{}^*)$, SC computes: $D_{knew}{}^* = h(id_{Ui} \|h(h(m \oplus PW_{new}{}^*)))$. Finally, the SC replaces D_i with $D_{knew}{}^*$.

6.5 Security and Performance Evaluation

6.5.1 Informal Security Analysis

In this subsection, we analyze the security of the proposed scheme and compare it with the other state-of-the-art schemes.

6.5.1.1 Mutual Authentication

The scheme proposed in this chapter provides mutual authentication, where all entities (i.e. user, FEN, and HEN) are mutually authenticating each other. More specifically, when the HEN receives the message $\{\mu$ and C_i α, $t1$, $t2$, $Kidx_{Ei}\}$, it can ensure that the user message $\{C_i, \alpha, t1, Kidx_{Ei}\}$ is included in the FEN message. When the FEN receives message $\{\beta, t3\}$, it ensures that this message is generated by the HEN node. Furthermore, when the user receives message $\{\pi, t4\}$, he/she can also confirm that this message is generated by the FEN. Hence, mutual authentication is achieved.

6.5.1.2 Session Key Establishment

The proposed scheme provides session key establishment after the authentication phase. A session key $(S_{Key} = h(h(id_{Ui})\|id_{HEN}\|id_{FEN}\|r\|t1\|t4))$ is set up between the user and the FEN for subsequent secure communication. The S_{Key} will be different for each session and cannot be replayed after the time expires. More importantly, the user and the FEN can securely execute encryptions and decryptions by using S_{Key} and hence, the proposed scheme achieves confidentiality for subsequent messages.

6.5.1.3 User Anonymity

In our scheme, user anonymity U_i is preserved at the registration phase by computing $B_i = h(id_{Ui} \| Id_{HENi} \| \$)$ and $C_i = E_{KE}(id_{Ui} \| \$)$. In addition, it is impossible to extract id_{Ui} from $E_M[h(id_{Ui}), id_{FEN}, r \| t1]$, which is encrypted using M, and it is also very difficult to revert the $h(id_{Ui})$. So, the proposed scheme can preserve user anonymity.

6.5.1.4 Replay Attacks

Our scheme is resistant to replay attacks, because the authenticity of messages $\{C_i,$ α, $t1$, $Kidx_{Ei}\}$, $\{\mu$ and C_i α, $t1$, $t2$, $Kidx_{Ei}\}$, $\{\beta$, $t3\}$ and $\{\pi$, $t4\}$ are validated by checking the freshness of four timestamps $((t2 - t1) \leq \Delta T$, $(t3 - t2) \leq \Delta T$, $(t4 - t3) \leq \Delta T$, and $(t5 - t4) \leq \Delta T)$. Let's assume an intruder intercepts a login request message $\{C_i$, α, $t1$, $Kidx_{Ei}\}$ and attempts to access the FEN by replaying the same message. The verification of this login attempt fails, because the time allowed expires (i.e. $(t2 - t1) \leq \Delta T$). Similarly, if an intruder intercepts a valid message $\{\mu$ and C_i α, $t1$, $t2$, $Kidx_{Ei}\}$ and attempts to replay it to the HEN node, the verification request will fail at the HEN because the time specified expires again (i.e. $(t3 - t2) \leq \Delta T$). Likewise, other messages are secured against the replay attack. Thus, our scheme is secure against the replaying of messages.

6.5.1.5 User Impersonation Attacks

An attacker cannot impersonate the user. Suppose an attacker forges a login message $\{C_i$, α, $t1$, $Kidx_{Ei}\}$. Now, he/she will again try to login into the system with the modified message $\{C_i^*$, α^*, $t1^*$, $Kidx_{Ei}\}$, since, the fake α^* will not be verified at the HEN, and the HEN cannot get the original sub-message $\{h(id_{Ui})\}$ by decrypting α^*. Therefore, it is hard to impersonate the user.

6.5.1.6 Gateway Impersonation Attacks

If an attacker does not possess the secret key Z, he/she cannot impersonate the HEN and cannot cheat the FEN. Hence, it frustrates attackers to generate the valid message.

6.5.1.7 Insider Attacks

It is possible in a real-world environment, when the HEN manager or system administrator can use the user password PW_{Ui} (e.g. weak password) to impersonate the user U_i through any other network, e.g. HEN. In this case, our scheme does not give any room for privileged insiders, since, in the registration phase, the user U_i is passing $h(m \oplus PW_{Ui})$ instead of the plain password. Thus, the insider of the HEN cannot get PW_{Ui} easily. Here, m is a sufficiently high entropy number, which is not revealed to the HEN.

6.5.1.8 Man-in-the-Middle Attacks

An attacker may attempt a man-in-the-middle (MITM) attack by modifying the login message $\{C_i$, α, $t1$, $Kidx_{Ei}\}$ into $\{C_i^*$, α^*, $t1^*$, $Kidx_{Ei}\}$. However, this malicious attempt will not work, as the false α^* will not be verified at the HEN and the HEN cannot obtain the original sub-message $\{h(id_{Ui}), id_{FEN}\}$ by decrypting α^*. Thus, MITM attacks may not applicable to the proposed scheme.

Table 6.2 Comparison of security features.

	Amor et al.'s [7]	Ibrahim's [8]	Hassan et al.'s [10]	Ours
S1	Yes	Yes	Yes	Yes
S2	Yes	Yes	Yes	Yes
S2	Yes	Yes	Yes	Yes
S4	Yes	Yes	Yes	Yes
S5	—	Yes	—	Yes
S6	—	—	—	Yes
S7	Yes	Yes	—	
S8	—	—	Yes	Yes

S1: Mutual authentication, S2: session key establishment, S3: User anonymity, S4: Replay Attacks, S5: Impersonation Attacks, S6: Insider Attacks, S7: Man-in-the-Middle, S8: Offline-Password/Key Guessing.

6.5.1.9 Offline-Password Guessing Attacks

Our scheme is free from any password verifier table, so password guessing attacks are not feasible. In the login phase, passwords are not simply transmitted, but they are transmitted with some other secret (i.e. $D_i^* = h(id_{Ui}\|h(m \oplus PW_{Ui}))$), which makes it difficult to guess the user's password.

Table 6.2 compares the security features of proposed scheme and the Amor et al.'s [7], Ibrahim's [8], and Hassan et al.'s [10] schemes. It can be noticed from Table 6.2, the proposed scheme provides more security services than the existing schemes.

6.5.2 Performance Analysis

This subsection discusses the energy prices of the cryptographic primitives of the proposed scheme. We simulated the user, the HEN, and the FEN on the on the Intel(R) Core(TM)i-7-7500, memory (RAM) 16GB, and 64-bit operating system. Then, we measured the energy prices for the advanced encryption standard (AES-128) encryption, SHA2 and HMAC operations. In our simulations, the AES-128 block cipher takes 916 *picojoule* for the encryption and decryption. A hash function SHA-2 and HMAC are consuming 526 *picojoule* and 603.45 *picojoule* of energy, respectively.

Table 6.3 shows the performance comparison results (in the terms of computation costs) of the proposed scheme and Amor et al.'s [7] and Ibrahim's [8]. The following notations are used: *Hash* is the operation of the one-way hash function; *Sym* is the

Table 6.3 Comparison of computation costs.

	User	FEN	HEN
Amor et al. [7]	—	—	1 Asym
Ibrahim's [8]	—	2 Sym	2 Asym
Ours	5Hash + 1Sym	4Hash + 1HMAC + 2Sym	1Hash + 1HMAC + 2Sym

operation of symmetric encryption/decryption; and *Asym* is the encryption/decryption operation or the signature operation by using the asymmetric cryptosystem. It can be seen from Table 6.3, the proposed scheme requires *5Hash* + *1Sysm* operations at the user end, *4Hash* + *1HMAC* + *2Sym* at HEN and *1Hash* + *1HMAC* + *2Sym* at FEN. Whereas the scheme proposed in [7] requires *1Asym*, and the scheme proposed in [8] takes *2Aysm* at HEN and *2Asym* at FEN. Note that in the Amor et al.'s [7] and Ibrahim's [8] schemes, the end users are not directly involved in their network model, therefore the user does not have to perform any operation. Nevertheless, our proposed scheme provides more security features than Amor et al.'s [7] and Ibrahim's [8], as shown in Table 6.2.

6.6 Conclusion

The MEC is an emerging paradigm in the IoT world. Mobile users can use MEC services in an inter-domain when they are on move. We designed a new, efficient and anonymous mobility supported mutual authentication scheme in MEC environments. The scheme utilized the two-factor approach in MEC. The proposed scheme can facilitate several services to the users such as user anonymity, mutual authentication, secure session key establishment, mobility support, and it also allows users to choose/update their password regularly.

References

1 Taleb, T., Samdanis, K., Mada, B. et al. (2017). On multi-access edge computing: a survey of the emerging 5G network edge cloud architecture and orchestration. *IEEE Communications Surveys & Tutorials* 19 (3): 1657–1681.

2 Porambage, P., Okawuibe, J., Liyanage, M. et al. (2018). Survey on multi-access edge computing for internet of things realization. *IEEE Communications Surveys & Tutorials* 20 (4): 2961–2991.

3 Rahmani, A., Gia, T.N., Negash, B. et al. (2018). Exploiting smart e-health gateway at the edge of health internet of things: a fog computing approach. *Future Generation Computer Systems* 78 (2): 641–658.

4 Salix, A. and Mancini, G. (2017). Making use of a smart fog hub to develop new services in airports. In: *European Conference on Parallel Processing (Euro-Par 2017)*. Santiago de Compostela, Spain (28–29 August 2017): Springer.

5 Hu, Q., Wu, C., Zhao, X. et al. (2018). Vehicular multi-access edge computing with licensed Sub-6 GHz, IEEE 802.11p and mmWave. *IEEE Access, Digital Object Identifier* https://doi.org/10.1109/ACCESS.2017.2781263.

6 Imine, Y., Kouicem, D.E., Lounis, A., and Bouabdallah, A. (2018). MASFOG: an efficient mutual authentication scheme for fog computing architecture. In: *17th IEEE International Conference on Trust, Security and Privacy in Computing and Communications/12th IEEE International Conference on Big Data Science and Engineering,*. New York, USA (31 July–3 August 2018). New York, USA: IEEE.

7 Amor, A.B., Abid, M., and Meddeb, A. (2017). A privacy-preserving authentication scheme in an edge-fog environment. In: *2017 IEEE/ACS 14th International Conference on Computer Systems and Applications, October 30th – November 3rd*

2017, Hammamet, Tunisia (30 October–3 November 2017). Hammamet, Tunisia: IEEE.

8 Ibrahim, M.H. (2016). Octopus: an edge-fog mutual authentication scheme. *International Journal of Network Security* 18 (6): 1089–1101.

9 Rios, R., Roman, R., Onieva, J.A., and Lopez, J. (2017). From smog to fog: a security perspective. In: *2nd IEEE International Conference on Fog and Edge Mobile Computing (FMEC 2017)*, Valencia, Spain (8–11 May 2017). Valencia, Spain: IEEE.

10 Hassan, A., Omala, A.A., Ali, M. et al. (2018). Identity-based user authenticated key agreement protocol for multi-server environment with anonymity. *Mobile Networks and Applications* https://doi.org/10.1007/s11036-018-1145-5.

11 Han, W. and Zhu, Z. (2012). An id-based mutual authentication with key agreement protocol for multi-server environment on elliptic curve cryptosystem (ECC). *International Journal of Communication Systems* 27 (8): 1173–1185.

12 Tseng, Y.M., Huang, S.S., and You, M.L. (2016). Strongly secure ID-based authenticated key agreement protocol for mobile multiserver environments. *International Journal of Communication Systems* 30 (11): e3251.

13 Irshad, A., Ahmad, H.F., Alzahrani, B.A. et al. (2016). An efficient and anonymous chaotic map based authenticated key agreement for multi-server architecture. *KSII Transactions on Internet and Information Systems* 10 (12): 5572–5595.

14 Wu, F., Li, X., Xu, L. et al. (2018). Authentication protocol for distributed cloud computing. *IEEE Consumer Electronics Magazine* 7: 38–44.

7

Biometric-Based Robust Access Control Model for Industrial Internet of Things Applications

Pardeep Kumar and Gurjot Singh Gaba

Abstract

Information and operational technologies are being used together and making the industrial Internet of Things (IIoT) happen in the Industry 4.0 paradigm. In this paradigm, smart devices (i.e. sensors) will be offered services and shared data to the user and so the cloud. As these devices will communicate with the users through the open network (i.e. Internet), user authentication is one of the most important security features to protect IIoT data access from unauthorized users. However, there exist traditional security techniques but these require heavy computational complexities. Therefore, such traditional schemes cannot be deployed directly to the smart devices in IIoT applications. This chapter proposes a biometric-based robust access control model (i.e. user authentication) that would perform a robust authentication and establish a session key between the user and smart devices. The effectiveness of the proposed scheme is demonstrated in terms of computation cost in the IIoT environment.

Keywords   *industrial internet of things; security; access control; biometric*

7.1   Introduction

The connection between industry and the advancement in computing, analytics, low-cost sensing and seamless connectivity of internet is full of promise [1]. The degree of transformation is emerging and there is reference to a "breakthrough" in terms of production and operational speed and efficiency. The new innovative constructs are all around the "data," which can now be gathered from plants, equipment, electrical and mechanical machines, thanks to low-cost smart sensors and other smart devices. These smart devices are equipped with processing and communication capabilities. Therefore, new and innovative technologies, concepts and platforms are significantly on the increase in the setting up of the industrial automation: industrial Internet, Industry 4.0, and IIoT [2]. In [3], the authors reported that the IIoT revolution will impact economic sectors that currently account for nearly two-thirds of global gross domestic product, changing the basis of competition and redrawing industry' boundaries.

IoT Security: Advances in Authentication, First Edition.
Edited by Madhusanka Liyanage, An Braeken, Pardeep Kumar, and Mika Ylianttila.
© 2020 John Wiley & Sons Ltd. Published 2020 by John Wiley & Sons Ltd.

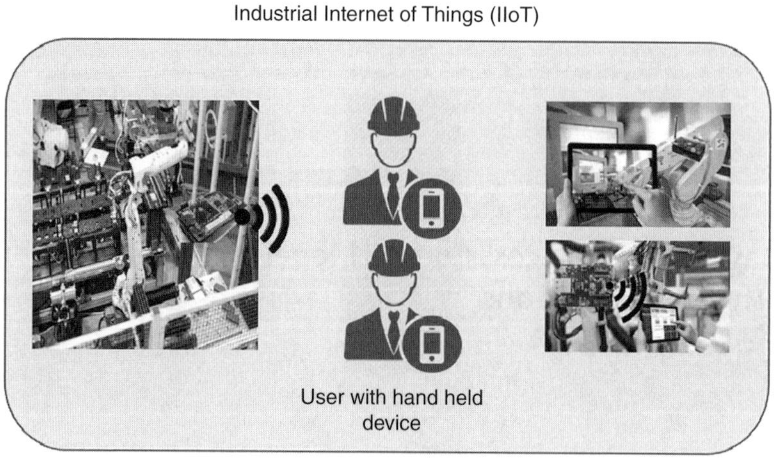

Industrial Internet of Things (IIoT)

User with hand held
device

Figure 7.1 Industrial internet of things applications.

In the smart factories or industries (i.e. manufacturing, assembly, etc.), IIoT makes best use of production and assembly processes producing more fine-grained data by integrating seamless connectivity and computing to various machines, assembly lines and tools. More precisely, during the working process, smart factories generate an enormous amount of data through "smart devices," i.e. devices with microprocessors onboard [2]. This data is transmitted to the users, control centers and other machines via a wireless communication network to maintain smooth and accurate operations in the factories. As smart devices are resource-constrained devices, the potential deployment of smart devices (i.e. sensors) for the real-world IIoT applications must deal with many challenges, including system architecture, availability, quality-of-services, etc.

Among these challenges, security is also one of the big concerns as the smart devices exchange data with other devices via insecure networks (e.g. Internet) [4]. Exploiting insecure networks, an attacker can trace and collect the data via eavesdropping and can redraw the profile of the process (i.e. production status) or other useful information of personal interest in a factory use-case. Moreover, in various applications, smart devices provide services to users directly or a user can directly access the smart devices via their own hand-held device. However, it is necessary to control who is accessing the smart device data as shown in Figure 7.1. Therefore, security services, i.e. access control (and/or authentication) is one of the core requirements for IIoT to protect the data access from unauthorized parties [4].

7.2 Related Work

Recently, several authentication schemes have been proposed which focus on IIoT applications. For instance, in [5], Ma et al. proposed a new certificateless searchable public key encryption method with multiple keywords (SCF-MCLPEKS) for the IIoT environment. The authors demonstrated that their proposed scheme is secure against two types of adversaries, e.g. Type 1 and Type 2. However, SCFMCLPEKS exploits the concept of a network-wide master, therefore, a leak of master key may lead to several attacks. In addition, the SCFMCLPEKS utilized the traditional public key primitives, such as scalar multiplication and bilinear pairing, therefore, it needs more energy for the smart devices attached with IIoT.

Gope et al. proposed a lightweight and physically secure anonymous mutual authentication protocol for real-time data access in an industrial wireless sensor network (WSN) [6]. The authors discussed three different application scenarios, environmental sensing, condition monitoring in body-area network, and process monitoring. The scheme makes use of the physically unclonable function and bitwise XOR operation. However, Katzenbeisser et al. [7] claimed that the main drawback of Physical Unclonable Function (PUF) is limited reproducibility and openness. In addition, raw PUF data is rarely available for subsequent research, which greatly hinders a fair comparison.

In [8], Das et al. proposed a new biometrics-based privacy-preserving user authentication scheme (BP2UA) for cloud-based IIoT deployment. BP2UA uses the user's smart card and biometric as two factors for authentication purpose. The scheme proposed in [8] uses bitwise Exclusive-OR and cryptographic hash operations at the smart devices's side, whereas the fuzzy extractor method is applied for biometric verification at the user side. The authors claimed that their proposed scheme is secure against many attacks, e.g. impersonation, man-in-the-middle, replay, insider, denial-of-service attacks, etc. The scheme does indeed cover many security properties; however, the scheme may be vulnerable to masquerade attack. In addition, the overall communication cost is still expensive as the packet length is high compared to other schemes mentioned by the authors. In [9], Bilal-Kang designed an authentication protocol in the future sensor network setting in which IoT can be embedded with WSN. In this scheme, a sensor node (a legitimate user) can establish multiple concurrent secure data sessions. They may be vulnerable to a parallel-session attack that can lead to other issues.

As shown above, several secure services and attacks have been addressed in the literature [5, 9]. However, several papers revealed that the most likely threat to information security is not the typical hacker, virus, or worm, but rather the malicious insider user [10]. In existing literature, the security-related all parameters (e.g. passwords, biometrics, plain identities, etc.) are stored onto corresponding smartcards. Therefore, security related parameters (especially row information) from the smartcards, are easy to retrieve via the power analysis tools [8] and that may lead to high risks of security breaches.

To address the above issues, this chapter proposes a biometric-based robust access control model for IIoT applications. The proposed scheme utilizes the biometric to perform robust authentication – because biometric identifiers are known to unique to individual's and more reliable in verifying identity than those of the sole password-based methods. The proposed model provides a robust mutual authentication and establishes a session key between the user and smart devices. To attain a low-computational overhead, we utilize elliptic curve cryptography (ECC), symmetric cryptosystem, and hash operation. Security analysis shows that the proposed model can defend popular attacks and also achieve efficiency.

The rest of the chapter is structured as follows: Section 7.2 discusses the network model, threat model, security requirements for the proposed model. Section 7.3 proposes our model in detail. Section 7.4 discusses the security analysis, efficiency evaluation and comparison with existing schemes for WSNs. Section 7.5's conclusions are drawn for the proposed access control model.

7.3 Network Model, Threat Model and Security Requirements

7.3.1 Network Model

Assume an IIoT network, consists of several low-cost smart sensor devices, which are deployed in the industrial environment. These sensors sense the environmental information and transmit them to the users for analysis. In a real-world IIoT, the sensory data is not only accessed through a gateway, but a user can also access it directly using a hand-held device (e.g. personal digital assistant (PDA)/smart-phone) over wireless communication. The basic network architecture is shown in Figure 7.1, where a user directly sends a data request to the smart sensor node. Upon receiving the data request, a sensor node first verifies user authenticity through the gateway node and then the user can access sensory data from the IIoT applications.

7.3.2 Threat Model

We consider the Dolev-Yao attack model [10], where an attacker can eavesdrop on the traffic, inject new messages, replay and change messages, or spoof other identities. In addition, the attacker may come from inside or outside the network. However, the unauthorized user's goals might be to obtain illegitimate data access, to control the smart devices, and to perform service degradation or denial of service (DoF) to disrupt the IIoT application.

7.3.3 Security Goals

In the IIoT network, a secure scheme should consider transparent security goals, as follows.

Mutual authentication: Every entity (user, gateway, and sensor) must be mutually authenticated; hence they can ensure the communication is only taking place between authentic entities.

Session key establishment: A session key should be established between a user and sensor node, so that subsequent communication could take place securely.

Confidentiality: It is desirable that a user authentication protocol facilitate confidentiality of messages; as a result, these confidential messages can only be used by authorized users.

Robust against popular attacks: Clearly, the scheme should defend against different popular attacks, such as impersonation, replay, and man-in-the-middle attacks. As a result, the scheme should be easily applicable to real-world applications.

7.4 Proposed Access Control Model in IIoT

To provide strong security to IIoT applications, this section presents a biometric-based robust access control model. In the proposed scheme, each user should perform a biometric-based registration with the gateway in a secure manner so that the sensory

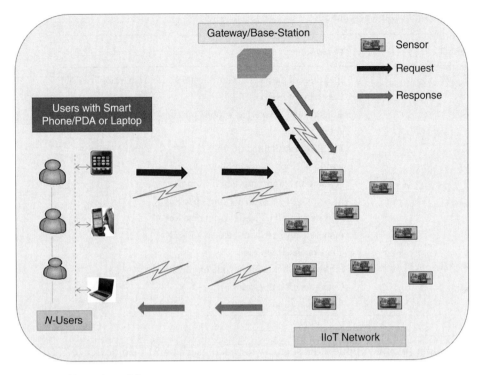

Figure 7.2 Network model.

data in IIoT applications can be accessed only by the registered users in a secure way, as shown in Figure 7.2. After user registration, the gateway node issues the security tokens for every registered user. Then, a user can submit his/her query in an authentic way and request the sensor data at any time within an administratively configurable period. The proposed scheme consists of two phases: system setup, and mutual authentication and key establishment.

Assumptions: Before starting the system, we assume that the gateway is a trustworthy entity. It is also assumed that the clocks of the user's mobile device, gateway, and smart sensor are synchronized in IIoT application [11]. Consider the elliptic curve discrete logarithm problem (ECDLP), to find an integer r, given an elliptic curve E defined over F_q, a point $P \in E(F_q)$ of order n, and a point $Q = r P$ where $0 \leq r \leq n - 1$, as shown in [12]. The notations and descriptions are shown in Table 7.1.

7.4.1 System Setup

To start the system setup, the user (Ui) and the GW need to perform the following steps to finish the system setup:

(i) Each user (U_i) generates own identity (id_U) and password (pw_U) and inputs id_U, pw_U and personal biometric (B_U) to the GW for the registration.

(ii) GW chooses two private keys y, $z \in$ Zp, and then computes two public keys, $Pub_1 = yP$ and $Pub_2 = zP$. It generates a long-term shared key (K_{GWSD}) and shares a private key (z) and a public key (Pub_1) with the SD. Now, the GW generates a random

Table 7.1 Symbols and descriptions.

Symbol	Description
id_U, id_{GW}, and id_{SD}	Identities of a user (U), Gateway (GW), and smart device (SD)
PWu	Password of user U.
HD	A hand-held device, e.g. mobile phone
F_q	A finite field
E	Elliptic curve defined on finite field F_q with prime order n
G	Group of elliptic curve points on E
P	A point on elliptic curve points on E
$E_K[M]$	M is encrypted (E) with symmetric Key K
$D_K[M]$	M is decrypted (E) with symmetric Key K
K_{GWSD}	A shared key between the GW and SD
$h()$	One-way hash function
$MAC\{M\}$	Message authentication code on message M
$\|,$	Concatenation operation
$\oplus$	Ex-or operation

number m and computes the security for each user's hand-held device (e.g. HD), as follows: a proxy key pair $S = mP$ and $\alpha = y + m\,h(h(S)\|id_U)$. In addition, the GW generates a unique token $(UT_U \in Zp)$ and computes $g_U = h(UT_U\|id_U\|pw_U\|B_U)$ for each user. The GW stores the proxy key pair (S, α) along with id_U, pw_U, B_U of all SDs. Then each SD's key pair (S, α), public key (pub_2), UT_U, g_U, id_{GW}, $h()$ are stored securely to each corresponding HD SIM card.

7.4.2 Authentication and Key Establishment

This phase invokes when the user wants to access the IIoT data locally. For this, user inputs id_U, pw_U and B_U and then the HD performs following steps:

(i) HD computes $g_U{}^* = h(UT_U\|id_U\|pw_U\|B_U)$ and verifies $g_U{}^* = g_U$. It generates a random integer o, computes $A = oP$, $\Phi = h(o.\,pub_2) \oplus (S,id_U,B_U, t1)$, $\mu = \alpha.\,h(h(A)\|\Phi\|id_{SD}\|\ id_{GW}) + o$, and $tag = MAC\ \{UT_U,(id_U\|B_U\|\ \alpha\|id_{SD}\|t1)\}$. Here $t1$ is the time stamp of the user's HD. Now, HD sends $\{A, \Phi, \mu, id_{SD}, tag, t1\}$ to SD.

(ii) SD checks if $(t2\text{-}t1) \geq \Delta T$ then SD aborts the operations. Here $t2$ is the current timestamp of the SD and ΔT is the expected transmission delay.

(iii) SD uses own private key (z) and computes $\Phi \oplus h\,(z\,A)$ to obtain S, id_U, B_U, and $t1^*$. It checks $t1^* = t1$, if not then aborts the request. It computes μP and $h(A\|\ \Phi\|\ id_U\|id_{SD}\|id_{GW})$. $(pub_1 + S\,h(h(S)\|id_U)) + A$ and then checks whether the computed values are identical. If yes, then goes to the next step. Generates random integer v, and computes $\beta 1 = E_{KGWSD}[v\|S\|id_U\|t1\|t2]$, here $t2$ is the timestamp of SD. Now, it sends $\{\beta 1, id_{GW}, id_{SD}, tag, t2\}$ to the GW. In addition, SD keeps store, id_U for the current session.

(iv) The GW verifies if $(t3 - t2) \geq \Delta T$ then GW aborts the further step. Here $t3$ is the current timestamp of the GW. Now the GW decrypts $\beta 1$ using D_{KGWSD} and obtains $v^*, S^*, id_U{}^*, t1^*, t2^*$. It verifies $t2^* = t2$, if yes then it retrieves the corresponding authentication token (UT_U) of $Id_U{}^*$, α of S^* from its table. Now the GW, computes $tag^* = MAC \{UTu^*(id_U||B_U|| \alpha||id_{SD}{}^*|| t1^*)\}$, and checks $(tag^* = tag)$ and $(id_{SD}{}^* = id_{SD})$, if it holds, then the user, HD and SD are authenticated entities. Now, the GW generates a random integer f, computes $S_{Key} = h(v||f||id_U||id_{SD}||\alpha||t1||t2 ||t3)$, $\beta2 = E_\alpha[v||f||id_{SD}||id_{GW}||t2||t3]$, and $\beta3 = E_{KGWSD}[\beta2||v||S_{Key}||id_U||id_{GW} || t3]$. Here, $t3$ is current timestamp of the GW and S_{Key} is the session key. Finally, it sends $\{\beta3, id_{GW}, id_{SD}, t3\}$ to the SD.

(v) The SD verifies if $(t4\text{-}t3) \geq \Delta T$ then aborts the system. Here $t4$ is the current timestamp of the SD. It decrypts $\beta3$ using K_{GWSD} and obtains $\beta2||v^*||S_{Key}|| id_U|| id_{GW}{}^*||t3^*$. It checks $(t3^* = t3)$, $(v^* = v)$, and $(id_{GW}{}^* = id_{GW})$. If all the conditions are true, then it sends $(\beta2, id_{SD}$, and $t4)$ to user's HD.

(vi) Upon receiving the message, HD verifies if $(t5\text{-}t4) \geq \Delta T$ then aborts the system. Here $t5$ is the timestamp of HD. Decrypts $\beta 2$ using α and gets $v||f||id_{SD}{}^*||id_{GW}{}^* || t2 || t3$. It checks $(id_{SD}{}^* = id_{SD})$, and $(id_{GW}{}^* = id_{GW})$, if yes then MD computes a session key $S_{Key} = h(v||f|| id_U||id_{SD}||\alpha||t1||t2|| t3)$.

7.5 Security and Performance Evaluations

7.5.1 Informal Security Analysis

We analyze the security of the proposed scheme under the Dolev-Yao attack model [10]. An adversary may intercept, modify, and insert any message over the public communication channels. The advantages of our scheme are explained as follows:

7.5.1.1 Save Against Masquerade Attack

(i) An adversary cannot masquerade GW to cheat SD, since he/she does not have knowledge of the secret key (K_{GWSD}). Hence, it is not easy for an adversary to compute the valid response, i.e. $\beta3 = E_{KGWSD}[\beta2||v||S_{Key}||id_U||id_{GW}||t3]$ to SD.

(ii) SD cannot masquerade GW to cheat HD. It can be noticed that the SD does not have any idea about a secret parameter α, and thus, SD cannot decrypt $\beta2 = E_\alpha[v||f ||id_{SD} ||id_{GW} ||t2||t3]$ as this message is encrypted by α. Here, α is shared between the GW and legitimate SD.

(iii) An adversary cannot masquerade HD as the user uses biometric (BU) to prove own legitimacy. In addition, if an adversary uses a fake identity (id_U') and false (α'), then the corresponding spurious $tag = MAC \{UT_U(id_U'||B_U||\alpha'||id_{SD}||t1)\})$ can be identified by HG, because HG cannot be verified (i.e. $tag' = tag$).

7.5.1.2 Safe Against Man-in-the-Middle (MITM) Attack

In this attempt, an attacker can eavesdrop on all the packet exchanges between the involved entities. Then he/she can resend these eavesdropped packets to make the other entities trust that they are rightfully exchanging information with each other. By doing so, an attacker can take over the whole communication and degrade the performance of IIoT. In the proposed scheme, it not easy for an ill-intentioned attacker to mount such a type of attack, successfully, since the adversary requires knowledge of the user's biometric (BU) and needs the secret value α. Without knowing all the secure parameters,

Table 7.2 Comparison of security features.

	Ma et al. [4]	Das et al. [7]	Proposed Scheme
Resist to masquerade attack	Partially	—	Yes
Resist to replay attack	—	Yes	Yes
Safe to MITM	—	Yes	Yes
Resist to DoS attack	Yes	—	Yes

the attacker would not be able to decrypt messages $\beta 2$ $(= E_{\alpha}[v||f||id_{SD}||id_{GW}||t2||t3])$ and $\beta 3$ $(= E_{KGWSD}[\beta 2||v||S_{Key}||id_U||id_{GW}||t3])$ to compute the session key. Therefore, the proposed scheme is resistant to the man-in-the-middle (MITM) attack.

7.5.1.3 Safe Against Denial-of-Service Attack

In the proposed protocol, the proxy key pair of the SIM card of HD and the database of GW do not require synchronous updates. Therefore, even if the attacker interferes with the transmitted authentic messages among HD, SD, and GW, they still cannot mount the DoS attack successfully.

7.5.1.4 Safe Against Replay Attack

An attacker can collect messages $\{A, \Phi, \mu, id_{SD}, tag, t1\}$, $\{\beta 1, id_{GW}, id_{SD}, tag, t2\}$, $\{\beta 3, id_{GW}, id_{SD}, t3\}$, and $(\beta 2, id_{SD}, and t4)$, which are sent among HD, GW, and SD. The attacker might replay these captured messages later to respective recipient. However, as can be seen, each message recipient checks the validity of the timestamp at first place as follows $(t2-t1) \geq \Delta T$ (at SD); $(t3-t2) \geq \Delta T$ (at GW); $(t4-t3) \geq \Delta T$ (at SD) and $(t5-t4) \geq \Delta T)$ (at HD). Therefore, the proposed scheme is safe against message replay attacks.

Table 7.2 lists the selected security features and makes a comparison between the proposed scheme and others, e.g. Ma et al. [4] and Das et al. [7].

7.5.2 Performance Analysis

We evaluate the performance of our proposed scheme and the scheme of Ma et al. [5], in terms of computation and communication costs.

The computational cost of the proposed model is analyzed. Let T_h by the time of performing a one-way hash $h (.)$, T_E and T_D by the time for performing a symmetric encryption and decryption, respectively, T_{PM} by the time for performing an ECC point multiplication operation, T_{MAC} by the time for performing a Message authentication code (MAC) operation, T_{SM} by the time for a scalar multiplication, T_{BP} by the time for performing a bilinear pairing operation, T_H by the time for a Hash-to-point operation, and T_{PA} by the time for performing a point addition operation. At HD device requires $6Th + 3T_{PM} + 1T_D + 1T_{MAC}$; the SD requires $4Th + 2T_{PM} + 1T_E + 1T_D$; and the GW incurs $1Th + 2T_E + 1T_D + 1T_{MAC}$. Whereas Ma et al. scheme requires $2TH + 4T_{SM}$ for KeyGen, $3TH + T_h + 4T_{SM} + 3T_{BP} + T_{PA}$ for certificateless public key encryption scheme, $T_H + T_{SM} + T_{PA}$ for a Trapdoor, and $2TH + T_h + T_{SM} + 2T_{PA} + T_{BP}$ for the test verification. Table 7.3 summarizes the overall computation cost between the proposed

Table 7.3 Comparison of computation costs.

	Ma et al. [4]	Proposed scheme
Overall computation cost	$6TH + 2Th + 10T_{SM} + 4T_{BP} + 5T_{PA}$	$11Th + 5T_{PM} + 3T_E + 3T_D + 2T_{MAC}$

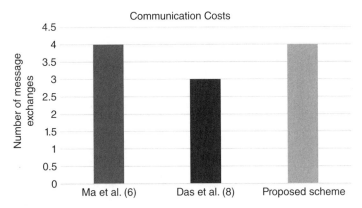

Figure 7.3 Comparison of communication costs.

scheme and Ma et al.'s [4] scheme. Note that we did not compare the computation cost of the proposed scheme with the Das et al.'s Scheme [8], as their scheme is based on the sole hashing and XoRing operations. Overall, the scheme from Das et al. requires an excessive hashing operation, e.g. 30 T_h (approx.).

We evaluate and compare communication costs in terms of the number of message exchanges for the proposed scheme, and Ma et al. [5] and Das et al.'s [8] schemes. To execute the whole scheme, Ma et al.'s scheme requires four rounds of message exchanges, Das et al.'s scheme takes three rounds of message exchanges and the proposed scheme requires four rounds of message exchanges, as shown in Figure 7.3. However, considering the security features (refer to Table 7.2), the proposed requires one more round of message exchanges than Das et al.'s scheme but provides more security features. Therefore, the proposed scheme can be a practical solution for such real-world IIoT applications.

7.6 Conclusions

IIoT is an emerging paradigm in the Industry 4.0 where smart devices (i.e. sensors) will play an important role and offer services and share data to the user. However, providing security to such time-critical applications is challenging. This chapter proposed a biometric-based robust access control model (i.e. user authentication) that would perform a robust authentication and establish a session key between the user and smart devices. The effectiveness of the proposed scheme has been demonstrated in terms of computation and communication costs in the IIoT environment.

References

1 Da Xu, L., He, W., and Li, S. (2014). Internet of things in industries: a survey. *IEEE Transactions on Industrial Informatics* 10 (4): 2233–2243.

2 Ferrari, P., Flammini, A., Sisinni, E. et al. (2018). Delay estimation of industrial IoT applications based on messaging protocols. *IEEE Transactions on Instrumentation and Measurement* 67 (9): 2188–2199.

3 Luvisotto, M., Tramarin, F., Vangelista, L., and Vitturi, S. (2018). On the use of LoRaWAN for indoor industrial IoT applications. *Wireless Communications and Mobile Computing* 2018: 1–11.

4 Gurtov, A., Liyanage, M., and Korzun, D. (2016). Secure communication and data processing challenges in the industrial internet. *Baltic Journal of Modern Computing (BJMC)* 4 (4): 1058–1073.

5 Ma, M., He, D., Kumar, N. et al. (2018). Certificateless searchable public key encryption scheme for industrial internet of things. *IEEE Transactions on Industrial Informatics* 14 (2): 759–767.

6 Gope, P., Das, A.K., Kumar, N., and Cheng, Y. (2016). Lightweight and physically secure anonymous mutual authentication protocol for real-time data access in industrial wireless sensor networks. *IEEE Transactions on Industrial Informatics* 63 (11): 1.

7 Katzenbeisser, S., Kocabas, U., Rozic, V. et al. (2013). PUFs: Muth, fact or busted? A security evolution of physically unclonable functions (PUFs) cast in silicon. In: *IEEE International Symposium on Hardware-Oriented Security and Trust (HOST)*, Texas, USA (2–3 June 2013). Texas, USA: IEEE.

8 Das, A.K., Wazid, M., Kumar, N. et al. (2018). Biometrics-based privacy-preserving user authentication scheme for cloud-based industrial internet of things deployment. *IEEE Internet of Things Journal* 5 (6): 4900–4491.

9 Bilal, M. and Kang, S.-G. (2017). An authentication protocol for future sensor networks. *Sensors* 17 (5): 1–29.

10 Dolev, D. and Yao, A. (1983). On the security of public key protocols. *IEEE Transactions on Information Theory* 29 (2): 198–208.

11 Neuman, B.C. and Stubblebine, S.G. (1993). A note on the use of timestamps as nonce. *ACM SIGOPS Operating System Review* 27 (2): 10–14.

12 Lee, H., Shin, K., and Lee, D.H. (2012). PACPs: practical access control protocols for wireless sensor networks. *IEEE Transactions on Consumer Electronics* 58 (2): 491–499.

8

Gadget Free Authentication

Madhusanka Liyanage, An Braeken, and Mika Ylianttila

Abstract

One of the major trends in the development of the Internet is the one of ambient Internet of Things (IoT) and even more ambient Internet of Everything (IoE). In this case, the user seamlessly interacts with various systems and devices, which automatically adapt to the user's needs based on the surrounding context. To even further enhance the user experience, a gadget-free or also called Naked-approach scenario is considered, where the user is able to experience the services without carrying any gadget. We present in this chapter a use case from the medical and healthcare sector using such a Naked approach, enabling an ambient IoE experience for the patient. Through direct interaction with the environment, using biometrics-based authentication and symmetric key-based operations, the patient is identified and is able to perform confidential and authenticated communication. A concrete and total security solution is presented, together with a discussion on the strength against well-known attacks. The performance analysis of the scheme shows the feasibility of the proposed approach.

8.1 Introduction to Gadget-Free World

The world is inevitably developing towards a society of hyper-connectivity, where each object has an identity and each person is multi-dimensionally connected to a network of digital services and information reservoirs. The growing number of connected devices offers huge potential for data collection and service business. However, there is a clear need for a change from the perspective of each individual user. Each citizen should have the means to control how their personal data is collected and utilized to grow new more personalized services around them. Many kinds of future scenarios can be seen for this development depending on the value basis of the developers, controllers and users (citizens). We need to make sure that human rights will be part of the future [11, 14].

In this framework of developing digitalism, the user perspective has not yet been fully appreciated. Currently, we are living in a post-industrial world, where our life patterns are more and more dominated by the electronic devices and gadgets that we use for documenting, planning, communicating and scheduling our daily activities. Even if those devices give us a powerful feeling of efficiency and expanded capabilities; in most cases

IoT Security: Advances in Authentication, First Edition.
Edited by Madhusanka Liyanage, An Braeken, Pardeep Kumar, and Mika Ylianttila.

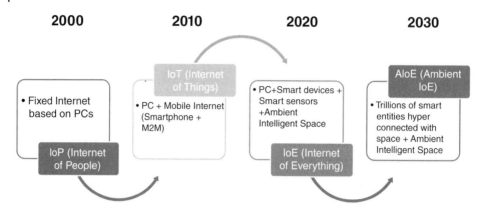

Figure 8.1 The Evolution of Internet

they are actually narrowing our senses. They are squeezing our attention and tunnelling our vision through a miniature high-definition display and forcing us to stiffen our joints to repeat micromotoric movements needed to interact with the device. We are sliding into a world of joint digital escapism. Figure 8.1 gives an overview of the evolution of Internet since 2000 [13, 16]. There must be an alternative.

Our guideline for the new-user experience and design is the Naked user approach. In this digital paradise any user - like me - can live Naked without carrying (or bumping into) gadgets. My services and interfaces will appear from the texture of the environment when I need them and disappear when not needed [1, 2]. If I want, I will be constantly but discretely connected to my communities - family, friends, work - and have instant access to my digital information sources and services. My surroundings - and the services behind them - will support my lifestyle, boosting my daily routines. The environment seems to know me better than I do. But I feel safe, because the use of the information is fully transparent to me. I own and have complete control over my data. And my data can be accessed only with my permission by trusted actors via trusted channels. Digital privacy and domestic peace legislation ensures I have the ability and right to disconnect - be unavailable, untraceable when I need privacy [12, 14]. Figure 8.2 illustrates this vision of a gadget-free world.

The physical devices connected to the higher order systemic functions (networks, data processing, services) will form a technology basis for the Surroundings as a Service. As an embodiment of the digital services, the surroundings will provide information, connections, tools, and guidance for the user as well as adjustments for living conditions (automation). The surroundings will collect data that is processed and bred to various kinds of digital information that can then be used by other services. This constantly growing personal or joint information property is one of the key components of future societies and businesses. Instead of growing wood or vegetables, we will be growing, breeding, processing, and using information [1].

The fundamental benefits for the user are:

- enabling the use of digital services through our surroundings, with a human sized, intuitive interface (like picking a fruit in paradise) without the need to carry or use any personal gadgets.

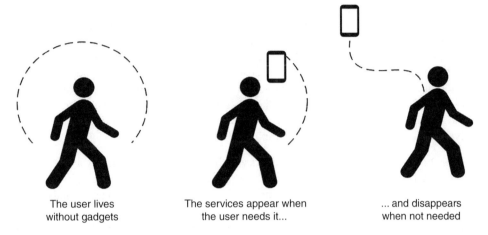

| The user lives without gadgets | The services appear when the user needs it... | ... and disappears when not needed |

Figure 8.2 The vision of Gadget-Free World

- providing the ambient connectivity and information in the texture of our surroundings without the need for our full concentration on our interfaces of our carriable gadgets. We will functionalize and enhance our surroundings instead of pulling our focus away from it.
- enabling the growth of users' personal digital information property by collecting and processing the data related to their environment, daily routines, communities, and enabling smart digital services based on this pool of information - leading to efficiency of daily tasks and advanced service and usage models in communications and lifestyle.
- enabling local harvesting of energy as an integral part of the system.

Benefits for society:

- enabling sustainable growth towards hyper-connected society via the paradigm change in the embodiment of digital services.
- enabling energy optimization on several levels - energy harvesting as a major functionality of the devices, reduced role of wireless cellular networks as the last-mile connectivity solution, decreased need for global telecommunications as a part of surroundings-as-a-service concept; tools for smarter automation and control.
- new knowledge and services through hyper-connected citizens, objects and spaces: better understanding of the daily connections, rhythms, routines and conditions of the society; stronger networking between citizens, helping in constant care, delivery of services, and new ways of working.

The Naked-user approach will compete with the current gadget-centric development trends such as mobile terminals and pads, the wearable gadgets (smart watches, wellness monitors, "google" glasses), the converging of building management systems (including HVAC, smart lighting, people-flow technologies), and the smart appliances and entertainment electronics. However, compared to these, our approach has its own niche, complementing or replacing these technologies. The Naked Approach provides an alternative where people live inside the digital reality instead of using it through gadgets.

The vision is challenging with respect to many of its major disciplines. Certain features for the user interface, such as identification, privacy control, and context recognition may require solutions like the carrying of ID chips or requests about activities from the user (biometric identification, keywords). The migration and adaptation of services to different surroundings may be limited at first. The realization of communications, energy solutions, and swarm-intelligent functionalities of the physical interaction layer may require extensive work. To overcome the risks and dodge the potential show-stoppers, we have built a team of highly co-operative top experts representing different disciplines as well as visionary cross-technological leadership.

The transition from gadget to the Naked World can easily be defined by three key phases i.e. bearable, wearable and nearable.

- **Bearables** are kind of hand-held gadgets which are the most common way of acquiring digital services nowadays. The most commonly used bearables are smartphones, laptops, tablets and PDAs. The current trend of bearables are getting declined due to wearable technology.
- **Wearables** are digital devices, which are worn by users to obtain digital services. It is the combination of smart sensors along with fashionable wearing items resulting in stylish wearable devices. Some of the most-used wearables are smart watches, smart clothes and google glasses among others. Wearables have a huge application in health-care because nowadays they are used frequently to monitor the fitness and health parameters of users like heart rate, temperature and blood pressure.
- **Nearables** are the final phase towards the Naked World where the user would have direct and seamless interaction with smart surroundings. The digital services appear to the user from the texture of the environment when needed.

During the transition from gadget to the Naked World, there are a number of factors that must evolve. For instance, the interaction of the user will be quite different from the current gadget-based interaction. Multi-modal interfaces will be required for the user's seamless interaction with the environment. Data sharing is another major thing which would also change as we move from gadget-centric to the user-centric world. Data is collected and shared by the environment, devices and systems present in the environment. Data is moved from local storages to storages in the infrastructure, such as servers or clouds. The identification of the user would also be different from the current username- and password-based solutions. The biometrics-based identification mechanisms seem to be an ideal candidate for the Naked environment, but require special attention with respect to theft and tamper resistance.

8.2   Introduction to Biometrics

The important challenge in this Naked environment is the identification of a valid user. Traditional password-based authentication mechanisms might not be suitable and efficient in this case as they require explicit input of the user at each point. Biometrics-based solutions, using either physiological or behavioral characteristics of the user, are much more promising. Moreover, they have several advantages over the classical passwords, which are are either too simple or too complex. It is not too difficult to break them. In cases where they are too simple, passwords are easy to crack. In cases of too difficult passwords, users often have the tendency to write down complex passwords in easily

Table 8.1 Comparison of Various Biometric Technologies

Characteristic	Universality	Distinctiveness	Permanence	Collectability	Performance	Acceptability	Circumvention
Face	H	L	M	H	L	H	H
Fingerprint	M	H	H	M	H	M	M
Hand geometry	M	M	M	H	M	M	M
Iris	H	H	H	M	H	L	L
Keystrokes	L	L	L	M	L	H	H
Voice	M	L	L	M	L	H	H

accessible locations. Moreover, passwords can be lost or forgotten by the user and are easy to share with different users.

Therefore, in order to enforce more or less seamless and ambient authentication, the use of biometrics seem to be a potential candidate for the identification and authentication of the users in such Naked environment. The most popular types of biometric characteristics are face, fingerprint, hand geometry, iris, keystrokes and voice. Table 8.1 [17] gives a short overview of these biometric characteristics with respect to the following features:

- Universality addresses the fact that all people possess the biometric.
- Distinctiveness corresponds to the possibility of being able to distinguish based on the characteristic.
- Permanence defines how permanent the identifier is with respect to time or other environmental conditions.
- Collectability gives an answer on how well the characteristic can be captured and quantified.
- Performance addresses both issues related to speed and accuracy.
- Acceptability corresponds to the willingness of people to use and share the characteristic.
- Circumvention addresses the possibilities of being capable of going wrong or being misused (foolproof).

The level of satisfaction of the biometric characteristic for each of these features is based on the author's perception [17] and is equal to High, Medium, and Low, which is represented in the table by H, M, and L, respectively.

As can be concluded, each biometric characteristic has advantages and disadvantages. There is no single optimal solution. The choice depends on the application. Moreover, in our case, the choice mainly depends on the accuracy of the characteristic. This is explicitly discussed in Section 7 of this chapter.

Note that there is currently also research to investigate the use of other biometric characteristics such as retina, infrared images of face and body parts, gait, odor, ear, and DNA. However, insufficient evidence of accuracy is currently available for these characteristics.

8.3   Gadget-Free Authentication

With the advancement in recent communication technologies, there is a clear need for enhancement in living environments for elderly and disabled persons. It is vital to have novel ways of multi-modal interaction and identification of the user with the environment for improved and easy access to services. Healthcare services are considered crucial for senior people and thus they need to be delivered on time. Therefore, it is required that the environment should be intelligent enough to identify the valid patients and provide digital services accordingly without any external intervention of a third person or device. This would make the identification process faster and reduce the overhead of checking/filling of registration forms. It is also essential in emergency situations in the hospital where critical patients do not have much time and cannot undertake physical efforts to register themselves. Hence, biometrics-based efficient authentication solution is needed for smart and ambient environments.

In this chapter, we propose an anonymous and biometrics-based authentication scheme for the Naked hospital environment. We consider a hospital scenario and show how patients can be authenticated without using gadgets to access services from their environment. Medical sensors embedded in the environment will provide them with the required digital services following secure authentication. Moreover, our proposed scheme can resist various well-known attacks such as insider attacks, replay attacks and identity privacy attacks. We have also compared our scheme with some of the already available remote biometric authentication schemes in terms of computational and communication costs [10].

We consider a potential hospital scenario, where an elderly person or critical patient can directly be authenticated by the smart environment in order to avail the required healthcare facilities as shown in Figure 8.3. These services may include the monitoring of the pulse rate, blood pressure and heartbeats of patients. The medical sensors are embedded in the environment to provide the services. We assume that when a patient enters a room in the hospital, he/she does not carry any gadget for identification, instead the camera/device placed in the environment can directly authenticate him/her. Biometric features are used for the authentication of the patient in this Naked environment. If the results of these services are presented to the patient on a screen or device present in the room, we need to assume that at a single time only one patient is available in the room (no other patient at one time, i.e. doctors or nurses may be in the room) in order to ensure the confidentiality of the information. On the other hand, if the information retrieved by the services is consulted afterwards by the patient by logging into a secure platform, this restriction can be dropped.

The central administration of the hospital (also called the registration center) is supposed to be a trusted party, which will have access to control and generate the required key material for the Access Point (AP). The AP has capabilities to capture the biometric features of the patient and can retrieve the required services of the user from the remote sensors, that is then sent to the medical server. This server contains all the medical information about patients. In particular, the AP and the remote sensors are vulnerable to security attacks.

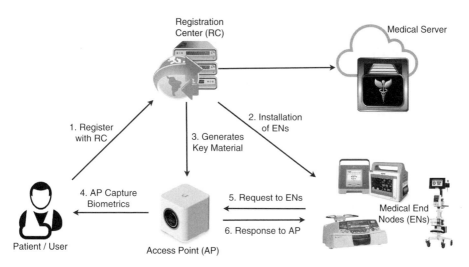

Figure 8.3 Problem scenario of authentication in the Naked Environment

8.4 Preliminary Aspects

8.4.1 Security Requirements

- **Confidentiality:** Information on the patient is kept secret from all unauthorized entities available in the surroundings.
- **Data Authentication:** Any authorized person should not be able to alter the original data.
- **Access Control:** The access to the data should be controlled by the central administration of the hospital and no other entity should be able to access it.
- **Identity Privacy:** It guarantees that the identity of the user cannot be revealed to any outsider.
- **Untraceability:** No outsider is able to link different messages to the same person.

8.4.2 Setting

We distinguish four different entities in the system, being the User (U), the Registration Center (RC), the Access Point (AP), and the Sensors or End Nodes (ENs) offering the services. Table 8.2 contains all the notations used for the proposed scheme.

The user first registers with the RC when requesting the services of the ENs corresponding to a particular AP. Then, the RC generates the appropriate key material for the AP. After this initialization, the user can now be authenticated by the AP, which will further process the request to the associated ENs. The AP is able to capture the biometric information of the user. Note that in case of a large number of users, the AP will serve as a gateway and an additional server will take over its role for user authentication and the construction of the secured response including the user's request to the ENs. For the sake of simplicity, we do not consider our explanation in this situation and limit ourselves to these four entities.

Table 8.2 Notation for Proposed Scheme

Notation	Description
RC	Registration Center
AP	Access Point
EN_j	End Node j
U_i	User i
x, y	Secret values by RC
b	Random number
ID_i	Identity of user i
P_{ref}^i	Reference biometrics of user i
$E_k(.)/D_k(.)$	Symmetric encryption/decryption with key k
$\parallel$	Concatenation operator
$\oplus$	XOR operator
R_j, s_j	Output and state of PRNG in step j

We assume that the communication between RC and user, RC and AP is secured using well-established mechanisms. The RC is considered a robust and secure entity, whereas the AP and ENs in the field might be more vulnerable to tampering. The explanation will focus on the interaction between user and AP on the one hand and AP and ENs on the other hand.

An outsider should not be able to derive the identity of the user in the whole process, nor to derive the content of the transmitted data. In addition, even if one of the devices, APs or ENs are tampered, an attacker might not be able to steal the biometric characteristics of the user or to perform other damaging actions. Only authenticated users are able to request services or access to the ENs.

The attackers may come from inside or outside the network. They are able to eavesdrop on the traffic, inject new messages, replay and change messages, or spoof other identities. Their goals might be to obtain illegitimate data access to the nodes, to perform service degradation or denial of service.

8.4.3 Notations

We represent a hash function by H. The encryption operation of message m under a key K to obtain the ciphertext c is denoted as $c = E_K(m)$, and the corresponding decryption operation as $m = D_K(c)$. Furthermore, the concatenation of values m_1 and m_2 is denoted by $m_1 \parallel m_2$ and the xor operation by $m_1 \oplus m_2$.

We denote by S the finite set of states of the PRNG. The initial state $s_0 \in S$ is obtained after mapping with seed z. The next state function is denoted by δ and the output function by ρ producing the value R_j after step j.

8.5 The System

Figure 8.1 presents the different phases that can be distinguished: the registration (2), the installation of APs (3) and ENs (1), the capturing of biometric data of the user (4), the actual request phase (5) and the corresponding response phase (6). We now discuss each of them in more detail. For ease in notations, we restrict the explication to only one AP and one end node with identity EN_j.

8.5.1 Registration Phase

The RC computes a biometric characteristic of the user and generates the corresponding reference template P^i_{ref}. As will be explained later, we choose to use an iris-based template, typically represented by 2048 bits. As studied in [5], on average half of all the bits differ for two templates of two different people, while a difference score of 0.32 statistically guarantees a positive match. We assume physical contact between user and RC in this phase in order to check the identification and authorization.

8.5.2 Installation Phase

Let x, y be two secrets chosen by the RC. The RC shares the parameters EN_j, $H(x\|y)$, $H(EN_j\|H(x))$ with each node EN_j in the network.

Before sharing secret material with the AP, RC first requests the current stage j of the PRNG at the AP. After this request, the AP updates the PRNG to the next state $j + 1$. We assume that both AP and RC have the same implementation of a PRNG with the same initial seed z. Consequently, the RC can compute $R_j = \rho^{j-k}(s_k)$, where k refers to the previous stage of the PRNG. Next, the following computations are made by the RC for the registration of user U_i. Let N be the number of registrations for a user with that identity.

$$A_i = H(y\|ID_i\|N)$$
$$B_i = H(x\|y) \oplus A_i$$
$$C_i = H(H(P^i_{ref})\|H(x)) \oplus H(A_i)$$
$$D_i = H(x) \oplus H(P^i_{ref})$$
$$E_i = P^i_{ref} \oplus R_j$$

The RC then sends the values $B_i, C_i, D_i, E_i, H(B_i\|C_i\|D_i\|E_i)$ to the AP, where they are stored. The RC also sends a list of the identities of end nodes EN_j, which are in its surrounding and can offer services for the involved user. Note that only B_i, T_e, with T_e a corresponding due date, are stored at the RC. All other intermediate computations and variables are securely erased from memory, except the latest state s_j of the PRNG.

8.5.3 Request Phase

We assume the PRNG, programmed on the side of the AP, is at state s_{j+1}, and has securely stored this state s_{j+1} together with the last output value R_j in its memory. When a patient enters a room, the AP captures the biometric characteristic P^*_{ref} of the user and computes $P^*_{ref} \oplus R_j = E^*_i$. It further computes $d(E^*_i, E_i)$ for all values E_i in the database and checks whether there is a potential candidate, meaning that the distance is lower than the predefined threshold of 0.32 [5]. If successful, the process can continue and an activation of EN_j can be initiated by sending a request message to it. The biometric recognition system proposed in [6, 7] highlights the importance of distance while acquiring and matching for biometric traits.

Denote a random nonce by N_i. Following computations are made:

$$P^i_{ref} = E_i \oplus R_j$$
$$H(x) = D_i \oplus H(P^i_{ref})$$
$$H(A_i) = C_i \oplus H(H(P^i_{ref})\|H(x))$$
$$C_1 = H(EN_j\|H(x)) \oplus H(H(P^i_{ref})\|N_i)$$
$$C_2 = H(A_i) \oplus N_i$$
$$CID_i = B_i \oplus H(H(EN_j\|H(x))\|H(H(P^i_{ref})\|N_i))$$
$$K_1 = H(N_i \oplus B_i)$$
$$V_1 = K_1 \oplus H(A_i))i$$

The following message is sent to EN_j.

$$CID_i\|C_1\|C_2\|V_1 \tag{8.1}$$

After this process, the PRNG and database of users is updated. This includes the following operations for the PRNG $R_{j+1} = \rho(s_{j+1})$, $s_{j+2} = \delta(s_{j+1})$. Also the value of E_i in the database for all users requires an update by replacing it to $E_i \oplus R_j \oplus R_{j+1}$, corresponding again to $E_i = P^i_{ref} \oplus R_{j+1}$. Finally, R_{j+1} and s_{j+2} must be securely stored and the other values securely erased from memory.

8.5.4 Answer Phase

Here, the EN_j executes the following operations using the values stored in the memory:

$$H(H(P^i_{ref})\|N_i) = H(EN_j\|H(x)) \oplus C_1$$
$$L_i = H(H(EN_j\|H(x))\|H(H(P^i_{ref})\|N_i))$$
$$B_i = CID_i \oplus L_i$$
$$A_i = B_i \oplus H(x\|y)$$
$$N_i = H(A_i) \oplus C_2$$
$$K_1^* = H(N_i \oplus B_i)$$
$$H(A_i)^* = K_1^* \oplus V_1$$

If $H(A_i)^*$ corresponds with the transmitted value V_1, the AP thus the user is authenticated. In case a response is required from the end node, the following computations are performed with N_j a random value:

$$C_3 = N_j \oplus H(H(P^i_{ref})\|N_i)$$
$$V_2 = N_i \oplus H(EN_j\|B_i\|N_j)$$
$$SK_{ij} = H(N_i\|N_j)$$

Finally, the message $C_3\|V_2\|E_{SK_{ij}}(M)$, with M the requested info (GET request) or a confirmation (POST, DELETE, PUT request) are sent to the AP.

The AP first derives $N_j = C_3 \oplus H(H(P^i_{ref})\|N_i)$. Next $N_i \oplus H(EN_j\|B_i\|N_j)$ is calculated and compared with the transmitted value V_2. If positive, mutual authentication is obtained and the shared symmetric key can be derived in order to decrypt the last part of the message.

8.5.5 Update Phase

This phase is only executed in case the user wants to update his security material. Therefore, he needs to register again with the RC, who computes his biometric characteristics.

Next, the RC generates $A_i = H(y\|ID_i\|N)$ and corresponding parameter $B_i = A_i \oplus H(x\|y)$ up to N times, until the computed B_i corresponds with the one in memory. Then, the new value for A_i is defined by $A_i = H(y\|ID_i\|N+1)$ and the corresponding updated values of B_i and C_i in memory are overwritten with their new values. The updated values of B_i, C_i are sent to the AP. Note that the value of N should be fixed depending on the system conditions.

8.6   Security Analysis

Let us discuss some security features and resistance against the most relevant attacks in literature.

8.6.1   Accountability

Note that a logging mechanism should be installed in each node. Each log contains the parameters $B_i\|N_i$. The parameter B_i gives no direct information to a certain identity. However, by keeping track of the same pseudonym, abnormal behavior leading to, for instance, service degradation and denial of service attacks, can be more easily detected. In case of doubt, the AP or RC will be contacted to derive the identity.

8.6.2   Replay Attacks

These type of attacks are avoided due to the usage of nonces. First on the side of the ENs, as logging is performed, replay attacks will be noticed. Second, on the side of the AP, replay attacks will not work either since the symmetric shared key with the EN is based on the first nonce, originally sent by the AP. Finally, the RC also keeps track of the number of registrations for a particular identity. One could also use timestamps, but this requires clock synchronization, which might be difficult to guarantee for low-cost nodes.

8.6.3   Insider Attacks

We distinguish the impact of two different situations, being a compromisd AP and EN.

8.6.3.1   Compromised AP

Let us assume that the attacker has physical access to the database of the AP, which stores the secret key material of all its users, being a list of valid combinations of $B_i, C_i, D_i, E_i, H(B_i\|C_i\|D_i\|E_i)$. Even in this situation, it is still impossible to formulate a valid request as the biometric information P^i_{ref} of a user U is required to retrieve the parameter $H(x)$ and the corresponding parameter $H(A_i)$. Note that we still assume that the AP has the capabilities to securely run its operations and to store the output and the next state of PRNG.

Also, a compromised AP is unable to derive information given by messages sent by the other APs as it is not aware of $H(x)$ and thus $H(EN_j\|H(x))$.

8.6.3.2 Compromised End Node

A compromised end node cannot take the role of AP as it does not know P^i_{ref} of the users or the parameter $H(x)$. Also, it does not learn anything from messages sent to another EN^*_j as it is not aware of the value $H(EN^*_j\|H(x))$.

In addition, it cannot release the critical identity related information of its users, since the received information is only indirectly associated with the user's identity. Finally, a compromised node does not have enough information to create users that can perform valid requests to other nodes, since the other secret key y is not known.

8.6.4 HW/SW Attacks

The system is based on a security protocol for tamper-resistant smart cards. The same ideas are applied on the side of the AP. Even with knowledge of the parameters $B_i\|C_i\|D_i\|E_i$, an attacker has no further advantage since the input of the biometric characteristic of the user is required, corresponding to the two-factor authentication feature.

8.6.5 Identity Privacy

Note that the user's request of the AP contains the parameter CID_i, which is a dynamic reference constructed by means of the nonce N_i, related to the pseudonym identity B_i of the user. Consequently, any outsider can never link the different requests to a particular user or to the same user. This also guarantees the location privacy of the user from any outside attacker.

From the request, the end node can derive the indirect link B_i with the user's identity. Only the RC is able to retrieve the real identity. Note that in contrast to an outsider, the end node has the possibility to link the requests to the same user. This feature is needed in order to easily detect abnormal behavior.

8.7 Performance Analysis

Here, we analyze the efficiency of the system, i.e., the cost and accuracy it needs to authenticate a person in order to gain access to the required services. The analysis is split between measuring the computational cost for cryptographic operations on the authentication protocol and the communication of the messages during the request and response phase.

8.7.1 Timing for Cryptographic/Computational Operation

In this section, we have computed the computational costs for two major phases of this authentication scheme i.e. the request phase and the answer phase (Table 8.3). The update phase is executed only when the user wants to modify the key material and thus it is not considered here. Suppose T_H represents the time required to execute one-way hash function SHA-1, T_S denotes a symmetric key encryption/decryption operation AES and T_M is the time required for an elliptic curve point multiplication [15]. The Elliptic Curve Cryptography (ECC) includes all necessary primitives of asymmetric cryptographic i.e., Elliptic Curve Digital Signature Algorithm (ECDSA), key exchange and agreement protocols. Point multiplication servers are considered as an elementary unit in all ECC and are computationally the most complex and expensive operations [3].

We have not included the computational cost for bitwise *XOR* and concatenation because these two operations take relatively less computational overhead. Based on results presented in [9], the computation times for T_H, T_S and T_M are 0.0023 *ms*, 0.0046 *ms* and 2.226 *ms* respectively. We have compared our results with existing remote biometric authentication schemes. The two schemes shown in [8] and [15] take higher execution time because of the need for elliptic curve point multiplication, which is not the case in our proposed scheme and the scheme of [4] as shown in Table III. The remote authentication scheme presented in [4] has quite similar computational cost compared with our scheme because it only uses hash functions. Our proposed scheme performs even better than [4], because it uses fewer numbers of hash functions and has relatively shorter execution time.

8.7.2 Communication Cost

Table 8.4 contains the communication costs for the request and answer phases of our proposed scheme and compared it with the other three available remote biometric-based schemes [3, 5, 15]. In order to evaluate the communication cost, we used *SHA-1* as the hash function having 160 bits as the message digest. For the symmetric encryption/decryption, we assume Advanced Encryption Standard (AES) having block sizes of 128 bits. Whereas, random nonce and timestamps each take 32 bits. In our scheme during the request phase the message $CID_i\|C_1\|C_2\|V_1$ needs $(160+160+160+128) = 608$ bits. The message at answer phase $C_3\|V_2\|E_{SK_{ij}}(M)$ requires

Table 8.3 Comparison of Computational Cost Using Our Scheme [10]

Scheme	Request	Answer	Total	Total Time
He [8]	$5T_H + 2T_S + T_M$	$12T_H + 4T_S + 7T_M$	$17T_H + 6T_S + 8T_M$	13.417*ms*
Baruah [4]	$6T_H$	$7T_H$	$13T_H$	0.0299*ms*
Odelu [15]	$4T_H + 2T_S + T_M$	$13T_H + 4T_S + 5T_M$	$17T_H + 6T_S + 6T_M$	17.847*ms*
Our Scheme	$6T_H$	$5T_H$	T_H	0.0253*ms*

Table 8.4 Comparison of Communication Cost Using Our Scheme [10]

Scheme	Request	Answer	Total Cost
He [8]	1920 bits	1620 bits	3520 bits
Baruah [4]	640 bits	320 bits	960 bits
Odelu [15]	864 bits	2080 bits	2944 bits
Our Scheme	608 bits	448 bits	1052 bits

(160+160+128) = 448 bits. Hence as a result, total communication overhead using our proposed scheme is 608+448 = 1052 bits. We can see that our scheme has significantly better communication costs in comparison with [8] and [15] which takes 2994 bits and 3520 bits respectively. However, our scheme possesses a slightly higher communication cost compared with the scheme of [4].

8.8 Conclusions

In this chapter we propose a complete security solution for a gadget-free patient to establish a secure and authenticated channel with medical sensors embedded in a hospital. Based on accuracy metrics of the iris biometrics, feasibility of the approach is proven. The scheme is shown to be resistant against replay attacks, insider attacks and even hardware/software attacks against the medical sensors in the hospital. Moreover, the identity of the patients is protected and accountability of the devices is obtained through logging. Thanks to the usage of solely symmetric key-based operations, the proposed scheme is very efficient, both with respect to computational and communication cost.

Acknowledgement

This work is supported by European Union RESPONSE 5G (Grant No: 789658) and Academy of Finland 6Genesis Flagship (grant no. 318927) projects.

References

1 Ahmad, I., Kumar, T., Liyanage, M. et al. (2018). Towards gadget-free internet services: A roadmap of the naked world. *Telematics and Informatics* 35 (1): 82–92.

2 Aikio, J., Pentikinen, V., Häikiö, J., et al. (2016). *On the road to digital paradise: The Naked Approach*. Lapland: University of Lapland.

3 Amara, M. and Siad, A. (2011). Elliptic curve cryptography and its applications. *Systems, Signal Processing and their Applications (WOSSPA), 2011 7th International Workshop*, Tipaza, Algeria (9–11 May 2011). IEEE.

4 Baruah, K.C., Banerjee, S., Dutta, M.P. and Bhunia, C.T. (2015). An improved biometric-based multi-server authentication scheme using smart card. *International Journal of Security and Its Applications* 9 (1): 397–408.

5 Daugman, J. (2009). How iris recognition works. In: *The essential guide to image processing* (ed. Alan Bovik), 715–739. New York: Elsevier.

6 Dong, W., Sun, Z. and Tan, T. (2009). A design of iris recognition system at a distance. *Chinese Conference on Pattern Recognition, 2009. CCPR 2009*, Nanjing, China (4–6 November 2009). IEEE.

7 Fancourt, C., Bogoni, L., Hanna, K. et al. (2005). Iris recognition at a distance. *International Conference on Audio-and Video-Based Biometric Person Authentication*, Hilton Rye Town, USA (20–22 July 2005). Springer.

8 He, D. and Wang, D. (2015). Robust biometrics-based authentication scheme for multiserver environment. *IEEE Systems Journal* 9 (3): 816–823.

9 Kilinc, H.H. and Yanik, T. (2014). A survey of sip authentication and key agreement schemes. *IEEE Communications Surveys & Tutorials* 16 (2): 1005–1023.

10 Kumar, T., Braeken, A., Liyanage, M. and Ylianttila, M. (2017). Identity privacy preserving biometric based authentication scheme for naked healthcare environment. *2017 IEEE International Conference on Communications (ICC)*, Paris, France (21–25 May 2017). IEEE.

11 Kumar, T., Liyanage, M., Ahmad, I. et al. (2018). User privacy, identity and trust in 5g. In: *A Comprehensive Guide to 5G Security* (eds. M. Liyanage, I. Ahmad and A.B. Abro), 267. New York: John Wiley & Sons.

12 Kumar, T., Liyanage, M., Braeken, A. et al. (2017). From gadget to gadget-free hyperconnected world: Conceptual analysis of user privacy challenges. *2017 European Conference on Networks and Communications (EuCNC)*, Oulu, Finland (12–15 June 2017). IEEE.

13 Liyanage, M., Ahmad, I., Abro, A.B. et al. (2018). *A Comprehensive Guide to 5G Security*. New York: John Wiley & Sons.

14 Liyanage, M., Salo, J., Braeken, A. et al. (2018). 5g privacy: Scenarios and solutions. *2018 IEEE 5G World Forum (5GWF)*, Santa Clara, USA (9–11 July 2018). IEEE.

15 Odelu, V., Das, A.K. and Goswami, A. (2015). A secure biometrics-based multi-server authentication protocol using smart cards. *IEEE Transactions on Information Forensics and Security* 10 (9): 1953–1966.

16 Porambage, P., Okwuibe, J., Liyanage, M. et al. (2018). Survey on multi-access edge computing for internet of things realization. *IEEE Communications Surveys & Tutorials* 20 (4): 2961–2991.

17 Uludag, U., Pankanti, S., Prabhakar, S. and Jain, A.K. (2004). Biometric cryptosystems: issues and challenges. *Proceedings of the IEEE* 92 (6): 948–9604.

9

WebMaDa 2.1 – A Web-Based Framework for Handling User Requests Automatically and Addressing Data Control in Parallel

Corinna Schmitt, Dominik Bünzli, and Burkhard Stiller

Abstract

Over the last decades the Internet of Things raised in its importance and became more and more part of everyone's life (e.g., smarthome, fitness tracking). In parallel different requests for privacy support, mobility support and flexible privilege handling raised. Thus, this book chapter summarizes the current situation and concerns of users and network owners in the IoT. Based on investigated and identified concerns and users' request, it categorizes and discusses the requirement design of WebMaDa (Web-based Management and Data Handling Framework) addressing the identified issues. WebMaDa supports the mobility request of users and at the same time place the total network and data control with the network owner, reducing the administrator or third-party involvement to a minimum. Thus, special focus in WebMaDa's design was in (i) automated request handling and (ii) addressing of data control with respect to privacy and transparency. The realized system is evaluated by a proof of operation.

Keywords   *Internet of Things (IoT); Wireless Sensor Network (WSN); WebMaDa; privilege management; data control; request handling*

9.1   Introduction

Today, different devices are connected forming small networks and being an integral part of the Internet of Things (IoT). Such networks are typically designed to provide individual solutions for a certain purpose (e.g., environmental monitoring or health monitoring) [1–3]. Devices used potentially show a large heterogeneity concerning hardware and software, thus, a linking to specialized systems allows for analysis and visualization of the data collected. While such an approach exists for IoT, it does not exist for an integrated handling of user requests and network owners changing over time in support of (i) mobility, (ii) ownership and controlling of data, and (iii) updating privileges granted immediately. Thus, the Web-based framework leads to the innovative and practical solution discussed here.

Many specific solutions exist to address mobility requests, while installing a dedicated application on the mobile device. This is considered to be a good solution, but

Corinna Schmitt: This work was performed during her employment at University of Zurich.

IoT Security: Advances in Authentication, First Edition.
Edited by Madhusanka Liyanage, An Braeken, Pardeep Kumar, and Mika Ylianttila.
© 2020 John Wiley & Sons Ltd. Published 2020 by John Wiley & Sons Ltd.

these solutions typically pose special requirements to the device's operating system and can quickly exhaust the device's resources, while being in operation. Integrating energy-saving mechanisms can solve the latter problem, but applications may still require memory. Thus, Web-based solutions are considered highly appropriate, as they only require Internet access and a browser on the controlling devices. Furthermore, the code base in use must only be updated in a single instance, which reduces maintenance costs.

The demand on handling ownership aspects and control of data increases due to wider offers of third-party services in support of analysis and visualization of data. Additionally, a possible misuse of unauthorized data access in IoT needs to be avoided. Thus, in combination with user demands, to be able to update privileges and to grant access to the data collected, challenges arise due to the situation that accesses granted to users can hardly be revoked or updated immediately, if at all.

WebMaDa, a **Web**-based **Ma**nagement and **Da**ta Handling Framework, addresses the three aforementioned aspects (i)–(iii) for sensor networks in an integrated manner. The development started in 2014 with basic support of mobile access to sensor networks owned by a single user, while allowing for the visualization of data in a flexible and hardware-independent manner [4]. WebMaDa was extended by addressing the general request of fine-grained access management and pulling data in emergency cases [5]. The drawback was that each request (e.g., create networks, access foreign networks, view, or pull data) required an interaction with a global administrator, thus, maintaining a central control and introducing unnecessary delay into the system. WebMaDa 2.0 solved this deficiency by automating the request and allowing for an immediate access grant handling without the involvement of a global administrator [6]. Additionally, WebMaDa 2.1 also addresses the demand for privacy and controlling data access besides the pure automated processing of requests by forwarding to the respective contact points using a mailing system. This ensures that network owners hold the full control of data collected and receive full transparency of when and by whom data was accessed and which rights had been granted.

This chapter summarizes the current situation and concerns of users and network owners in IoT. In turn, it categorizes and discusses the requirements design of WebMaDa. Consequently, WebMaDa is presented in detail with the special focus on (i) the automated request handling and (ii) the addressing of data control with respect to privacy and transparency. The evaluation provides a proof of operation.

9.2 IoT-Related Concerns

The IoT today includes many types of devices ranging from resource-rich devices (e.g., Tablets, Notebooks, Servers) to devices with fewer resources (e.g., Smartwatches, Smartphones, Sensors). Both types collect a lot of data with high variety and, partially publish them using different technologies as shown in Figure 9.1. At first glance, this does not seem to be a big problem, but over time this has changed as the number of devices with fewer resources increased and can be placed everywhere, and as well as devices, owners become more and more mobile and travel around, but still want to stay in control of deployed devices out of their current range. Thus, the first request on mobility support arose.

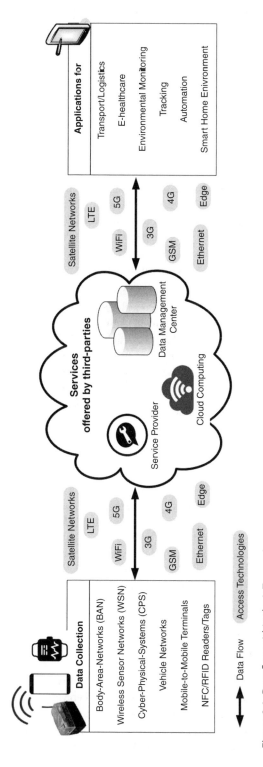

Figure 9.1 Data flow within the IoT.

Furthermore, users became aware of the fact that collected data can support profiling and predict habits. Both usually occur in the data flow as soon as third-parties are involved (e.g., displaying data or storing data in the Cloud) leading to a loss of data control. This is contradictable to the expectations and definition of ownership, which was strengthened by the release of the General Data Protection Regulation (GDPR) [7] in May 2018 in Europe. Thus, the second request of IoT-device users focuses on ownership support.

The third concern about immediate privilege update support follows the first two concerns directly. As users are more and more mobile, they have no direct contact to the deployed devices and also want to access them remotely, perhaps even giving specific people access to them and update granted privileges. In theory, this can happen easily, but here again a control-loss happens, because the owners must trust a third-party offering such an access management service and require an involvement of unknown persons (e.g., administrator) to react on requests in time. Where the timing issue might be critical, in some cases, like emergency cases or losing trust in people having access. Further, owners want to be kept informed about privilege changes and access to data that depend on the third-party if it is done in time or even at all.

All these concerns can partially be addressed by third-party services, but at the same time require full trustability that the service follows rules like those mentioned in the GDPR, for example, strengthen data subject rights, data protection officer in place, privacy-by-design and by-default, and data breach notification [7]. Thus, WebMaDa was developed to address the aforementioned three concerns step-wise until processes were automated sufficiently to reduce the involvement of administrators and giving data owners full control of their collected and published data. The upcoming section presents the design decisions taken leading to the establishment of the WebMaDa framework.

9.3 Design Decisions

Being aware of users' concerns and the devices they used (e.g., Smartphones, tablets) when traveling it was decided to develop a Web-based solution to support network monitoring when absent from the place of deployment. This approach is highly recommended, as it will be operating system independent, will not require additional Apps to be installed on the mobile devices, and will also support code maintenance in one place only. In order to link the deployed network to the Web-based framework, the data owner will first be required to register, providing unique credentials to a special framework CoMaDa (**Co**nfiguration, **Ma**nagement, and **Da**ta handling) [8, 9] which functions as a bridge between the deployed network and WebMaDa. The main responsibility of CoMaDa is to transmit collected data to a backend infrastructure providing the data to WebMaDa and storing the data and all access, as well as any data request placed in direction to the nodes in the network and received answers.

It was decided to have a MySQL database, called WebMaDa-DB, in the backend infrastructure to log all kinds of information. Each user of WebMaDa is stored with contact information, as well as whether a network is owned or not. For each registered network having an unique identifier information of the network's topology, the individual nodes, the received data, and the network owner are stored. As also foreigners should get access to special networks granted by owners, the WebMaDa-DB underwent an extension. New tables were integrated to store individual access rights to foreign networks and, to address the controlling request of owners, an exhaustive logging system was established storing all access including timestamps and the requested data.

In order to address the third user concern on immediate handling of requests, especially for privilege updates, an automated request handling was designed. This requires an Email system informing owners of placed requests for access on the one hand while, on the other hand, immediately notifying requestors about grants and denials. Thus, the administrator is only involved at the beginning when users request access to WebMaDa. When this has been granted, the users themselves can interact without any involvement of the administrator with the owners of another network. As a result, delays are reduced and owners can immediately take privilege decisions on their own. In order to be able to document Email communication the WebMaDa-DB was required to be extended again to log the placed requests, the answers sent out, and the changed privileges.

The resulting data flow in an abstract way as shown in Figure 9.2 including four main components having the following purpose [5]:

- The network collecting data (here a Wireless Sensor Network (WSN)) is connected to CoMaDa.
- CoMaDa periodically processes data received from the network to forward the data to WebMaDa's backend. Therefore, it links the data to the unique credentials of the network, which in return, links the data in the backend to the corresponding tables. Furthermore, CoMaDa supports communication to the network received from Web-MaDa to request immediate sensor readings.
- WebMaDa's backend stores all data received from CoMaDa, handles user and network registration, offers privilege management, supports automated request handling, logs all kinds of information, and offers the code and infrastructure for WebMaDa's frontend support. The backend itself is build by for items: (i) An Apache HTTP Server, called entry server, where all traffic to WebMaDa passes through working as a proxy server and terminating TLS connections before the traffic is forwarded to WebMaDa's components. (ii) A server hosting database of WebMaDa (WebMaDa-DB) and (iii) an Apache HTTP Server, called Web server, hosting the final front-end of WebMaDa. And (iv) an Apache Tomcat Server hosting the WebSocket end point.
- WebMaDa's frontend offers a user-friendly graphical user interface (GUI) offering different functionalities like registration, data viewing, privilege request and adjustment, data request placements, and data filtering. The frontend is privilege-sensitive constructed; meaning different functions are available depending on whether someone is a user or an owner.

9.4 WebMaDa's History

As mentioned in Section 9.2 users requested a solution supporting mobile access and interaction with their deployed networks. Thus, WebMaDa was developed offering user's the possibility to monitor their deployed network during their absence by using manifold devices. Until that time existing solutions to access sensor networks required

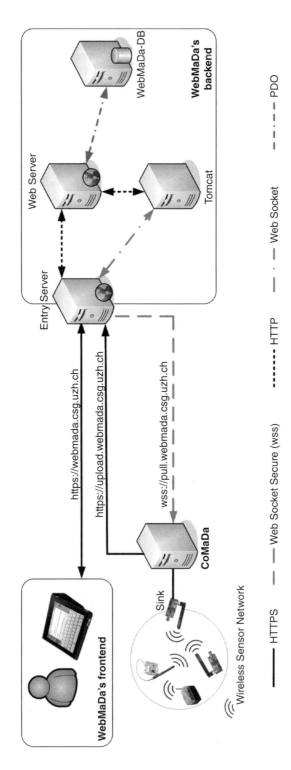

Figure 9.2 WebMaDa's architecture and data flows.

a SmartCard solution issued by the administrator of each network [12–15]. This in turn brought several drawbacks: (1) User had to have carry additional card and special terminals were required to insert the card in order to authenticate and authorize the user before accessing the network. (2) Additional cost occurred due to SmartCard production and special terminals required. (3) For each network one administrator was required, needed to be available to issue SmartCards. And (4) the network owner lost control of the data, as they do not know why someone requested access and if it was gained assuming they did not function of administrator issuing the SmartCards themselves. All these drawbacks inspired the design and development of WebMaDa in order to develop a user-friendly solution to access networks when being far away with less involvement of an administrator, staying control of own data and granted privileges, as well as being flexible to extend WebMaDa with further features continuously.

WebMaDa 1.0 was released in 2014 [4] having a similar data flow in place as shown in Figure 9.2 with the only exception that the backend existed only of a Web-Server and a database. Also, the communication ways between CoMaDa and WebMaDa instances were not secured at all. Included services allowing registration of users, monitoring the network by owners, viewing data received in raw format and in line charts, viewing network topology, and requesting access to foreign networks. WebMaDa's administrator was highly involved in the total system, as he/she was in charge when people wanted to use WebMaDa requesting access to the network by default and, on the other hand, when users request access to foreign networks. This situation introduced delays to the reaction time on handling requests in general, and the structure of the WebMaDa-DB was limited to the essential tables logging access grants, network identifiers, network topology, data received, and information about link to the line charts provided by a third-party service.

Due to the increasing awareness on security and privacy [1], WebMaDa underwent some modifications to address this over the next few years. The first security update was integrated by changing the visualization of data received from the network. Here the third-party involvement was replayed by graph creation using Google Charts [9]. Here, the library is used to draw line charts without transmitting data to externals and, thus, the network owner stays in full control of his/her data collected. Furthermore, additional visualization functionalities became available like layering data and zoom in for a specific time period. The biggest security changes were performed in 2016 [5, 11] by updating the backend infrastructure completely and securing the communication ways as illustrated in Figure 9.2. This secure communication includes (i) an user interface (cf. https://webmada.csg.uzh.ch) used to maintain and access WSNs after successful authentication, (ii) an upload interface (cf. https://upload.webmada.csg.uzh.ch) that can be accessed using HTTP POST and uses a set of well-defined JSON messages to upload received data from CoMaDa, and (iii) an secure pull request interface (cf. wss://pull.webmada.csg.uzh.ch) used by WebMaDa to send pull queries to CoMaDa receiving immediate sensor reading. For further details to these interfaces it is referred to [5, 11]. This infrastructural update was mainly caused by the user's wish to request data immediately instead of periodically as well as the greater security awareness arising about communication in general. As a result, WebMaDa itself was extended with a fine-grained access management solution, more tables in the WebMaDa-DB logging privileges in greater detail, and pull-support in the frontend realizing the immediately data request. All these modifications resulted in WebMaDa 2.0 addressing the users' request on ownership and controlling of data, as well as granting privileges on their own and requesting immediate data [5, 10]. However, the administrator is still involved in different stages:

- Privilege requests must be handled, as no direct contact possibility between owner and user exists.
- Network or data deletion is requested by owners as there exists no right to cause it directly.
- Controlling requests must be addressed manually, i.e. when an owner wants to know how to access the data and when.

Thus, WebMaDa 2.1 was developed addressing the concerns mentioned in Section 9.3 and reducing the involvement of the administrator as much as possible to improve reaction times of requests and to address the owner's requests on control and immediate privilege updating. The resulting solution includes an automated user request handling solution, mailing and notification solution, and an extension of the WebMaDa-DB for controlling purposes.

9.5 WebMaDa 2.1

In order to address a data-controlling request from a network owner, WebMaDa distinguishes between two stakeholder roles, namely "network owner" and "user." Initially, everyone registering with WebMaDa is considered to be a "user." As soon as a user registers his/her own sensor network, this role changes to "network owner" and unique credentials are received for this network identifying it within CoMaDa and WebMaDa. Automatically, corresponding tables are created in the backend linked to these unique credentials for the network and the network owner takes over the administrator activity for his/her network. The tables include, on the one hand, general information (e.g., topology, nodes, data received) and, on the other hand, special information like access privileges, placed requests, and accessed data. Up until now, the administrator was required to grant potential users access to WebMaDa and to handle request from owners to receive controlling information of their network. In previous versions of WebMaDa, the administrator stays now in charge of forwarding requests from users to network owners, including access requests and rights for foreign networks, as well as accessing the WebMaDa-DB to query controlling questions and forward the results. All these steps introduced delays into the workflow as requests placed required immediate reaction from network owners and were time consuming for the administrator in cases many requests were placed at the same time. To overcome this, WebMaDa's 2.1 missions are to automate processes and to shift workload from the administrator directly to the respective entities (e.g., owners, database management).

9.5.1 Email Notifications

Including a mailing solution into WebMaDa solved the designed automation process of handling request. This was possible, as users needed to provide Email contact information to obtain access to WebMaDa. WebMaDa 2.1 includes Email notification in the process of handling (i) access request to WebMaDa, (ii) access request to foreign networks, and (iii) request for password reset. As PHPMailer[1] offers a number of advantages (e.g., active community, good documentation, and extensive error handling)

---

1 https://github.com/PHPMailer/PHPMailer (last access October 29, 2018).

compared to PHP's built-in feature; it was decided to use it here. Another advantage of PHPMailer was that most of WebMaDa was already implemented in PHP and, thus, the integration of the library can be considered a logical step on the one hand, while on the other, it also pays attention to security standards like Secure Sockets Layer (SSL)/Transport Layer Security (TLS) that were already included in WebMaDa 2.0 and should further be supported.

9.5.1.1 Access Request Handling to WebMaDa

It was decided to start the Emailing notification in the early stages of WebMaDa contact. This means, until version 2.0, potential users had to contact WebMaDa's administrator via mail or in person to receive invitation codes and credentials to create an account for WebMaDa. This situation required the administrator to periodically check his/her mail for an access request, then logging into a special admin interface and creating an invitation code that is manually sent back to the potential user, along with registration instructions. In turn, the potential user could use this invitation code to finalize the registration process and become an user of WebMaDa. With the now integrated Email notification solution, the potential user just needs to click a button within WebMaDa to start the registration process by announcing his/her mail contact and the reason for using WebMaDa. Figure 9.3 illustrates the complete process where envelopes stand for automated Email notifications.

As soon as a request was placed, the administrator received a notification via Email with the required information. After login to the admin interface, the request is either denied or accepted by clicking buttons, causing the corresponding notification to be sent out. In case of declination, it is a negative information (red marked envelope) and the potential user can issue a new attempt. In case of agreement, a positive answer (green marked envelope) is sent out including the automatic created invitation code and the instructions on how to continue with the registration. WebMaDa-DB receives an entry on generated invitation-code linked to the corresponding Email placing the request and a timestamp until it is valid. In both cases, new entries appear in the tables logging the Email traffic for controlling purposes. As soon the potential user receives a positive validation, the registration can be completed by using the invitation-code and providing the required additional information (e.g., username, real name, affiliation, password) to create a unique account. In WebMaDa's backend, it is automatically checked if the invitation-code is still valid. If yes, the information is stored in the WebMaDa-DB where the password is stored for security reasons in a hashed version. In return, a notification is sent out to the requestor that the registration process was successfully completed and WebMaDa can be used. From this point on he/she holds the role of a "user." Else the registration process must be initialized and performed again.

Now the user can register own networks following existing workflows from WebMaDa 2.0, which are already automated and do not involve actions by WebMaDa's administrator anymore [5]. Further, the user is able to place access requests to foreign networks that up to now would involve WebMaDa's administrator again by forwarding the request to the network owner. To erase this involvement, automated Email notifications are placed in this process as well.

9.5.1.2 Access Request Handling to Foreign Networks

In order to receive access to a foreign network the workflow shown in Figure 9.4 is processed. It is initiated by an authenticated user selecting a public available WSN of interest after login to WebMaDa 2.1. As each WSN is represented in WebMaDa-DB

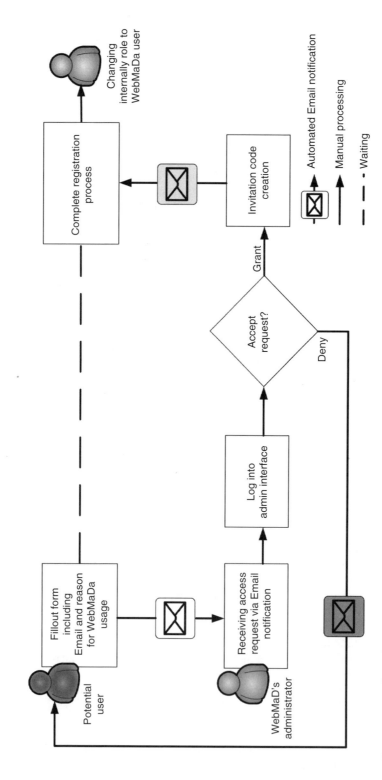

Figure 9.3 Handling of access request to WebMaDa 2.1.

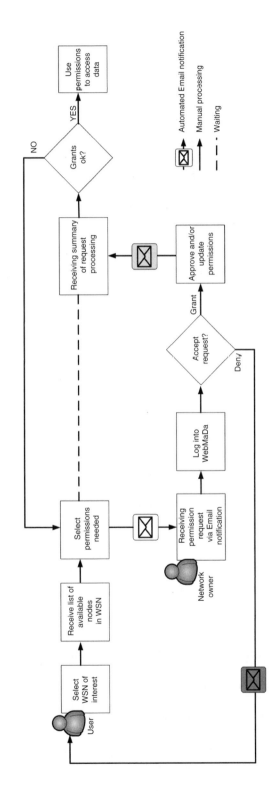

Figure 9.4 Handling of access requests to foreign networks.

with a unique identifier – the WsnId – each placed request from now on is linked to this unique identifier. In return, a list of available nodes in the selected WSN is presented in the frontend, where the user now selects the permissions he/she wants to request. As soon as the request is placed, an Email is automatically created including the information about the request and is directed to the network owner. The owner's Email contact is received from the WebMaDa-DB by placing an automated processed query to look up the contact information via the unique WsnId of the selected WSN. The network owner receives a notification in his/her Email account that a permission request was placed to the owned WSN. After logging into WebMaDa, the owner can deny or grant the request. In the first case, an Email is automatically sent out to the requestor with the information that the permission request was denied. In the other case, the network owner can either directly approve the placed request or modify the permissions requested (e.g., adding further rights or limit request rights). Next, an Email is automatically generated and sent back to the user, who placed the request. In both cases, entries in the WebMaDa-DB are made including information of changed permissions to the respective WSN and about the sent Emails. In return, the user receives a summary of the request processing, can agree on it, when the request satisfies his/her initial request continuing with accessing the data or start the process from the beginning again to ask for modified permission.

As it can be seen by the flow shown in Figure 9.4 and with the above description interaction by WebMaDa's administrator is no longer required for handling any access requests for foreign networks. Thus, the time required for receiving access is only triggered by the reaction times of the network owner his/herself. He/she is also able to update given rights on his/her own without involving the administrator at all and is able to inform the permission holder automatically by Email when granted rights are modified or revoked at any time.

9.5.1.3 Password Reset

In older versions of WebMaDa, the administrator was required to reset a password. Using the Email notification by meeting both security and usability requirements at the same time, now also optimizes this process. Due to security reasons, no standard password can be set and sent to the user via Email. A simple link with a randomized token to reset the password would also not suffice, as it would be possible to intercept this message and use the link. Thus, it was decided that the automated process includes the following steps:

1. A user requests a password reset link for a user account. In order to receive a password reset link, the user has to enter the Email address that is assigned to his/her user account.
2. After submitting the form, the user is informed that if the address is linked to an account, he/she will receive an Email.
3. Within WebMaDa's backend, it is checked if the address matches a user account registered.
   a. If not, no further action is done, not even an Email is sent as a response to the request.
   b. If yes, a randomized token is created and sent to the user. This token is only valid for one reset.
4. For the reset, the user is requested to enter the username and a new password. The username is requested because it is never mentioned together with the Email address

and can, therefore, be used as a secure combination that is known only to the user itself.

5. The updated information is stored in WebMaDa-DB and the user can log into Web-MaDa with his/her new credentials.

9.5.2 Data Control Support

As the number of users is rapidly growing over time, the request for data control that is also supported by the GDPR has increased. WebMaDa 2.1 received an update to address the request. Data control implies support for privacy and transparency in parallel. Both aspects and their realization are described in the upcoming sections.

9.5.2.1 Privacy Support

The understanding of the term "privacy" is that restrictions to the access of an owned network exist. As soon as a potential user of WebMaDa passed the process for Web-MaDa access described in Section 9.5.1.1, he/she becomes a WebMaDa user and can own networks. During the registration process of a new network WebMaDa 2.1 now requires the specification of the network as "private" or "public." If a network is marked as "private," only the network owner can have access to it and see all data. As soon as he/she wants to grant access to the owned network it has to be marked as "public." This updated specification results in the fact that a user can see the network in the list of available WSNs and start the process described in Section 9.5.1.2. The owner of the network can still change the specification of the network whenever necessary.

9.5.2.2 Transparency Support

In order to fulfill the data-controlling request of network owners in full, transparency must be addressed in parallel with privacy. The simplest way to do this is to log all interactions with WebMaDa including any placed access request, data queries, and granted/revoked rights, as well as mails sent out. In order to do this, WebMaDa 2.1 includes a logging system that extends the existing database structure in the backend. Thus, the following log tables were integrated into WebMaDa-DB [6]:

- *ActiveWSN_Log* stores information about changes made to a WSN. This includes privacy settings, reset, deletion, and creation of a WSN.
- *Admin_Log* is required if WebMaDa is newly set up and no users are created that could potentially be an administrator; a root account is needed to enter the backend. This root user is stored in a separate table called Admin and therefore separately logged. If the password of the root user is changed, it will be visible in this log table.
- *User_Log* stores all changes that affect a user. This includes password resets, alteration of the administrator status, as well as creation and deletion of a user.
- *InvitationCode_Log* receives an entry if a new invitation code is sent, deleted, or used by a potential new user to create his account.
- *InvitationRequests_Log* receives an entry as soon as a potential new user requests an invitation link and an entry in the corresponding log table is created. Furthermore, it is logged if the request is approved or declined by an administrator.
- *Mail_Log* stores every outgoing mail which is logged together with its subject and the timestamp.

- *Rights_Log* stores all changes that affect the effective permissions of a user to a certain WSN which are logged in this table. This includes newly granted permissions but also access revocation and changes to existing rights. Additionally, it is mentioned if it impacts push or pull permissions.
- *PermissionRequests_Log* stores all permission requests to a certain WSN which are stored in this table. This includes the creation of the request, as well as an accept, deletion, or alteration by the administrator.

For the sake of completeness, it must be mentioned here that this detailed logging solution might affect the privacy of users and/or owners at the same time. The purpose of such a logging system is to provide a history of the changes that were made to the system – here WebMaDa and network access. This issue becomes highly relevant as soon as a WSN is reset (data deleted but WsnId still exists) or deleted (data and WsnId deleted) by the network owner. Based on the understanding of transparency mentioned above, it was decided that even if a WSN is deleted, the logs of this network are kept in the WebMaDa-DB. This should not have any impact on privacy related topics as there is only meta data stored in the WebMaDa-DB. For instance, no actual data like temperature or humidity is logged, but only data on administration activities on the network (e.g., access requests, changed rights, and network privacy settings). In order to obtain logging information for such deleted or reset networks the administrator must be contacted and the requestor must prove the ownership. [6]

Further, only network owners have access to see the logs of changed rights and permission requests of their owned networks. Additionally, they can see logged data of their active WSN, for example, when it was created, reset, deleted, or when the privacy settings were changed. On the other hand, administrators can see different Emails that were sent, high-level meta data of the active WSNs (e.g., reset, deletion, creation, and privacy), invitation requests and codes, as well as changes to the root account and the different users. [6]

Within WebMaDa 2.0, a filtering option was included allowing the network owners to stay in control of their owned networks [10]. This filtering system uses the exhausting logging system to present the network owner with requested information. In order to ensure that only network owners can perform filtering, this option is only activated as soon as the credential check for ownership is successful. This is done directly as soon as a user logs into WebMaDa and clicks on the button "Filtering" in the frontend of WebMaDa. If the validation fails, the user is informed about not owning the networks. Otherwise, he/she can proceed and adjust the filter to his/her requests and filter each attribute of the stored tables. The following attributes can be used to specify the filter [10]:

- Username,
- Type (push or pull or push&pull),
- SensorName (e.g. Voltage, NodeTime, Temperature, Humidity),
- Action (grant or remove), and
- Time.

Depending on the configuration of the filter and the size of the dataset handling, the query can take time. Thus, an efficient and scalable frontend solution must be in place to simplify usability. Therefore, it was decided to use DataTable[2] plug-in as it is explicitly designed for bootstrap styling and prevails negative aspects.

---

2 https://datatables.net (last access: October 29, 2018).

9.6   Implementation

This section presents implementation details for the notification process, the logging, and the filtering in WebMaDa 2.1. All relevant scripts and files are fully integrated into the previous version of WebMaDa, both in the front- and backend. With these enhancements and modifications, WebMaDa 2.1 can now fulfill the wishes and requests of current users mentioned in Section 9.2. A corresponding proof of functionality is given in Section 9.7.

9.6.1   Mailing Functionality

As described throughout Section 9.5.1 WebMaDa 2.1 includes an automated notification solution to reduce the workload of the administrator and allow direct contact between involved stakeholders (here: users, owners). In order to realize this, a mailing solution was designed having the requirement of guaranteed maintainability of the implementation. Therefore, it was decided to implement only one script containing Email templates and to be able to send the corresponding messages matching the different answering types required (cf. Figures 9.3 and 9.4). As a result, all activities that a user can trigger (e.g., permission request, new user request) to deliver an Email, must have their requests sent to this central location and changes in the mail template are only required in one script and, thus, lower the maintenance costs. To solidify the mailing solution requires two PHP files: (i) *mailConfig.php* and (ii) *mailFactory.php*. [6]

The first file – *mailConfig.php* – contains all necessary configurations for the PHP-Mailer. It also includes the sender's Email account and all basic configurations like the host and port of the mail server used, as well as the login information and the *Mail From* name. By intention, this file should only be included once, namely in the mail factory script.

The second file – *mailFactory.php* – is responsible for routing different requests to the correct method and sends the corresponding mails. This means, that if a user wants to call this script, sending an Ajax POST request to the factory along with a purpose is required. This request includes all required information to route the request to the correct method via a switch case statement. If necessary, some additional data can be added to the request. In most cases, this is an object that contains the recipient's Email address and the body of the message.

Sending mails is triggered by user interactions via an input as described throughout Section 9.5.1. This means, the process starts with a JQuery action (e.g., click or change) by the user caused in WebMaDa's frontend. In the following, an example of requesting the deletion of an access request by a network owner is assumed to describe the implemented workflow. Figure 9.5 [6] illustrates the linkage of scripts and gathered inputs for this process.

After initiating this request handling by clicking the corresponding button in the GUI the required data is gathered, namely Username and WsnId. In the next step, this data is sent by an Ajax request to the corresponding PHP script, which performs the desired operation. For details about the deleting operation see [5]. If the deletion is successful a true statement (=1) is returned by the script, or alternatively, a false statement (=0). Assuming the deletion was successful the method *sendEmail(purpose, emailData)* is

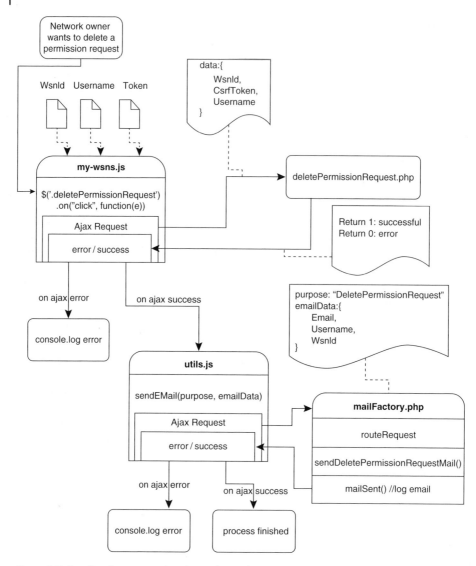

Figure 9.5 Dataflow between scripts for mailing solution in WebMaDa 2.1.

called located in the *util.js*, because it is a method that will be utilized by different scripts. In order to create the necessary mail for notification, the required data (here: recipient Email and information for the mail body) is gathered. Next, a new Ajax request is sent to the mail factory including the gathered data. In the mail factory, the request is then routed based on the provided purpose (here: DeletePermissionRequest). Finally, the method *sendDeletePermissionRequestMail(additionalData)* is called, which fills the provided data into the mail template and then sends it via the function *sendMail()*. At the end of the process, the method *mailSent()* is called creating a log entry for the recently sent mail in the corresponding table in WebMaDa-DB.

9.6.2 Logging Functionality

The logging solution follows the same principles as implemented for the mailing functionality, meaning that only one single PHP script should be required to create logs, but at the same time staying flexible and allowing creation of new logging methods was required. As a result, the *logFactory.php* was implemented responsible for routing requests and writing log entries. One of the most important attributes of the logs by definition is the possibility to have a historical view on the data gathered. As for the usage of a network owner or administrator, it should be possible to trace every change (e.g., permission requests granted or denied, accesses modified or revoked) that has been made within the network. This already implies that for each table that should be logged, a separate table has to be available in WebMaDa-DB.

For each table defined in Section 9.5.2.2 a separate table *<Tablename>_Log* is created. This means for example that table *Rights* contains the access permissions for the different networks, is accompanied with a table called *Rights_Log*. All attributes present in the original table are also present in the log table. The assumption ensures that entry changes cause direct writing to the corresponding log tables without further processing. Additionally, each log table includes the attributes *action*, *byUser*, and *date* [6]:

- The attribute *action* contains the action executed on the dataset. For example, if a new permission is granted, the string "granted" is written into the WebMaDa-DB. Other actions can be deletion, creation, or alteration. If the "alteration" is written in a field, it is indicated that an attribute has changed in comparison to the last record. An example of this is the changing of the privacy status. Setting the *IsPrivate* attribute from 0 to 1 or vice versa changes the status.
- The attribute *byUser* indicates who caused the corresponding action. In case of accepting or rejecting an invitation request to WebMaDa the administrator is set for the attribute *byUser* and a corresponding entry is made in the *InvitationRequest_Log* table. In case of a new potential user for WebMaDa no username exists yet and, thus, the Email address from the request form is considered for the attribute *byUser*.
- The last attribute *date* includes a timestamp of the action supporting a timing sorting to create a history within the logs.

These three additional attributes together with the original dataset suffice to trace which user took what action at which time and how the current situation could possibly be reverted.

Similar to the mailing process was the process of adding a new log entry implemented. If an action is triggered via JQuery, the operation is executed and based on the return value an Ajax POST request is sent to the script *logFactory.php*. Here as parameters considered are the purpose of the request and the additional data containing the elements that need to be logged (e.g., username, WsnId, and the sensor that access is requested for). Next the *routeLogRequest* method is called where a switch statement is used to decide where the request should be routed. The route is determined by the provided purpose in the Ajax POST request. Due to the broad variety of different tables and actions that must be considered, an individual method must be created for each type of event that should be logged. Finally, the data is written into WebMaDa-DB using PHP Data Objects (PDO) calling a storing procedure consisting of an insert statement.

9.6.3 Filtering Functionality

The main goal of the filtering functionality is to provide a simple, yet powerful way to allow a network owner or administrator to view all changes that have been made in a network including when who accessed data. Achieving this goal, it was defined that a network owner can see all information logged related to his network owned. This includes the rights and permission requests, as well as the changes made to his network. Here, users can only see owned networks in a filtering of the table *ActiveWSN*, including all registered networks in WebMaDa. On the other hand, the administrators should see system relevant logging tables such as the Emails that have been sent by the framework, the *User* and *Admin* tables, the *Invitation Requests* and *Codes*, as well as an overview of all registered WSNs.

Based on these assumptions the filtering process is implemented as follows: Each filtering form contains fields that a logging table could possibly be filtered with (cf. Section 9.5.2.2). Further it contains the name of the assigned stored procedure as a hidden input field, which is annotated with the class *pdoName*. This design decision was taken in order to reduce code duplication in the corresponding JQuery script resulting in one filtering method calling dynamically selected stored procedures in it.

As the frontend of WebMaDa is designed with HTML-files itself and the filtering should also support HTML, the fields in the HTML form have to be provided in the correct order as defined in the store procedure in the backend. This means, if the filtering procedure expects the username as first parameter, this has to be the first input element in the form, too. For each input field that should be considered as a filter, the class *logFilter* has to be added to the corresponding HTML element. The reason for this design request comes from the JQuery script scanning the HTML container for all elements that are annotated with this class and adds their value as filter criteria. Only the limit operator is handled separately, as it is used in all forms. The limit filter is annotated with the class *limitFilter* and is not included in the *logFilter* array.

As soon as the network owner specified in the frontend the filter criteria in a satisfying manner, the function *openLog* is called to receive the data of the corresponding log table. As a first step, the name of the PDO, all filters from the user interface, the Cross-Site-Request-Forgery (csrf) token, as well as the container in which the log table is located have to be gathered. Next, the parameters are passed to the PHP script *generateLogTable* fetching the data from WebMaDa-DB. Therefore, a customized PDO is generated with the name of the passed stored procedure and all filter parameters that have been sent with the request. First the stored procedure shown in Listing 9.1 [6] contains all inputs present in the HTML form and then the corresponding log table used as the FROM clause (line 15). The WHERE clause (lines 16–18) is built to either ignore the input values if they are null or check if they have any values given. Due to this, it is possible to use the same procedure to search for various combinations of filter inputs and not creating a separate one for each condition. Finally, the PHP script generates a HTML table containing all obtained records. The generated string is returned to the JQuery method *openLog*, where it is set as the HTML content of the table container.

9.7 Proof of Operability

In this section a proof of operability for the implemented features Email notification, logging, and filtering is presented showing that the concerns listed in Section 9.2 are

```
1   CREATE DEFINER= `admin `@ `localhost ` PROCEDURE ` GetLogs_PermissionRequest `(
2           IN p_limit INT
3           , IN p_wsnid VARCHAR(200)
4           , IN p_username VARCHAR(200)
5           , IN p_sensorname VARCHAR(200)
6
7   BEGIN
8           SELECT WsnId
9                   ,Username
10                  ,SensorName
11                  , action
12                  , byUser
13                  , date
14
15          FROM PermissionRequest_Log
16          WHERE (p_wsnid IS NULL OR WsnId = p_wsnid) AND
17                  (p_username IS NULL OR Username = p_username) AND
18                  (p_sensorname IS NULL OR SensorName = p_sensorname)
19
20                  ORDER BY date DESC
21          LIMIT p_limit;
22  END
```

Listing 9.1 Stored procedure getting the logged entries of the selected network.

addressed by WebMaDa 2.1, namely (i) mobility support, (ii) ownership control, and (iii) immediate privilege handling.

9.7.1 Automated Request Handling

As described throughout Section 9.6.1 WebMaDa 2.1 received an extension for notifications per Email if requests are placed in order to support administrators, network owners, and users to received immediate notifications about received requests and update status. Proving the operation of the functionality is done in the following with two examples: (i) A potential user requests access to WebMaDa (cf. Figure 9.3) and (ii) a user requests access to a foreign network. In both cases screenshots are presented illustrating the GUI from requestor's perspective and from the respective answering party (e.g., administrator, network owner).

9.7.1.1 Example for Automated Handling Access Request for WebMaDa

For case (i) Figure 9.6 illustrates the form a potential user must fill out to requesting access to WebMaDa. Here the contact information and the purpose for the request must be provided. As soon as the button "Send Request" is pushed the automated notification system in the backend starts its work generation three mails automatically:

- Email to WebMaDa's administrator to inform that an invitation request was placed, triggering the him/her to log into the administrator page to access or grant the request (cf. Figure 9.7).
- Email to the requestor that the request was received by WebMaDa (cf. Figure 9.8, mail no. 1).
- Email to the requestor that the request was open (cf. Figure 9.8, mail no. 2).

As soon as WebMaDa's administrator receives the notification and logged into the admin view of WebMaDa he/she can see in the users menu which requests for invitation wait for an accept or a revoke decision and pending invites (cf. Figure 9.9a). The latter

Figure 9.6 Invitation request for in the frontend of WebMaDa.

means that the request was granted, but the registration process not yet concluded. Independent of accepting or denying the placed request the requestor receives an Email about the status update. Here Figure 9.8, mail no. 3 shows the mail sent out when the request was granted. Automatically, the admin's view is updated and the granted request is shifted to the section pending invites waiting for the requestor concluding the registration (cf. Figure 9.9b).

As this process shows, the interaction with the administrator is very limited and everyone receives notification about status and actions. Furthermore, there is no longer any need to publish an Email address, which prevents the administrators Email address from ending up in a spam mailing list. Additionally, with slight adaptations in the backend WebMaDa's administrator can now give authenticated users the right to act as an administrator in parallel. This means work can now be distributed to several people and, thus, it is guaranteed that the creation of users can remain operative even in case of illness or other absence of a key person. Since the process has been simplified and unnecessary intermediate steps (e.g., opening and writing an Email) have been omitted, the goal can be considered of intermediate and automated request handling is fulfilled.

9.7.1.2 Example for Automated Handling Access Request to Foreign Network

As soon as the user has access to WebMaDa he/she can request access to foreign networks assuming the network owners specified them as "public." If not set to "public" they would not appear in the available list presented to the user when specifying a request. The presented list includes only the names of the networks, meaning the user needs to know that (e.g., timWSN, G4T9R04PYS, Buenzli2), which is linked to an unique WsnId in WebMaDa-DB. The user follows the steps shown in Figure 9.4. First, he/she logs into WebMaDa and navigates to the submenu "Other WSNs" where granted access to foreign networks is shown and new requests can be placed (cf. Figure 9.10). Here the user can see, where privileges to foreign network(s) were already granted including details (here: access to network G4T9R04PYS). In this view also granted access can be given back by pressing the red button causing automatic notification sent out to the corresponding partners and an entry in the corresponding logs and tables

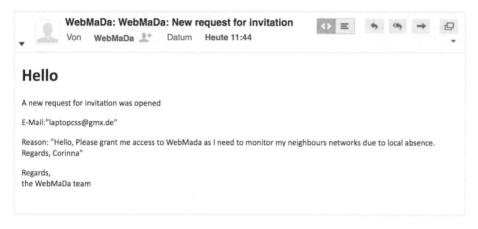

Figure 9.7 Automated Email notification at the administrator's side.

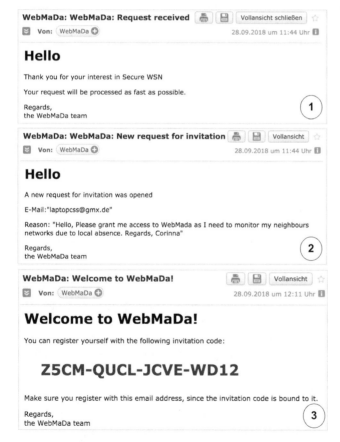

Figure 9.8 Automated Email notification at the requestors side.

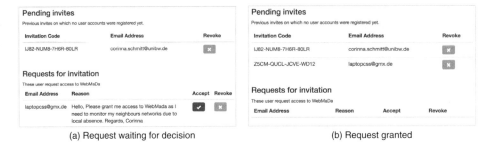

(a) Request waiting for decision   (b) Request granted

Figure 9.9 Administrators view showing placed requests and pending invites.

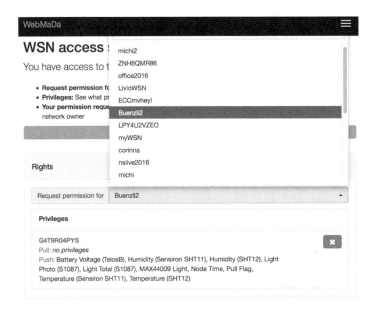

Figure 9.10 View of current access permissions to foreign networks and selection opportunities.

in the database. When now querying a new permission request the user selects the foreign network in a dropdown menu (here: Buenzli2, cf. Figure 9.10). Immediately after selecting the network of interest, a window pops up where the user can place the detailed request (cf. Figure 9.11; here: requesting push for Humidity-TelosB and pull for Temperature-TelosB), where A0LDUV5L6O equals the WsnId for the selected foreign network "Buenzli2" in WebMaDa-DB. As soon as the user pushes the button "Open Request" the automated notification system becomes activated, creating corresponding Email to the involved people that raised the request. Additionally, all information is written in the WebMaDa-DB. Visual feedback is given to the user in the GUI by listing pending permission requests (cf. Figure 9.12a).

When the network owner receives the notification per Email he/she logs into WebMaDa and can see in the submenu "My WSNs" pending requests as shown in Figure 9.12b. Now he/she can either directly grant it without changes, deny it or update it. In the last option the network owner can modify the placed request before granting it. This means he/she can remove requested permissions, grant additional ones or grant

Figure 9.11 Popup to specify request.

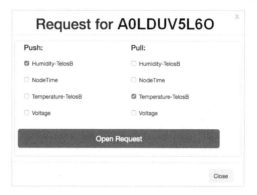

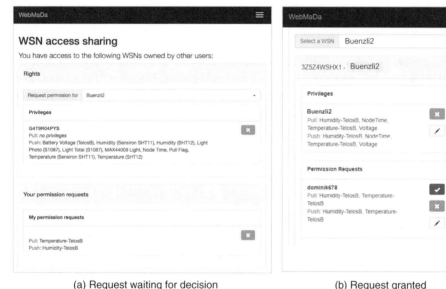

(a) Request waiting for decision   (b) Request granted

Figure 9.12 GUI for permission requests.

others. When the settings satisfy the network owner he/she pushes the button "grant" and in the backend notifications are created to inform the requestor and network owner about changes, as well as entries are created in WebMaDa-DB. Automatically the requestors view is updated by moving the request from section "My permission requests" into the section "Privileges" in the GUI and in parallel a notification is send per Email to the requestor to inform about the status change.

With this solution design the network owner stays in full control of gathered data and can grant, update and/or revoke privileges whenever necessary without involving the administrator anymore, because the process automatically matches the selected WSN to the linked unique WsnId in the database and checks the corresponding owner and its mail. This information is then used to construct the initial notification. The answering process works in a similar way. The delay in handling the request is only influences by the network owner, who needs to handle the request.

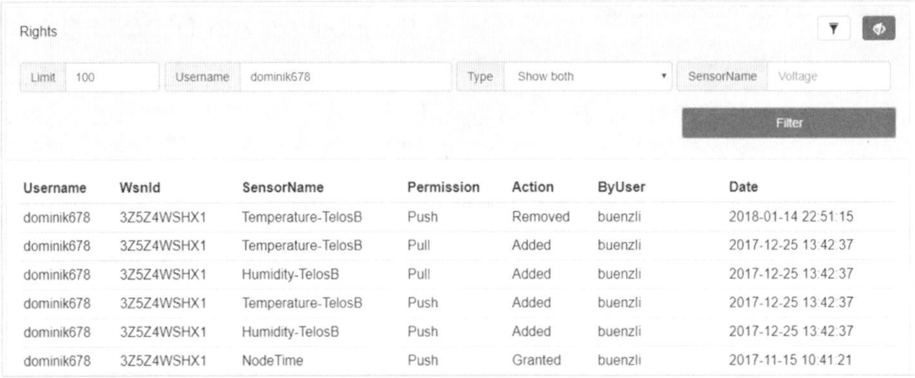

Username	WsnId	SensorName	Permission	Action	ByUser	Date
dominik678	3Z5Z4WSHX1	Temperature-TelosB	Push	Removed	buenzli	2018-01-14 22:51:15
dominik678	3Z5Z4WSHX1	Temperature-TelosB	Pull	Added	buenzli	2017-12-25 13:42:37
dominik678	3Z5Z4WSHX1	Humidity-TelosB	Pull	Added	buenzli	2017-12-25 13:42:37
dominik678	3Z5Z4WSHX1	Temperature-TelosB	Push	Added	buenzli	2017-12-25 13:42:37
dominik678	3Z5Z4WSHX1	Humidity-TelosB	Push	Added	buenzli	2017-12-25 13:42:37
dominik678	3Z5Z4WSHX1	NodeTime	Push	Granted	buenzli	2017-11-15 10:41:21

Figure 9.13 Filtering result for network owner.

The option for the network owners to mark the network as "private" serves as an important means of data security. This option is enabled by default and prevents other users from seeing the network or making access requests. However, this setting does not affect any existing access rights, instead it just controls the visibility of the network under "Other WSNs" in WebMaDa's submenu.

9.7.2 Filtering Functionality Using Logging Solution

In order to address the transparency request and the control request of network owners a logging system was integrated into WebMaDa 2.1 as described in Section 9.6.2. This logging system is automatically filling tables in WebMaDa-DB with corresponding information if actions are triggered (e.g., access granted/revoked/updated, queries placed). Access to all tables in the WebMaDa-DB is reserved for the administrator only. Network owners can access part of the information by using the filtering solution without having the administrator in the queue to process the request. This filtering solution becomes only available if the user is a network owner for the selected network. This decision was taken by purpose to prevent misuse of information. The required credential check is already done when a user logs into WebMaDa meaning that under the submenu "My WSNs" he/she can see owned networks and has the option for filtering available. In the filter option the network owner can specify the query handled in the backend and receiving in the GUI the result. An example is shown in Figure 9.13 for access performed by user dominik678.

9.8 Summary and Conclusions

Within this book chapter a user-friendly Web-based framework for handling user requests automatically was presented by addressing user concerns for mobility support, ownership support, and immediate privilege update having the goal (i) to limit involvement of third-parties in the process chain and (ii) to inform involved parties immediately about status changes.

The first concern on mobility support was the driving force to develop WebMaDa itself, by offering a Web-based framework with functionalities such as registering network, grant/revoke/update privileges, and viewing gathered data. WebMaDa 2.1 extends now these basic functionalities by offering network owners filtering functionality, allowing seeing data access and granted rights also when being physically absence from the deployed network and the instance CoMaDa. This functionality also addresses the second concern on ownership support, as in WebMaDa 2.1 network owners can query requests to WebMaDa-DB directly accessing the exhausting logging system. The third concern about immediate privilege handling is now possible by WebMaDa 2.1 as no third-party like the administrator is involved anymore when access requests are placed. This became possible with WebMaDa 2.1 by including the automated notification system by informing involved parties directly about placed requests, status about the request, and updated that were processed. Everything is automatically in the backend logged to address the ownership support additionally.

Overall it can be stated that with the new included functionalities – automated request handling and addressing data control in parallel – WebMaDa 2.1 now addresses current concerns in IoT and strengthens the ownership with minimal involvement of third parties. For the future it is envisioned to offer more visualization opportunities via Web-MaDa 2.X, like dynamic graph creation, and optimizing the logging solution to improve scalability of tables by including compression for archiving purposes. Further WebMaDa should be linked to other types of networks to collected data and on the other side link WebMaDa to actor systems such as climate control systems in order to trigger events when being abort (e.g., closing/opening window, turning off/on heating).

References

1 Porambage, P., Ylianttila, M., Schmitt, C. et al. (2016). The quest for privacy in the Internet of Things. *IEEE Computer Society* 3: 34–43, April.

2 Schmitt, C. and Carle, G. (2010). Applications for Wireless Sensor Networks (Chapter 46); *Handbook of Research on P2P and Grid Systems for Service-Oriented Computing: Models, Methodologies and Applications*, Edited by. N. Antonopulus, G. Exarchakos, M. Li and A. Liotto, pp. 1076–1091, ISBN: 1-61520-686-8, Information Science Publishing, January.

3 Atzori, L., Iera, A., and Morabito, G. (2018). The Internet of Things: A survey; *Journal Computer Networks* 54 (15): 2787–2805, October.

4 Keller, M. (2014). Design and Implementation of a Mobile App to Access and Manage Wireless Sensor Networks; Master Thesis, Communication Systems Group, Department of Informatics, University of Zurich, Zurich, Switzerland, November.

5 Schmitt, C., Anliker, C., and Stiller, B. (2016). Pull Support for IoT Applications Using Mobile Access Framework WebMaDa; *IEEE 3rd World Forum on Internet of Things (WF-IoT)*. New York, NY, USA December, pp. 377–382.

6 Bünzli, D. (2018). Efficient and User-friendly Handling of Access Requests in Web-MaDa; Bachelor Thesis, Communication Systems Group, Department of Informatics, University of Zurich, Zurich, Switzerland, January.

7 European Parliament (2016). Regulation (EU) 2016/679 of the European Parliament and of the Council of 27 April 2016 on the protection of natural persons with regard

to the processing of personal data and on the free movement of such data, and repealing Directive 95/46/EC (General Data Protection Regulation); Document ID 32016R0679, Brussels, Belgium, April, https://eur-lex.europa.eu/eli/reg/2016/679/oj (last access: September 13, 2018).

8 Schmitt, C., Freitag, A., and Carle, G. (2013). CoMaDa: An Adaptive Framework with Graphical Support for Configuration, Management, and Data Handling Tasks for Wireless Sensor Networks; *9th International Conference on Network and Service Management (CNSM), IFIP*, Zurich, Switzerland, ISBN: 978-3-901882-53-1, pp. 211–218, October 2013.

9 Schmitt, C., Strasser, T., and Stiller, B. (2016). Third-party-independent Data Visualization of Sensor Data in CoMaDa; *12th IEEE International Conference on Wireless and Mobile Computing, Networking and Communications*, New York, NY, USA, pp. 1–8, October 2016.

10 Silvestri, N. (2017). WebMaDa Extension Addressing Transparency Request for Data Owners; Assignment, Communication Systems Group, Department of Informatics, University of Zurich, Zurich, Switzerland, July 2017.

11 Schmitt, C., Anliker, C., Stiller, B. (2017). *Efficient and Secure Pull Requests for Emergency Cases Using a Mobile Access Framework; Managing the Web of Things: Linking the Real World to the Web*, In: M. Sheng, Y. Qin, L. Yao, and B. Benatallah (Eds.), Morgen Kaufmann (imprint of Elsevier), Chapter 8, pp. 229–247, ISBN: 978-0-12-809764-9, February.

12 Das, A.M. (2009). Two-factor user authentication in wireless sensor networks. *IEEE Transactions on Wireless Communications* 8 (3): 1086–1090, March.

13 Chen, T.H., and Shih, W.K. (2010). A robust mutual authentication protocol for wireless sensor networks. *ETRI Journal* 32 (5), October.

14 Turkanovic, M., Brumen, B., and Hölbl, M. (2014). A novel user authentication and key agreement scheme for heterogeneous ad-hoc wireless sensor networks, based on the Internet of Things notion. *Ad Hoc Networks* 20: 96–112, April.

15 Amin, R., and Biswas, G.P. (2016). A secure light weight scheme for user authentication and key agreement in multi-gateway based wireless sensor networks. *Ad Hoc Networks* 36: 58–80, January.

Part IV

IoT Device Level Authentication

10

PUF-Based Authentication and Key Exchange for Internet of Things

An Braeken

Abstract

Key agreement between two constrained IoT devices that have never met each other is an essential feature to provide in order to establish trust among its users. Physical Unclonable Functions (PUFs) on a device represent a low-cost primitive exploiting the unique random patterns in the device allowing it to generate a unique response for a given challenge. These so-called challenge-response pairs (CRPs) are first shared with the verifier and later used in the authentication process. The advantage of a PUF at the IoT is that even when the key material is extracted, an attacker cannot take over the identity of the tampered device. However, in practical applications, the verifier, orchestrating the authentication among the two IoT nodes, represents a cluster node in the field, who might be vulnerable for corruption or attacks, leading to the leakage of the CRPs. Possessing a huge number of CRPs allows its usage in machine learning algorithms reveal the behaviour of the PUF.

Therefore, in this chapter we propose a very efficient method to provide authentication between two IoT devices using PUFs and a common trusted cluster node, where the CRPs are not stored in an explicit way. Even when the attacker is able to get access to the database, the stored information related to the CRPs will not be usable input for any type of learning algorithm. The proposed scheme uses only elliptic curve multiplications and additions, instead of the compute intensive pairing operations as an alternative scheme recently proposed in the literature.

10.1   Introduction

Authentication between two IoT devices that have never met before is a frequently occurring problem. For instance, consider the following healthcare situation of a patient in a hospital room [1]. Suppose the patient is wearing a bracelet for blood pressure. In case of measurements above a certain threshold, the alarm button in the room should send a signal to the nurses. The communication between the bracelet and alarm button over the Internet and on local networks needs to be secured to gain trust and acceptance and to avoid direct physical harm to the patient, even loss of life. Both the bracelet and the alarm button are not authenticated before, but should be able to generate in a very

IoT Security: Advances in Authentication, First Edition.
Edited by Madhusanka Liyanage, An Braeken, Pardeep Kumar, and Mika Ylianttila.
© 2020 John Wiley & Sons Ltd. Published 2020 by John Wiley & Sons Ltd.

efficient way trusted security material with the help of a common trusted verifier node, acting as a gateway.

In 2001, Physical Unclonable Functions (PUFs) were introduced as an interesting cryptographic primitive [2] and can be seen as the hardware equivalent of a one-way function. A silicon PUF is a physical entity embodied in a physical structure that is easy to fabricate but practically impossible to clone, duplicate or predict, even if the exact manufacturing process is produced again. Instead of using a private key that is linked to the device identity, the authentication of PUFs is based on the usage of so-called challenge-response pairs (CRPs). The electrical stimulus, called the challenge (C), is applied to the physical structure in order to react, called the response (R), in an unpredictable manner due to the complex interaction of the stimulus with the physical micro-structure of the device, which is dependent on physical factors introduced during the manufacturing process in an unpredictable and uncloneable manner. PUFs have relatively low-hardware overhead and are thus very interesting in an IoT context. In [3, 4] a list is provided of process parameter variations potentially impacting the delay and leakage characteristics of Complementary Metal Oxide Semiconductor (CMOS)-based digital circuits, which can be used as PUF.

PUFs on devices have already been applied for device identification and authentication, binding hardware to software platforms, secure key storage, keyless secure communication, etc. Figure 10.1 illustrates a straightforward usage of PUFs for device identification and authentication as the PUFs can be seen as the unique fingerprint of the device. A security protocol is called a PUF-based protocol when at least one of the entities is able to construct PUF challenges and responses, which are used in the rest of the protocol.

We assume for our construction the existence of a PUF mechanism in both devices willing to authenticate each other. Moreover, instead of the devices communicating directly with a common authentication server storing all CRPs of the devices, we also consider the existence of local cluster nodes, acting as verifier nodes for the two IoT devices in the field. The cluster nodes are responsible for requesting the required security information of the authentication server, where the security information does not include explicit CRP data which can potentially be abused later on in any kind of learning algorithm. We show that our proposed PUF-based protocol is very efficient, compared to state-of-the-art schemes presented in literature.

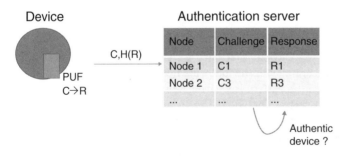

Figure 10.1 Example of PUF usage for device identification (C = Challenge, R = Response)

10.2   Related Work

Here, we focus the related work on the key agreement protocols in IoT as this is the part where PUFs are involved. For the secure communication phase, we can refer to the multiple identity-based signcryption schemes [5, 6] in the literature, which are cryptographic primitives performing both signature generation and encryption in one single phase.

Two main categories of key agreement schemes in IoT can be distinguished, these being key agreement from device to server and key agreement between two different devices. Note that we do not involve the discussion IoT devices have with user access, where two-factor and three-factor authentication is possible. We refer to the survey of [7] for these types of schemes and [8] for an anonymous symmetric key-based key-agreement protocol. The most important trends and schemes in each of these categories are now discussed.

10.2.1   Key Agreement from IoT Device to Server

For the first category of key agreement schemes, for simplicity, we assume that both device and server are registered with the same trusted third party (TTP). If not, the negotiation is further dealt with on the level of the TTPs. In the registration phase, there are two possible types of key material to be shared at the device side in order to guarantee that the identity of the device is linked with the claimed key material. The first type is through classical identity-based public key cryptography. This can be arranged using different types of key establishment schemes, being the typical identity-based schemes [9], certificateless [10], certificate-based schemes [11] and proxy-based schemes[12]. The identity-based key establishment schemes require the usage of computationally intensive pairing operations and have inherent key escrow. Moreover, the schemes are vulnerable for a curious and honest TTP, which is a TTP that is honest in the sense that it performs all the required steps but curious in retrieving the data for own purposes (e.g., selling to third parties). Both the key escrow problem and the vulnerability for a curious and honest TTP can be avoided in the other two types of key establishment schemes since there, the resulting private key is generated by means of secret information coming both from the device and the TTP. The main difference between certificateless and certificate-based schemes is that, in the first one, a secure channel is required for sharing the secret information of the device. Consequently, the certificate-based key establishment schemes offer the most interesting features and the Elliptic Curve Qu-VanStone (ECQV) [13] certificates mechanism is considered as the most efficient one as it is based on Elliptic Curve Cryptography (ECC). In the ECQV scheme, the public key of the user is generated by means of the identity of the user and the certificate generated by the TTP. As a consequence, instead of sharing the public key and identity of the user, it is sufficient to share the certificate and the identity of the user. Any other party, knowing the public key of the TTP, is then able to generate the public key of the user. Note that our proposed key agreement scheme is based on this principle. A key agreement mechanism based on the ECQV scheme has been proposed in [14]. In [14], Porambage et al. have proposed a key agreement protocol between sensor and server using the ECQV principle to be applied in the context of wireless sensor networks in distributed IoT applications.

The second type of key material, which can be used in these key agreement protocols, are the PUF-based challenges and responses. The main advantage with PUF-based key material is that the attacker cannot take over the identity of a tampered device, whose key material has been extracted. There exists multiple PUF-based key agreement protocols for device to server in the literature. In [15], 21 server-device/token key agreement protocols have been classified with respect to the features, device authenticity, server authenticity, device privacy, leakage resilience, number of authentications, resistance to noise, modelling resistance, denial of service (DoS) prevention, and scalability. It has been shown that only a very limited number are able to satisfy these features at a reasonable level. The main problems were vulnerability for DoS attacks, replay attacks, impersonation attacks, and synchronization problems. In the lightweight category of proposals are the Slender PUF, [16] noise bifurcation [17] and PUF-lockdown protocol [18] retained, while in the non-lightweight category only Reference protocol II-A [15] and the protocol proposed by Sadegi et al. [19]. The main difference between these protocols [15–19] and our PUF-based protocol is that these protocols take the noisiness of the PUF into account, while our protocol considers the usage of a strong and controlled PUF. Moreover, [16, 17] also take countermeasures to offer resistance against machine-learning attacks, although this cannot be completely excluded [15]. The proposed protocol in [18] prevents an attacker from querying a token for CRPs that has not yet been disclosed during a previous protocol run. The main weakness of [19] is that it does not scale well, as the server needs to exhaustively evaluate a pseudo-random function for each registered token.

Another method for key agreement with the usage of PUFs is described in [20]. Here, the private key of the device is securely stored using a PUF construction. During the first communication message, the certificate issued by the manufacturer needs to be included. This approach is interesting, but strongly relies on the trustworthiness of the manufacturer, which is, in many cases, not verifiable by the device. In [21], the concept has been explained how PUFs, in combination with Blockchains, are able to establish authentication for IoT devices. Although the idea is promising, the impact of Blockchains on the performance of IoT devices is not fully clear for the moment.

10.2.2 Key Agreement between Two IoT Devices

For the key agreement between two IoT devices, e.g., in the case of sensor nodes in automobiles, smart home and smart cities, we assume that both devices are registered with the same TTP. During the key agreement process, the TTP can be involved or not. In the case where TTP is not involved, the most efficient identity-based authentication and key agreement protocol proposed in literature can be found in [22], which only requires two phases and is also based on the ECQV key establishment protocol. We also mention in this context the standard key agreement scheme in IoT, called the Datagram Transport Layer Security (DTLS) protocol [23] with raw public keys, which has as the main difference, usage of less efficient certificates. In [24], the authors have replaced in the DTLS protocol these more compute intensive certificates by the ECQV certificates and evaluated the impact of this on the DTLS protocol.

However, without involvement of the TTP, a revoked device cannot be detected by the other party, without storage of a revocation list, which is very memory demanding and not advisable in a constrained IoT device. Therefore, it makes sense to also consider key agreement protocols with an active involvement of the TTP. Such protocol using classical public key cryptography mechanisms is evident, assuming that the TTP stores the list of valid identities and corresponding public keys of the participating IoT devices. Note that this scheme will be used as a benchmark to also compare the efficiency of our proposed scheme.

For the PUF-based key agreement protocols between two IoT nodes with the aid of a common server (taking the role of TTP), who has stored the challenge-response pairs of the PUFs from the different nodes, Chatterjee et al. recently proposed a protocol in [25]. The public keys of the devices are generated using the PUF results, followed by an identity-based encryption mechanism for the actual secure communication. In [27], we show that their protocol is not resistant against man-in-the-middle, impersonation, and replay attacks. In addition, we present an alternative PUF-based protocol for the key agreement phase, which is even more efficient. In order to overcome the weakness of CRP leakage at the authentication server in case an attacker manages to get access to the database, Chatterjee et al propose in [26] a new version of the authentication protocol. In this chapter, we build further on the architectural model proposed in [26], but develop a slight modification of the scheme from [27] in order to also enable resistance against CRP leakage. Our proposed scheme is very efficient compared to [26], as we do not require compute-intensive pairing operations.

10.3 Preliminaries

10.3.1 System Architecture

In [26], Chatterjee et al. introduced a cluster node in the field to perform the authentication between the two IoT devices, instead of the central authentication server supervising the different IoT nodes as in [25, 27]. This is a more realistic scenario given the storage requirements of the authentication server. Therefore, we consider the similar architecture and protocol phases as in [26]. Consequently, the scheme consists of a registration phase, security association and authentication and key agreement phase. Note that we do not discuss the final secure communication phase, as this process is similar as in [27] and only relies on a signcryption scheme based on the ECQV scheme. Figure 10.2 depicts the system architecture and the different phases in the system.

- In the registration phase, both IoT device and cluster node register with the authentication server, which is considered to be a TTP and fully trusted. This phase is assumed to be performed in a trusted and secure environment, e.g., by physical contact between the trusted authentication manager and the devices. For the IoT node, the identity ID and a set of challenges C with corresponding hash of the responses $H(R)$ are shared with the TTP. The cluster node shares its identity ID_S, public key P_S and a common shared secret K_S with the TTP. The corresponding secret key d_S is securely stored in the cluster node.
- Upon request of the cluster node ID_S for authentication of a certain IoT node ID_A, the security association process is activated. Here, the Secure Association Provider (SAP) requests an entry of the TTP involving both ID_A, ID_S, containing an implicit CRP, which is uniquely coded to become usable for ID_S in the authentication process and which is not leaking explicit information on the behavior of the PUF.

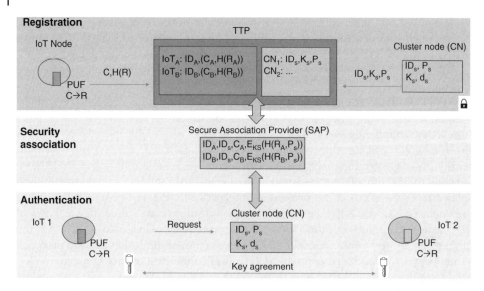

Figure 10.2 System architecture

- In the authentication and key agreement phase, two nodes, registered to the same TTP and in the same communication range of a CN, are able to generate a private and public key pair. The public key of the one node is shared with the other with the help of the CN to guarantee the authenticity.

10.3.2 Assumptions

Taking into account the described system architecture, we consider the following assumptions in the system.

- The registration phase of both the IoT nodes and the CNs happens in a trusted and secure environment. The TTP is completely honest and stores the data with the highest security standards, being regularly audited by third parties. Consequently, no attacker is able to derive information at this stage or at the level of the TTP.
- The communication between TTP, SAP, CN and IoT can run over insecure channels.
- The CN possesses non-volatile memory to store at least its private key d_s and its common shared key K_S with the TTP. No attacker is able to retrieve this information.
- The CN is considered to be honest but curious. In this scenario, the CN performs all the required actions, but might be interested in abusing some of the information in order to sell it to third parties.
- The PUF designed in the IoT is inaccessible and cannot be predicted by the attacker as it is an implicit property of the device.

10.3.3 Attack Model

The attacker has two main goals. First, he can be interested in collecting CRPs of an IoT device in order to predict the behavior of the device by using it as input of a machine-learning algorithm. This would allow impersonation of the IoT device at a later stage. The second goal of the attacker can be to directly impersonate a registered and legitimate IoT node without possessing the node and thus to derive a key with another unaware registered and legitimate IoT device. As the CN is considered to be honest but curious, the attacker can also corrupt the CN in order to derive information on either the CRPs or the key derived by the IoT nodes.

We consider the Dolev-Yao attack model [29]. Here, the attacker is able to eavesdrop on the traffic, inject new messages, replay and change messages, or spoof other identities. Consequently, the attack is only limited by the constraints of the cryptographic methods used. We may assume that the attacker carries the message.

In practice, eavesdropping, intercepting and modifying data are activities that can be launched at any point where the traffic passes, from the IoT device to the authentication server. Some examples of how this can happen are as follows:

- In a local network:
  - Anyone who is connected to the same Wi-Fi is able to read the traffic.
  - If the router (or some other part of the network) is hacked, the hacker can read and modify the traffic.
  - The person that legitimately controls the network, e.g., the person responsible for the authentication of the server, can read and modify the traffic without even having to hack anything.
- Over the internet:
  - The Internet Service Provider (ISP) is able to read and modify all the traffic, since it controls the hardware it passes through. The traffic can also go through other networks owned by unknown companies, eventually from different countries, and those who can also read and modify the traffic.
  - When a nation state hands over to a court one of these companies passing the traffic, it can also read and modify the data (e.g., NSA).

10.3.4 Cryptographic Operations

First, we briefly explain the two most important building blocks to be used in the scheme. Next, some other required operations are also briefly mentioned.

10.3.4.1 PUFs

There are two types of PUFs, strong and weak. The difference is related to the number of responses that can be generated. A strong PUF is able to generate a large amount of challenge-response pairs, while for a weak PUF, the number is limited, often to one.

The practical realisation of a strong PUF is challenging. PUFs also have problems for stabilizing the noise when generating the responses. In order to solve this issue, error correcting codes and assisting computed helper data are required. A good construction of both is essential to avoid leakage of information and resistance against fault and reliability attacks.

Recently, [28] proposes the construction of PUF-FSM, a controlled strong PUF without the need for error correcting codes and helper data by only using the error-free responses, which are fed in a finite state machine. In [26], a 5-4 double arbiter PUF, consisting of five 64-bit Arbiter PUF instances, together with a BCH encoder and decoder has been designed. Here, besides the challenge also helper data needed to be accompanied.

We do not focus on the design of a PUF in this paper, but assume the usage of such type of PUF-FSM in our protocol. So, we are able to generate a large amount of challenges and responses in a controlled way. When using the PUF-FSM, we can assume that the PUF evaluation behaves as a random oracle. Consequently, in our protocol, we do not integrate the helper data. In case a PUF requiring helper data like in [26] is used, the helper data should be dealt with in the same way as the challenge.

10.3.4.2 Public Key-Related Operations

The public key-related operations in our proposed scheme rely on ECC [30], offering more light-weight public-key cryptographic operations than the classical discrete logarithms or RSA-based systems. For example, 1024-bit RSA corresponds with a 160-bit ECC scheme from a security point of view.

Let us denote the elliptic curve (EC) $E_{p(a,b)}$ to be used in our scheme by $y^2 = x^3 + ax + b$ with a and b two constants in F_p and $\Delta = 4a^3 + 27b^2 \neq 0$. In [31, 32], standardized curve parameters are described for p between 113 and 571 bits. The base-point generator P of $E_{p(a,b)}$ is of prime order q. All points on $E_{p(a,b)}$, together with the infinite point form an additive group. There are two elementary operations related to ECC resulting in another point of the EC, the EC multiplication $R = rP$ with $r \in F_q$ and the EC addition $R_1 + R_2$. ECC relies on two computational hard problems.

- The Elliptic Curve Discrete Logarithm Problem (ECDLP). This problem states that given two EC points R and Q of $E_{p(a,b)}$), it is computationally difficult for any polynomial-time bounded algorithm to determine a parameter $x \in F_q^*$, such that $Q = xR$.
- The Elliptic Curve Diffie Hellman Problem (ECDHP). Given two EC points $R = xP, Q = yP$ with two unknown parameters $x, y \in F_q^*$ and P the base-point generator of the defined EC, it is computationally difficult for any polynomial-time bounded algorithm to determine the EC point xyP.

10.3.4.3 Other Operations

Furthermore, we denote the operation H as the one-way cryptographic hash function (e.g., SHA2 or SHA3) that results in a number of F_q. As encryption algorithm AES or even a lightweight crypto algorithm can be used. The encryption of a message M with key K is denoted by $C = E_K(M)$ and the corresponding decryption operation on the ciphertext C by $M = D_K(C)$. The concatenation of two messages M_1 and M_2 is denoted by $M_1 \| M_2$ and the xoring of two messages by $M_1 \oplus M_2$.

We assume that these functions, the EC parameters together with the EC operations, are implemented in each entitiy participating the scheme. In addition, the IoT nodes possess a reliable and secure PUF.

10.4 Proposed System

We now discuss the different operations to be executed in each of the phases in more detail.

10.4.1 Registration Phase

For the IoT node, a number of random challenges (C) and corresponding responses (R) are generated by the IoT node. These pairs $(C, H(R))$, together with the identity of the node, are sent to the TTP during the registration process of the IoT. The TTP is storing the pairs for each IoT node in a highly secure environment. Also the CN registers with the TTP by first deriving a common shared key K_S with the TTP and by sending its identity ID_S and public key P_S. The tuple (ID_S, P_S, K_S) is also stored securely by the TTP.

10.4.2 Security Association Phase

When an IoT node ID_A launches a key establishment request with another IoT node ID_B on timestamp TS_1, both in the same communication range of CN, the CN forwards the request ID_A, ID_B, ID_S, TS_1 to the SAP. The SAP in turn queries the TTP for two valid entries to be used in the authentication and key agreement phase. As a result, the TTP first verifies if ID_A, ID_B are registered. If so, it then derives four messages, using two CRPs of each node randomly selected from its database. Denote $i = \{A1, A2, B1, B2\}$, then the following key material is computed.

$$R_i' = H(H(R_i) \| TS_1 \| P_S)$$
$$V_i = E_{K_S}(R_i')$$

Finally, the four tuples (ID_i, C_i, V_i, TS_1) are sent by the TTP to the SAP and further forwarded to the CN.

10.4.3 Authentication and Key Agreement Phase

Due to the symmetry of the protocol, it is suffice to explain the steps from the side of Node 1 after the second step. Figure 10.3 illustrates the complete overview of the different steps between Node 1 and the server. A similar parallel process is executed between server and Node 2.

- To start the key agreement phase, Node 1 shares with the server the identities ID_A, ID_B of Nodes 1 and 2 respectively, together with the current timestamp TS_1. Node 1 opens a new session in its memory and stores the three parameters ID_A, ID_B, TS_1 in it.
- Upon arrival of the message, the CN forwards to the SAP the message (ID_A, ID_B, TS_1). As explained before, the Security Association phase is launched and results in the four tuples (ID_i, C_i, V_i, TS_1) with $i = \{A1, A2, B1, B2\}$. Using its common shared key K_S with the TTP, the CN is able to derive $R_i' = H(H(R_i) \| TS_1 \| P_S)$ by decrypting V_i.
- Based on this information, the CN now computes both:

$$C_{A2}' = C_{A2} \oplus H(R_{A1}' \| TS_1 \| ID_B)$$
$$C_{B2}' = C_{B2} \oplus H(R_{B1}' \| TS_1 \| ID_A)$$

Also, a hash value to ensure the integrity and authentication of the server needs to be included. This value for Nodes 1 and 2 respectively is computed as follows:

$$h_{1A} = H(C_{A1} \| C_{A2} \| TS_1)$$
$$h_{1B} = H(C_{B1} \| C_{B2} \| TS_1)$$

Node 1	CN

$$ID_A \| ID_B \| TS_1 \longrightarrow$$

Contact SAP and get info
$$C'_{A2} = C_{A2} \oplus H(R'_{A1} \| TS_1 \| ID_B)$$
$$h_{1A} = H(C_{A1} \| C_{A2} \| TS_1)$$
$$C'_{B2} = C_{B2} \oplus H(R'_{B1} \| TS_1 \| ID_A)$$
$$h_{1B} = H(C_{B1} \| C_{B2} \| TS_1)$$

$$\longleftarrow C_{A1} \| C'_{A2} \| h_{1A} \| P_S$$

Derive R'_{A1}
$$C_{A2} = C'_{A2} \oplus H(R'_{A1} \| TS_1 \| ID_B)$$
Check h_{1A}
Derive R'_{A2}
$$q_A \in F_q, Q_A = q_A P$$
$$h_{2A} = H(R'_{A1} \| R'_{A2} \| TS_1 \| ID_B \| Q_A)$$

$$h_{2A} \| Q_A \longrightarrow$$

Check h_{2A}, h_{2B}
$$q_S \in F_q, Q_S = q_S P$$
$$c_A = Q_A + Q_S$$
$$c_B = Q_B + Q_S$$
$$r_A = H(c_A \| TS_1 \| ID_A \| ID_B)q_S + d_S$$
$$r_B = H(c_B \| TS_1 \| ID_A \| ID_B)q_S + d_S$$
$$P_A = H(c_A \| TS_1 \| ID_A \| ID_B)c_A + P_s$$
$$P_B = H(c_B \| TS_1 \| ID_A \| ID_B)c_B + P_s$$
$$h_{3A} = H(R'_{A1} \| R'_{A2} \| r_A \| c_A \| c_B \| P_A \| P_B)$$
$$h_{3B} = H(R'_{B1} \| R'_{B2} \| r_B \| c_B \| c_A \| P_A \| P_B)$$

$$\longleftarrow r_A \| c_A \| c_B \| h_{3A}$$

$$d_A = H(c_A \| TS_1 \| ID_A \| ID_B)q_A + r_A$$
$$P_A = d_A P$$
$$P_B = H(c_B \| TS_1 \| ID_A \| ID_B)c_B + P_s$$
Check h_{3A}

Figure 10.3 Proposed key agreement scheme.

The message $C_{A1}, C'_{A2}, h_{1A}, P_S$ is sent to Node 1 and $ID_A, TS_1, C_{B1},$ C'_{B2}, h_{1B}, P_S is sent to Node 2. The server opens a new session in which the parameters $ID_A, ID_B, R'_{A1}, R'_{A2}, R'_{B1}, R'_{B2}, TS_1$ are stored.

- Upon arrival of the message, Node 1 first checks if the message consists of three parameters of the expected length. If so, it opens the stored session and retrieves the corresponding parameters ID_1, ID_2, TS_1 of that session. Then, Node 1 computes R_{A1} for C_{A1} in order to find $R'_{A1} = H(H(R_{A1}) \| TS_1 \| P_S)$ and to derive C_{A2} from the received C'_{A2}. This allows verification of h_{11}. If correct, Node 1 also derives R_{A2} and the corresponding R'_{A2}. Next, Node 1 randomly chooses $q_A \in F_q$ to compute $Q_A = q_A P$. Then, it computes $h_{2A} = H(R'_{A1} \| R'_{A2} \| TS_1 \| ID_B \| Q_A)$. The message (h_{2A}, Q_A) is sent to the server and R'_{A1}, R'_{A2} are additionally stored in the session. Similarly, Node 2 generates the message (h_{2B}, Q_B) and the parameters $ID_A, ID_B, TS_1, R_{B1}, R_{B2}$ are stored in a new session at Node 2.
- Upon arrival of messages consisting of two parameters of the expected length, the server first verifies h_{2A}, h_{2B} for the identities involved in any of the open sessions stored at the server side, where the current timestamp is in a reasonable timeframe

with the corresponding stored timestamp. The verification of h_{2A}, h_{2B} ensures the integrity and authenticity of both nodes. If it is correct, it starts with the derivation of the certificates for both nodes. Therefore, it randomly chooses a variable $q_s \in F_q$ and computes $Q_S = q_s P$. Next, the two certificates of the nodes are computed as c_A and c_B:

$$c_A = Q_A + Q_s$$
$$c_B = Q_B + Q_s$$

These certificates are used to compute the auxiliary information r_A, r_B for Node 1 and Node 2 to compute their private and public key pair respectively. Recall, that (d_s, P_s) is the private-public key pair of the server, where P_s has been initially already transmitted by the CN.

$$r_A = H(c_A \| TS_1 \| ID_A \| ID_B)q_s + d_s$$
$$r_B = H(c_B \| TS_1 \| ID_A \| ID_B)q_s + d_s$$

The corresponding public keys P_A, B_B of the nodes 1 and 2 respectively are derived as follows:

$$P_A = H(c_A \| TS_1 \| ID_A \| ID_B)c_A + P_s$$
$$P_B = H(c_B \| TS_1 \| ID_A \| ID_B)c_B + P_s$$

In addition, to guarantee the integrity of the communication, the values h_{3A}, h_{3B} are also computed.

$$h_{3A} = H(R'_{A1} \| R'_{A2} \| r_A \| c_A \| c_B \| P_A \| P_B)$$
$$h_{3B} = H(R'_{B1} \| R'_{B2} \| r_B \| c_B \| c_A \| P_A \| P_B)$$

The values r_A, c_A, c_B, h_{3A} are sent to Node 1. In the same way, the values r_B, c_B, c_A, h_{3B} are sent to Node 2. The stored session is now closed at the server side.

- When a message of four parameters of expected length is received, Node 1 opens the stored session(s) where the current timestamp is in a reasonable timeframe with the stored timestamp. It starts with the computation of its private key $d_A = H(c_A \| TS_1 \| ID_A \| ID_B)q_A + r_A$. Its public key P_A equals to $d_A P$, but also to

$$P_A = H(c_A \| TS_1 \| ID_A \| ID_B)c_A + P_s$$

as derived by the CN. The private key is only known by the node itself as the random number q_a derived in the previous step is required. Using c_B, Node 1 is able to compute the public key P_B of Node 2. Note that this mechanism is based on the ECQV Implicit Certificate Scheme [13]. Finally, Node 1 checks the hash value h_{3A}. If OK, both nodes close the stored session and open a new session storing its private and public key pair together with the public key and identity of the other node.

10.5 Security Evaluation

We now discuss the security features, established in the proposed scheme. The focus is on the key agreement scheme, as the registration phase is assumed to be performed in a completely trusted and secure environment.

- *Integrity.* Integrity is obtained in every step of the protocol since every message contains a hash, which includes the other parts (or derivations of it) of the message.
- *Authentication of the Node.* The nodes are authenticated if the check on the values h_{2A}, h_{2B} are correct. These hash values contain the responses on the challenges, which can only be generated by the node possessing the PUF.
- *Authentication of the Server.* The server is authenticated by the node in both received messages. In the first received message by Node 1, the value h_{1A}, computed by the server, includes C_{2A}, while this value is not part of the received message. Instead, C_{2A} is masked by a hash value, which includes information on the response on the first challenge. Consequently, as only the CN is aware of the information of the response, Node 1 is ensured that the CN has sent the message. Moreover, the validity of the public key of the CN is also checked as it is included in the value R'_{A1}, containing the information of the response and generated by the TTP. If the computation of R'_{A1}, R'_{B1} by Node 1 and Node 2 respectively is similar to the stored value with the CN, the nodes are sure that the transmitted public key corresponds to the authentic one validated by the TTP.

  In the second received message, the hash value h_{3A} is included. This hash value contains the responses on the two challenges, which is only known by the server. In addition, by adding the public keys of the two nodes into the hash, it also allows the nodes to verify the success of the process. Due to the symmetry of the protocol, the same reasoning holds for Node 2.
- *Resistance against man-in-the-middle attacks.* As the authenticity of the sender is verified by the receiver in each step of the protocol, resistance against man-in-the middle attacks is guaranteed.
- *Resistance against impersonation attacks.* Even if Node 1 is malicious, it is not possible for the node to derive information on the challenge-response pairs of the other node from the messages exchanged between the CN and Node 2. This follows on from the fact that only hash values on the responses are included in the messages, which do not leak any information.

  Also the CN cannot be impersonated as both keys K_S, d_S are required to successfully perform the process. These keys are stored in the tamper-resistant part of the memory. The key K_S is necessary to derive the information of the CRPs. The key d_S is required to construct the private and public key pair of each node. Moreover, the public key P_S cannot be repudiated or changed as it is also embedded in the responses computed by the TTP during the security association phase. If d_S were not used, the node cannot verify the authentication of that particular CN as it could also be another registered CN.
- *Resistance against replay attacks.* Replay attacks are avoided, due to the usage of the timestamp TS_1, which is included in all the hashes of the different transmitted messages and in the information of the responses on the selected challenges.
- *Protection against denial of service attacks.* Besides the first message, both the nodes and server can check at each step of the key agreement scheme the integrity and authentication by verifying the hash included in the messages. Consequently, in case a huge amount of false messages flow over the network, it will be very quickly discovered by each entity. Note that this feature is not present in [26] as in their scheme only in the last step is the IoT able to verify the validity. Consequently, in [26] an attacker

can create a large amount of open sessions at each entity, where it is impossible to filter out the legitimate messages without finalizing the complete process.

10.6 Performance

Both the computational and communication costs are considered. We compare the efficiency of our solution with [26] and [27], as these are the only secure PUF-based protocols in the literature with comparable architecture, i.e., deriving a common secret key between two IoT devices. Note that [26] has the additional advantage that no information on CRPs is leaked, compared to [27], and thus [26] offers the same security features as our proposed scheme.

For the key agreement scheme, which is based on the existence of PUFs, we utilize a straightforward public key variant as a benchmark, taking into account a similar architecture as this scheme and [26]. The different steps to be performed in the benchmark scheme are presented in Figure 10.4. In this scheme, we assume that the TTP stores the identity and public key of each of the registered IoT devices, instead of the CRPs. The IoT devices possess the public key P_{TTP} of the TTP. Denote the private-public key pairs of Node 1, Node 2, and CN by $(d_A, P_A), (d_B, P_B)$ and (d_s, P_s) respectively. The signature operation on a message M with private key d is denoted by $sig_d(M)$, while the verification of the signature S with public key P is denoted by $ver_P(S)$. We assume the application of the Schnorr signature scheme for the complexity analysis of the next paragraph.

10.6.1 Computational Cost

We only focus on the most compute intensive operations, being pairings (P), Hash-ToPoint operations (HP), EC multiplications (EM), EC additions (EA), symmetric key encryptions/decryptions (S) and hashes (H). We did not include the performance of the

Node 1	CN	Node 2
Choose $r_A \in F_q$		
$R_A = r_A P$		
$\xrightarrow{ID_A\|ID_B\|R_A\|TS}$		
	Check TTP and receive	
	$P_A, P_B, s = sig_{d_{TTP}}(P_s\|TS)$	
	$\xrightarrow{Com.Req.\|ID_A\|TS}$	
		Choose $r_B \in F_q$
		$R_B = r_B P$
	$\xleftarrow{R_B\|ID_A\|TS}$	
	$K_A = H(d_s R_A\|TS)$	
	$K_B = H(d_s R_B\|TS)$	
	$C_A = E_{K_A}(P_B\|TS)$	
	$C_B = E_{K_B}(P_A\|TS)$	
$\xleftarrow{C_A\|TS\|s}$		$\xrightarrow{C_B\|TS\|s}$
$ver_{P_{TTP}}(s)$		$ver_{P_{TTP}}(s)$
$K_A = H(r_A P_s\|TS)$		$K_A = H(r_B P_s\|TS)$
$P_B\|TS = D_{K_A}(C_A)$		$P_A\|TS = D_{K_B}(C_B)$

Figure 10.4 Key agreement scheme based on public key cryptography used as benchmark

Table 10.1 Comparison of computational cost in key agreement phase for IoT device

Operation	This	[26]	[27]	BM
P	0	1	0	0
HP	0	1	0	0
EM	3	2	3	2
EA	1	2	1	0
S	0	0	0	2
H	10	6	10	2
Total (µs)	2.97	33.28	2.97	1.98

PUF operation as this is highly dependent on the underlying system. We note that in our scheme and [27], two PUF evaluations are used in order to validate the authentication of the request after each step. In [26], only one PUF evaluation is included, having the disadvantage that the authentication of the request can only be verified at the very end of the scheme. Without loss of security (besides intermediate authentication validation), we can easily remove the parameters C'_{A2}, C'_{B2}, also resulting in one PUF evaluation at the IoT node. However, as we believe that this additional check is an added value of the scheme, we will not use this adaptation in the discussion of the performance.

Tables 10.1 and 10.2 compare the number of operations and the corresponding resulting time between our scheme and the ones of [26, 27] and the benchmark scheme (BM) for the key agreement scheme in the IoT node and CN respectively, excluding the PUF evaluations. For the computation of the timings of both protocols, we have considered the numbers derived in [33], where all the operations have been evaluated on a personal computer with an Intel I5-3210M 2.5 GHz CPU and Windows 7 operating system. The cryptographic operations have been implemented using the MIRACL cryptographic library. We have also assumed that there is only one stored session at the nodes and server side, similar to that in [26].

As can be concluded from the table, our proposed protocol is considerably faster from the node and the server side, compared to [26]. The main difference between our scheme and [26] is that we do not need to compute the intense hash-to-point and pairing operations. With respect to [27], the performance at the node side is approximately similar, but there is a small degradation in time at the server side. With respect to the public key-based benchmark scheme, we can conclude that the difference in timing at the node side is acceptable, and there is even a small win for the timing at the server side. This is the price to pay for being resistant against hacking.

10.6.2 Communication Cost

For the determination of the communication cost, we take into account that the nodes' identity and the timestamp correspond to 32 bits. The challenges are represented by 64 bits and the responses by 48 bits, as in [26]. An EC with $q = p = 160$ bits is used, which

Table 10.2 Comparison of computational cost in key agreement phase for CN

Operation	This	[26]	[27]	BM
P	0	2	0	0
HP	0	2	0	0
EM	3	6	1	4
EA	4	10	2	1
S	0	0	0	1
H	10	6	10	2
Total (µs)	2.98	68.55	1.00	3.95

Table 10.3 Comparison of computational cost in communication phase
(Node 1 is the sending node and Node 2 the receiving node)

Entity	This	[26]	[27]	BM
Node 1	416	704	416	224
Server	1280	1152	1120	704

is also the length of the result of the hash value. As a consequence, the resulting sizes of the transmitted messages by both the node and the server in the key agreement protocol are enumerated in Table 10.3.

It can be concluded that the communication cost from the IoT node is slightly worse than the benchmark algorithm and equal to the system of [27]. It is almost twice as efficient compared to [26]. From the server side, again the benchmark algorithm outperforms all other schemes. Here, our scheme is the least optimal. However, if we remove one PUF evaluation, as discussed before, we are able to reach the same efficiency as [26]. Our scheme and the ones of [26, 27] have the same structure and thus the same number of exchanged messages, being four between Node 1 and Server and three between Node 2 and server. These numbers are one lower in case of the benchmark key agreement scheme.

10.7 Conclusions

We have presented in this chapter a highly efficient authentication algorithm for two IoT devices containing a PUF implementation. The main advantage of including a PUF mechanism is that the security of the devices is guaranteed even if they become compromised as there are no secret keys stored. An interesting feature of our system is that neither the attacker eavesdropping the channel, nor the SAP or CN, is able to learn more about the structure of the CRPs in order to collect pairs for building a learning algorithm to reveal the behavior of the PUF.

We have compared the efficiency of the scheme with a straightforward public key-based mechanism and could conclude that the performance difference between both is small. We have also shown that the scheme outperforms similar systems in literature.

References

1 Kumar, T.; Braeken, A.; Liyanage, M.; Ylianttila, M. (2017). May. Identity privacy preserving biometric based authentication scheme for Naked healthcare environment. *2017 IEEE International Conference on Communications (ICC)*, Paris, France (21–25 May 2017). IEEE.

2 Pappu, S.R. (2001). Physical One-Way Functions. Ph.D. Thesis. Massachusetts Institute of Technology.

3 Blaauw, D., Chopra, K. Srivastava, A. and Scheffer, L. (2008). Statistical timing analysis: From basic principles to state of the art. *IEEE Transactions Computer Aided Design Integrated Circuits Systems* 27 (4): 589–607.

4 Abu-Rahma, M.H. and Anis, M. (2007). Variability in VLSI circuits: Sources and digns considerations. *Proc. IEEE International Symposium Circuits Systems*, Los Alamitos, USA (27–30 May 2007). IEEE.

5 Zheng, Y. (1997). Digital Signcryption or How to Achieve Cost (Signature & Encryption) ≪ Cost (Signature) + Cost (Encryption). *Annual International Cryptology Conference*, Berlin, Germany (17–21 August 1997). Springer.

6 Braeken, A., Shabisha, P., Touhafi, A. and Steenhaut, K. Pairing free and implicit certificate based signcryption scheme with proxy re-encryption for secure cloud data storage. *CloudTech; IEEE*, Rabat, Morocco (24–26 October 2016).

7 Tashi, J.J. (2014). Comparative analysis of smart card authentication schemes. *IOSR Journal of Computer Engineering* (16): 91–97.

8 Braeken, A. (2015). Efficient anonymous smart card-based authentication scheme for multi-server architecture. *International Journal of Smart Homes* 9: 177–184.

9 Shamir, A. (1984). Identity-Based Cryptosystems and Signature Schemes. *Workshop on the Theory and Application of Cryptographic Techniques*, Paris, France (9–11 April 1984). Springer.

10 Al-Riyami, S.S. and Paterson, K.G. (2003). Certificateless Public Key Cryptography. *International Conference on the Theory and Application of Cryptology and Information Security*, Taipei, Taiwan (30 November–4 December 2003). Springer.

11 Gentry, C. (2003). Certificate-Based Encryption and the Certificate Revocation Problem. *International Conference on the Theory and Applications of Cryptographic Techniques*, Warsaw, Poland (4–8 May 2003). Springer.

12 Braeken, A., Liyanage, M. and Jurcut, A.D. (2019). Anonymous Lightweight Proxy Based Key Agreement for IoT (ALPKA). *Wireless Personal Communications*, pp.1–20.

13 Certicom Research. (2013). SEC4: Elliptic Curve Qu-Vanstone Implicit Certificate Scheme, Standards for Efficient Cryptography Group. Version 1.0. http://secg.org/sec4-1.0.pdf (accessed 25 June 2019).

14 Porambage, P., Schmitt, C., Kumar, P. et al. (2014). Two-phase Authentication Protocol for Wireless Sensor Networks in Distributed IoT Applications. *Proceedings of the 2014 IEEE Wireless Communications and Networking Conference (WCNC)*, Istanbul, Turkey (6–9 April 2014) IEEE.

15 Delvaux, J. (2017). Security Analysis of PUF-Based Key Generation and Entity Authentication. Ph.D. Thesis, Katholieke Universiteit Leuven.

16 Rostami, M., Majzoobi, M., Koushanfar, F. et al. (2014). Robust and reverse-engineering resilient PUF authentication and key-exchange by substring matching. *IEEE Transactions Emerging Topics in Computing* 2 (1): 37–49.

17 Yu, M.D. Verbauwhede, I. Devadas, S. and M'Raihi, D. (2014). A noise bi-furcation architecture for linear additive physical functions. *Proceedings of the 2014 IEEE International Symposium on Hardware-Oriented Security and Trust (HOST)*, Arlington, USA (6–7 May 2014). IEEE.

18 Yu, M.D., Hiller, M., Delvaux, J. et al. (2016). A lockdown technique to prevent machine learning on PUFs for lightweight authentication. *IEEE Transactions Multiscale Computer Systems* 2: 146–59.

19 Sadeghi, A.R., Visconti, I. and Wachsmann, C. (2010). Enhancing RFID security and privacy by physically unclonable functions. In: *Towards Hardware Intrinsic Security –Foundations and Practice* (eds. A.R Sadeghi and D. Naccache), 281–305. Berlin, Germany: Springer.

20 Tuyls, P. and Batina, L. (2006). RFID-Tags for Anti-counterfeiting. *Topics in Cryptology|CT-RSA 2006*, San Jose, USA (13–17 February 2005). Springer.

21 Guartime and Intrinsic ID. (2017). Internet of Things Authentication: A Blockchain solution using SRAM Physical Unclonable Functions. https://www.intrinsic-id.com/wp-content/uploads/2017/05/gt_KSI-PUF-web-1611.pdf (accessed 25 June 2019).

22 Simplicio, M.A., Jr., Silva, M.V., Alves, R.C. and Shibata, T.K. (2017). Lightweight and escrow-less authenticated key agreement for the internet of things. *Computer Communications* 98: 43–51.

23 Wouters, P., Tschofenig, H., Gilmore, J. et al. (2014). T. RFC 7250|Using Raw Public Keys in Transport Layer Security (TLS) and Datagram Transport Layer Security (DTLS), June 2014. https://www.rfc-editor.org/rfc/rfc7250.txt (accessed 25 June 2019).

24 Ha, D.A., Nguyen, K.T. and Zao, J.K. (2016). Efficient authentication of resource-constrained IoT devices based on ECQV implicit certificates and datagram transport layer security protocol. *Proceedings of the Seventh Symposium on Information and Communication Technology*, Ho Chi Minh City, Vietnam (8–9 December 2016).

25 Chatterjee, U., Chakraborty, R.S. and Mukhopadhyay, D. (2017). A PUF-Based Secure Communication Protocol for IoT. *ACM Transactions on Embedded Computing Systems* 16: 67.

26 Chatterjee, U., Govindan, V., Sadhukhan, R. et al. (2018). Building PUF based Authentication and Key Exchange Protocol for IoT without Explicit CRPs in Verifier Database. *IEEE Transactions on Dependable and Secure Computing (TDSC)* 99: 1–1.

27 Braeken, A. (2018). PUF Based Authentication Protocol for IoT. *Symmetry* 10 (8): 352.

28 Gao, Y., Ma, H., Al-Sarawi, S.F. et al. (2017). PUF-FSM: A Controlled Strong PUF. *IEEE Transactions on Computer-Aided Design of Integrated Circuits Systems* 30 (5): 99.

29 Dolev, D. and Yao, A. (1983) On the security of public key protocols. *IEEE Transactions on Information Theory* 29: 198–208.

30 Hankerson, D., Menezes, A.J. and Vanstone, S. (2003). *Guide to Elliptic Curve Cryptography*. New York, NY: Springer.

31 SEC 2: Recommended Elliptic Curve Domain Parameters, Certicom Research, Standards for Efficient Cryptography Version 1.0, September 2000. http://secg.org/sec2-v2.pdf (accessed 25 June 2019).

32 Recommended Elliptic Curves for Federal Government Use, National Institute of Standards and Technology, August 1999. http://csrc.nist.gov/groups/ST/toolkit/documents/dss/NISTReCur.pdf (accessed 20 November 2018).

33 He, D., Zeadally, S., Wang, H. and Liu, Q. (2017). Lightweight Data Aggregation Scheme against Internal Attackers in Smart Grid Using Elliptic Curve Cryptography. *Wireless Communications and Mobile Computing* 2017: 1–11. doi:10.1155/2017/3194845.

11

Hardware-Based Encryption via Generalized Synchronization of Complex Networks
Lars Keuninckx and Guy Van der Sande

Abstract

We present an encryption and authentication scheme suitable for ASIC or FPGA hardware implementation, which is based on the generalized synchronization of systems showing chaotic dynamical behavior. The scheme consists of a single-driver system, which provides two identical driven systems with a complex waveform. The driven and driving systems synchronize with correlation no higher than 1.2%. A bit-stream derived from their outputs is then used as an encryption or authentication key. The security of the scheme is based on the fact that it is easy to generate the response of the complex systems, given an input, but hard to do a system identification. Furthermore, the spectrum of the signals of the driver and driven system reveals no information. We show that regardless of their initial state, the distant receivers synchronize within a short time, relative to their internal timescale. We validate the bit-streams generated by the driver using the NIST test suite for randomness and have found no deviations. Finally, we provide pointers as to the practical implementation and application of the presented scheme.

11.1 Introduction

Although the Internet of Things (IoT) promises to bring enormous benefits, the risks associated with it cannot be ignored or underestimated. Take for example, the security risks associated when hackers are able to read a home's thermostat settings, identifying exactly if the occupants are at home or not. Or possibly disastrous: a hacker having access to the braking system of a (self-driving) car. The potential risks associated with wireless enabled pacemakers and other implantable medical devices do not even need to be explained. While offering the huge benefit of remote diagnostics, at the same time, malignant access goes far beyond just a privacy breech and is potentially life-threatening. Many current IoT devices have relatively weak security capabilities and are easy entry points for hackers. The reasons for this exacerbated cybersecurity risks are plenty. First of all, there is no real form of standardization, let alone a legal framework to adhere to, yet. Each device still speaks its own 'language', following its own custom protocols, making it very hard for the network to know which devices to really trust. Also, the cost of building extra security into these basic devices is often too

IoT Security: Advances in Authentication, First Edition.
Edited by Madhusanka Liyanage, An Braeken, Pardeep Kumar, and Mika Ylianttila.
© 2020 John Wiley & Sons Ltd. Published 2020 by John Wiley & Sons Ltd.

high for commercial purposes. In an industrial context, a big increase in the number of sensors and devices that are being connected can be seen, creating a huge potential attack surface. Typically, such industrial IoT consists of decades-old equipment and control systems that were never designed for exposure to the Internet and to its security risks. Together with an insufficient budget for implementing cybersecurity awareness, monitoring, and prevention technology, this leads to an ever-growing number and type of attacks. Therefore, the future of successful deployment and acceptance of IoT-enabled applications depends critically on the availability of fast, robust and low-power encryption and authentication methods. Fast in this context equals low power by implementing short sleep/wakeup/sleep cycles. The encryption and authentication methods used should not require much real-estate on the chip to be low cost. Hardware-based methods will be preferred over software-based to offload the main processor and provide encryption that is transparent at the application level.

In this Chapter, we will therefore propose a cost-effective lightweight hardware-entangled security solution for IoT systems. However, our proposed solution can also be employed in other security contexts. The development of new strategies to protect sensitive information from interception and eavesdropping has been receiving significant attention, also in our present-day worldwide communication networks. Generally speaking, the aim of our work is the development and implementation of a novel random key distribution system based on the concept of generalized synchronization between distant elements in large networks. Such a random key synchronization system can have a significant impact in the field of physical layer-based encryption techniques, offering not only high confidentiality but also potentially high-speed real-time encryption and decryption. Here, we put forward a scheme fit for a fully digitalized implementation.

Nowadays, confidentiality and the authenticity of information are mostly ensured through mathematical algorithms. Algorithmic key-based encryption systems usually take a digital data stream and convolute it with a given binary pattern, which we refer to as the key. The resulting encrypted binary string can then be transmitted through a public communication channel. A classic example of this type of encryption is the Vernam cipher [1], where the recipient decodes the message using the same key-string code as used for encryption. In this case, the key is agreed via another secure channel. This algorithm has been mathematically proven to be totally secure if the key is fully random, has the same length as the message and is used only once. This so-called one-time pad cryptography is, however, not suited for secure communications between two parties who have not been able to exchange encryption keys beforehand. To circumvent this drawback, other software cryptosystems relying on asymmetric-key algorithms (public-key cryptography such as RSA) have been developed. However, asymmetric algorithms use significant computational resources in comparison with their symmetric counterparts and therefore are not generally used to encrypt bulk data streams. Also, the effectiveness of these encryption techniques relies on the fact that it is computationally hard (but not impossible) to decrypt a message knowing only the public key. Therefore, the growing computational power and the fact that a key is used more than once remains a latent threat for current algorithmic cryptography. In order to strengthen the process of securely exchanging a private key other hardware-oriented approaches have been proposed such as Physically Uncloneable Functions (PUFs).

PUFs have gained extensive research attention since they were first proposed in 2002 [2]. Their operational principle is based on inherent and unavoidable differences within every physical object that appear during the manufacturing process. This gives each physical instance a unique identity and, as such, could be used as a lightweight authentication scheme as well as anti-counterfeiting solution. A first proposal was based on light scattering [2]. In Ref. [3], Gassend *et al.* proposed the first electrical PUF. They have gained a lot of attention as a secure key generation and storage mechanism. Their "unclonability" property is key: it should be impossible to build two physical instances of a PUF having exactly the same properties or characteristics.

PUFs are usually divided into two categories: weak PUFs and strong PUFs. A strong PUF can be queried with an exponential number of challenges to receive an exponential number of responses. They can be used in authentication protocols as well as for key generation and storage. Weak PUFs, on the other hand, only have a very limited challenge space and can only be used for key generation and storage.

PUFs suffer from some significant drawbacks. They are susceptible to environmental conditions and noise. Therefore, if a PUF is challenged twice by the same signal it can lead to two different and obviously unreliable responses. Also, existing electrical strong PUFs can be emulated in software. By relying on machine-learning techniques, it is possible to approximate the required parameters sufficiently to have a software model which is in fact a very good clone of the real PUF. This is particularly true for the Arbiter PUF. Typically, some non-linearity is added to make machine-learning attacks more difficult. XOR PUFs are widely assumed to be secure against machine-learning attacks if enough XORs are used [4–6]. Recently however, in Ref. [7] an attack method was demonstrated that scales favorably with the number of XORs.

While our approach resembles PUFs, the main idea originates from improving the confidentiality of an encrypted message by additional encoding at the physical layer using chaotic carriers. Chaos-based encryption systems rely on two spatially separated chaotic systems to synchronize with each other. Once the two systems are synchronized, the chaotic output of the sender can be used as the carrier in which the message is hidden as a small modulation [8]. The receiver can extract the message by comparing the incoming signal with the synchronized one. Multi-gigabit information transmission in real installed optical networks over several tens of kilometers have been demonstrated using this paradigm [9]. However, the necessity of sharing a chaotic carrier signal over a public channel reveals information on the specifics of the system used. Therefore, these chaos-based communication systems offer confidentiality but cannot, for the moment, guarantee security. Such chaos-based encryption schemes could augment the security by operating in a bidirectional fashion, whereby the modulating messages from both communicating parties involved disturbing the shared synchronization inducing signal. Since both parties have exact *a priori* knowledge of their own respective modulating signal, they are able to deduce the message imposed on the carrier by the other party [9, 10]. However, an optimized hardware solution (compatible with software methods) for confidential data transmission is currently lacking and highly desired. Recently, we have demonstrated a system which can encrypt data in a new way, with a high level of security and which can be built using current off-the-shelf components [11]. This secure key distribution scheme was based on synchronized random bit generation and relied on the synchronization between a transmitter and a distant receiver through an uncorrelated chaotic driver signal. From the synchronized chaotic signals, a random key can be distilled that would be extremely difficult to reconstruct from the information shared in the public channel. This system does not suffer from the drawbacks of PUFs nor from the drawbacks of standard vanilla chaos encryption. In this chapter, we revisit this work and not only translate it to a fully digital implementation but also demonstrate a novel easy-to-implement lightweight random bit generator, that is based on a delay dynamical system.

In the next section, we introduce our general scheme and the chaotic driver, of which, the dynamical behavior is investigated in detail. Then, in Section 11.3, we show how the chaotic followers are built, giving pointers for further practical use. In Section 11.4 we show how the drivers and followers can be combined to form a complete encryption or authentication solution. Furthermore, we reason that a brute-force attack on this system is futile. We wrap up our conclusions in Section 11.5, discussing ways in which this line of research may go further.

11.2   System Scheme: Synchronization without Correlation

In [11], we have presented an encryption-scheme based on generalized synchronization of driven chaotic delay systems. Usually, the dynamical signals observed from systems that operate synchronously, by virtue of a (uni- or bi-directional) coupling, exhibit some similarity. A simple example is found in identical pendulum clocks, which, when placed in each others, vicinity, from any given initial state tend toward swinging in unison after a while. This was first observed by the Christiaan Huygens in the 17th century [13]. Other examples include flashing fireflies, electrical oscillators and interpersonal brain activity as seen in mirror neurons [14]. Later, it was discovered that not only regular oscillators but also chaotic systems can synchronize, given the right circumstances [8]. Deterministic chaos is characterized as irregular motion that has a sensitive dependence on initial conditions: two identical chaotical systems brought into very close (yet unequal) initial states, will, after a while, show vastly diverging dynamics. The interested reader is refered to [15] for an excellent introduction to chaos theory. Generalized synchronization expands the common notion of synchronization to all causally related dynamics. An example of synchronization showing regular oscillatory dynamics is found in electronic oscillators that lock on a subharmonic of an injected signal. Inferring generalized synchronization from the observation of the system's dynamics is in general difficult and still an open problem [16]. This is especially true for driven chaotic systems and this is what is exploited in [11] to build an encryption system. Figure 11.1 shows the scheme of this system, which is also the basis of the scheme presented in this chapter. At the transmitter side, a chaotic driver generates a complex signal. This signal feeds into a chaotic follower or responder, which thus generates, in a deterministic way, its own output signal. Then, from the followers' signal a deterministic but by all accounts apparently random bitstream is derived. Naturally, this is only possible if the follower's dynamical behavior allows for this. This bitstream is then used as a one-time-pad key for encoding a

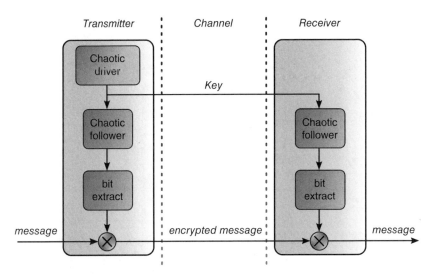

Figure 11.1 Encryption scheme based on generalized synchronization between two driven chaotic systems.

binary plaintext, via an XOR-operation. The probability of seeing a '1'-bit in the ensuing encrypted message is:

$$P\{\text{encrypted} = 1\} = P\{\text{message} = 1 \text{ and key} = 0\} + P\{\text{message} = 0 \text{ and key} = 1\}$$
$$= P\{\text{message} = 1\}P\{\text{key} = 0\} + P\{\text{message} = 0\}P\{\text{key} = 1\}$$
$$= P\{\text{message} = 1\}\frac{1}{2} + P\{\text{message} = 0\}\frac{1}{2}$$
$$= \frac{1}{2}, \tag{11.1}$$

since the bits of the message and the key are independent random variables. On the receiver side, the signal of the driver causes a second follower (identical to the one on the transmitter side) to generate the exact same signal. Thus, the bitstream derived from this second follower exactly equals the bitstream of the follower in the transmitter. The message is then recovered via an XOR-operation of the receiver-side bitstream and the cyphertext. Understandibly, a two-way link can be built in the same way. A qualitative requirement for such a scheme to be succesful is that it is computationally hard to identify the relation between the driver signal and the key bitstream.

In [11] this scheme was demonstrated using an analog electronics approach, as shown in Figure 11.2a. The chaotic driver consisted of a series of nonlinear filter blocks (NLBs), Figure 11.2b, with a non-monotonic response, Figure 11.2c, placed in a delay ring. Each responder used four similar nonlinear filters in a chain. The main purpose of that work was to illustrate how this scheme could be viable, yet there, as a final implementation, a fast photonic link was envisioned. It is instructional to explain why delay is used in the chaotic driver in Figure 11.2. First, in a photonic setting delay appears naturally by virtue of the finite propagation speed optical signals have when travelling along an optic fiber. It has long been known that delay can cause instabilities and even chaos in control systems. This is also the case for semiconductor lasers subjected to delayed feedback [17], thus

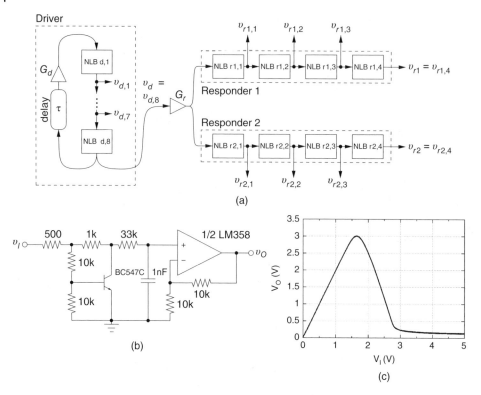

Figure 11.2 a) System diagram of the driver and responders or followers of [11]. Each nonlinear block (NLB) contains the subcircuit of Figure 11.2b. Pairwise NLB blocks in the responder chains were built using matched components. b) A single nonlinear block. The transistor circuit provides a non-monotonic "bump-like" function and is followed by a first-order filter and buffer. c) Response of a NLB when slowly scanned.

offering a straightforward way to produce optical chaos. Therefore using delay in an electronic system mimics this feature. Second, and more importantly, what is special about delay dynamics is that they live in an infinite phase space. To determine how the state of a delay system evolves in time, one has to provide the initial condition of the delay line. This initial condition is defined by a function, having in theory an infinite number of points along the delay line. Then, at any further time, the state itself is a function of all values along the delay line. In contrast, the state of a finite-order chaotic system (such as the archetypical Lorentz system) is fully defined by no more than the amount of numbers equalling the order of the differential equation that describes it. For the Lorentz system, the state is described at any time by only three numbers, hence its phase space is three-dimensional. The consequence of having an infinite-dimensional phase state is that delay-based chaotic systems offer what is often loosely referred to as "richer" dynamical behavior, meaning they behave in a more erratic way. By using functionally the same nonlinear blocks in both the driver and responders, we were able to obtain similar spectral signatures in their signals. This is an important safeguard, since attackers are likely to investigate the spectral characteristics of the signals involved.

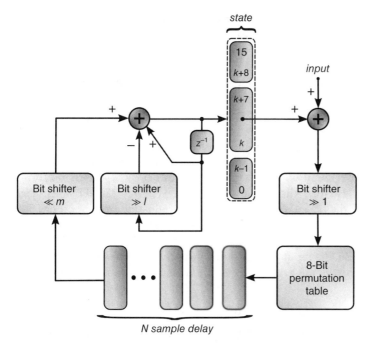

Figure 11.3 A Delay-Filter-Permute (DFP) block. For autonomous operation, the input and associated bit shifter in front of the permutation table are omitted.

Here, we illustrate how the same scheme can readily be adopted to a low-power and resource limited digital setting, suitable for both encryption and authentication. We begin by showing how a chaotic delay-based driver may be implemented in digital hardware.

11.2.1 The Delay-Filter-Permute Block

In Figure 11.3, we show the diagram of a digitally implementable "chaotic" delay system that we call a Delay-Filter-Permute block (DFP). Understandibly, a discrete-time and discrete-valued system cannot show deterministic chaos in the true sense, since the number of discrete states in the phase space is finite and therefore the dynamical behavior must eventually revisit these states. In contrast, a real-valued chaotic system has infinitely dense orbits, which nevertheless take up a finite volume in its phase space. Such a collection of dense orbits is called a "strange attractor". However, if the number of possible states in a discrete-valued system is high, then very complex dynamical behavior could be observed if the state update rules allow for it. An example of this is found in feedback digital filters which can exhibit chaos-like behaviors due to overflow, underflow or rounding [18]. In the scheme shown in Figure 11.3, this large phase space is obtained via a delay of N 8-bit values, having in total 2^{8N} possible states. The values in the delay originate from an 8-bit in/8-bit out permutation table, hence a bijective function Q of the form:

$$Q : \{0,1\}^8 \mapsto \{0,1\}^8 : x \mapsto Q(x), \tag{11.2}$$

which is chosen at random, such that for $x \neq y$ we have $Q(x) \neq Q(y)$ and $\forall y \in \{0, 1\}^8$ $\exists x : Q(x) = y$. We did not need to place any special requirements on the construction of the random permutation table to reach the results reported below. If we exclude mapping values onto themselfves $Q(x) = x$ (because this would make the dynamical behavior less diverse), then it turns out there exists a staggering $(2^8 - 1)! \approx 3.35 \times 10^{504}$ choices to build this function. The delayed and permuted 8-bit values are then bit-shifted to the left by m positions, i.e., multiplied by 2^m. These are added to a 16-bit "state" variable. By using an l-bit right-shift operation and a subtraction, a recursive lowpass filtering is performed on this state variable. The recursive nature of this filter gives it some memory, allowing it to mix the states of several delayed values. The input to the permutation table is given by the bitfield $(k, k + 7)$ of the 16-bit state variable, and these also form the key values of Figure 11.1. Several DFP blocks can be coupled by averaging the $(k, k + 7)$ bitfield with that originating from another DPF. Another method is to simply to let the sum of the $(k, k + 7)$ bitfields from two coupled DPFs overflow, and use the least significant eight bits as input to the permutation table. For autonomous operation, the input and divide-by-two shifter as shown in Figure 11.3 are omitted. In summary, the behavior of an autonomous DFP block answers to the following equation:

$$s_{n+1} = s_n - (s_n \gg l) + Q[(s_{n-N} \gg k) \text{ and } 255] \ll m, \tag{11.3}$$

where s denotes the 16-bit state variable and the operation $a \gg b$ $(a \ll b)$ means shifting the binary representation of a to the right (left) over b positions, which is equivalent to dividing (multiplying) a by 2^b, followed by rounding downwards to the nearest integer. In Eq. (11.3), N is the length of the delay line and "and" stands for the bitwise and-ing operation. Clearly, the DFP is readily integrateable in an ASIC, FPGA or as embedded code. We note that a similar system, executed in two steps:

$$s_{n+1} = s_n + Q[(s_{n-N} \gg k) \text{ and } 255] \ll m, \tag{11.4}$$

$$s_{n+2} = s_{n+1} - (s_{n+1} \gg l), \tag{11.5}$$

delivers qualitatively similar results as Eq. (11.3), while being slightly more easy to implement in hardware. Also, switching the permutation table and delay, does not alter the operation, apart from the entries in the delay line itself.

In Figure 11.4, we present some exploratory data analysis on the dynamical behavior of a single DPF, Eq. (11.3), with the parameters set to $(k, l, m, N) = (3, 4, 4, 1024)$. We emphasize that these values are merely exemplary. During our simulations we found that many other choices work just as well. In Figure 11.4a, we show the timetrace of the DFP. Here, time is shown in units of the delay line length, $N = 1024$ samples. Initially, the delay is filled with zeros and the state variable is also set to zero. After a startup transient of about $10 \times N$ samples, an irregular signal is seen. Of course, to make sure the system starts generating, it is required that $Q(0) \neq 0$. By not allowing self mappings $Q(x) = x$, we ensure a short transient-to-steady state and what is loosely called a "rich" dynamical behavior. The autocorrelation, Figure 11.4b, has a single central peak with a width (at half maximum) of 22 samples, and a maximum height outside of the central peak, of 0.015. Clearly, there is no indication of a periodic component. Figure 11.4c shows a return map from one state value s_n to the next, s_{n+1}. The wide central band indicates that one specific state value is in general preceded by many others. Often for chaotic

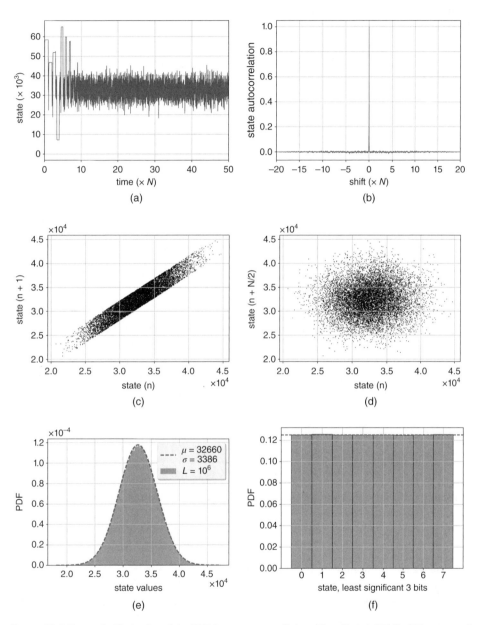

Figure 11.4 Dynamical behavior of the DFP for parameters $(k, l, m, N) = (3, 4, 4, 1024)$. a) Timetrace of the state variable s of Eq. (11.3) at startup. Time is in units of the delay line length N. b) Autocorrelation of the state. c) Return map $s_n \mapsto s_{n+1}$. d) Return map $s_n \mapsto s_{n+N/2}$. e) Histogram of the state and fit to a normal distribution from one million samples. f) Histogram of the values obtained from one million samples of the three least significant bits of the state. The dotted line indicates $P = 1/8$, the bin value for a theoretical uniform distribution.

delay systems, a suitable delay embedding reveals some structure reminiscent of the nonlinearity employed [19, 20]. We have found this is not the case here, as for example the half-delay return map $s_n \mapsto s_{n+N/2}$, shown in Figure 11.4d, reveals no recognizable structure. A histogram of the state derived from $L = 10^6$ samples, Figure 11.4e, shows a near-perfect fit to a normal distribution with location $\hat{\mu}_s = 32660$ and root-variance $\hat{\sigma}_s = 3386$.

11.2.2 Steady-State Dynamics of the DFP

We can estimate these values and gain some insight into the DFP by rewriting Eq. (11.3). Using $\alpha = 1 - 2^{-l}$, $\beta = 2^m$, thus ignoring flooring effects introduced by the bit shifters and finite 16-bit precision leads to:

$$s_{n+1} = \alpha s_n + \beta x, \tag{11.6}$$

where we have written the delayed output of the permutation table as a variable x. Taking the expectation of both sides:

$$E\{s_{n+1}\} = E\{\alpha s_n + \beta x\}, \tag{11.7}$$

and setting $E\{s_{n+1}\} = E\{s_n\}$, yields:

$$\mu_s = 2^{l+m} \mu_x \tag{11.8}$$

A histogram of the output of the permutation table (not shown) reveals that these values, interpreted as the realizations of a random variable X, closely follow a uniform distribution with $X \sim U[a = 0, b = 255]$. Therefore $E\{x\} = \mu_x = (a+b)/2 = 127.5$ and $\text{Var}\{x\} = \sigma_x^2 = (b-a)^2/12 = 5418.75$. Thus Eq. (11.8) yields $\mu_s = 32640$, very close to the experimental result of 32660. In a similar fashion, we can calculate the variance of the state variable:

$$\begin{aligned} \text{Var}\{s_{n+1}\} &= \text{Var}\{\alpha s_n + \beta x\} \\ &= E\{(\alpha s_n - \alpha \mu_s + \beta x - \beta \mu_x)^2\} \\ &= \alpha^2 \text{Var}\{s\} + \beta^2 \text{Var}\{x\} + 2\alpha\beta \text{Cov}\{s, x\}. \end{aligned} \tag{11.9}$$

Assuming state variable s and output of the permutation table x are uncorrelated, results in:

$$\sigma_s^2 = \frac{2^{2m}}{2^{1-l} - 2^{-2l}} \ \sigma_x^2. \tag{11.10}$$

After filling in the $l = m = 4$, we obtain $\sigma_s \approx 3384.6$, again very close to the experimental result (3386), shown in Figure 11.4e. By virtue of the central limit theorem, many averaged (or low-pass filtered) uniformly distributed random variables add up to a normal distribution. This explains the shape of the state variable distribution as shown in Figure 11.4e.

11.2.3 DFP-Bitstream Generation

In [11], we have introduced a method for generating an unbiased bitstream from an analog chaotic circuit. There, the bit value was chosen based on the comparison beween two values of the subsampled chaotic timeseries. This method could be applied here too.

Table 11.1 Results from the NIST randomness test suite, for the testing of 50 million bits obtained as the LSBs from the state variable of a DFP with parameters $(k, l, m, N) = (3, 4, 4, 1024)$.

Uniformity P-value	Pass Ratio	Test
0.616305	49/50	Frequency
0.911413	49/50	BlockFrequency
0.350485	49/50	CumulativeSums (min)
0.574903	49/50	CumulativeSums (max)
0.455937	49/50	Runs
0.699313	49/50	LongestRun
0.455937	50/50	Rank
0.613305	50/50	FFT
0.008879	48/50	NonOverlappingTemplate (min)
0.996335	50/50	NonOverlappingTemplate (max)
0.085587	49/50	OverlappingTemplate
0.419021	49/50	Universal
0.851383	49/50	ApproximateEntropy
0.054199	34/34	RandomExcursions (min)
0.911413	34/34	RandomExcursions (max)
0.000153	34/34	RandomExcursionsVariant (min)
0.911413	34/34	RandomExcursionsVariant (max)
0.030806	50/50	Serial (min)
0.096578	50/50	Serial (max)
0.319084	50/50	LinearComplexity

However, due to the highly irregular dynamical behavior of the DFP, a simpler method to generate random bits works as well. Here, we take the least significant bit of each sample of the state variable. In Table 11.1, we show the results of the National Institute of Standards test suite for random bit streams [21] on 50 million bits generated by this method. The random bits were divided in 50 sequences of 1 million bits each. The test suite was used with the default settings. Where a test has more than one result, both the highest and lowest results are shown. The results file states that the minimum pass rate for each statistical test, with the exception of the random excursion (variant) test, is approximately 47 for a sample size of 50 binary sequences. The minimum pass rate for the random excursion (variant) test is approximately 31 for a sample size of 34 binary sequences. We conclude that the bitstream thus generated by the LSB of the DFP state variable shows no sign of deviation from randomness.

11.2.4 Sensitivity to Changes in the Permutation Table

It is interesting to examine how sensitive the chaotic timeseries obtained from the DFP are with respect to changes in the permutation table. In Figure 11.5a, we show two signals of almost identical DFPs, again with parameters $(k, l, m, N) = (3, 4, 4, 1024)$.

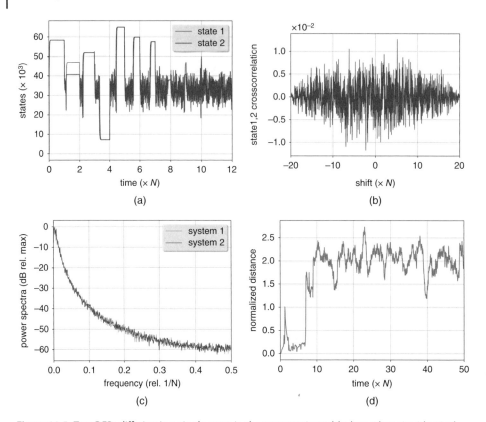

Figure 11.5 Two DFPs differing in a single swap in the permutation table, but otherwise identical. a) Transient at start up. b) Crosscorrelation. c) Power spectrum. d) Normalized distance of entries in the delay line.

The only difference beween them consists of two values that are swapped in the permutation table of one DFP. After an initial transient, the timeseries diverge completely. Their maximal crosscorrelation, shown in Figure 11.5b is 0.013, revealing no relation between the signals or the nearly identical systems they originated from. The power spectra, shown in Figure 11.5c, appear similar in nature. In Figure 11.5d, we show the normalized Euclidian distance of the N-component vectors X_1, X_2 formed by the delay line entries. This distance is calculated as:

$$d(X_1, X_2) = \frac{\sum\limits_{i=1}^{N} (x_{2,i} - \overline{x_2} - x_{1,i} + \overline{x_1})^2}{\sqrt{\sum\limits_{i=1}^{N} (x_{1,i} - \overline{x_1})^2 \sum\limits_{1}^{N} (x_{2,i} - \overline{x_2})^2}} \approx \frac{\mathrm{Var}\{X1 - X2\}}{\sqrt{\mathrm{Var}\{X1\}\mathrm{Var}\{X2\}}}, \qquad (11.11)$$

where $\overline{x_1}$ and $\overline{x_2}$ represent the mean values. This illustrates that, after the transients the system states stay separated. In Figure 11.5d, the mean value of $d(X_1, X_2) \approx 2$ coincides with what is expected for two independent identically distributed normal random variables. From 1 million bits derived from the LSBs of both DFPs, we found that the fraction of equal bits was 0.500423. This is in line with what can be expected for two unrelated random bit generators. In addition, we found that negating even one single bit in the permutation table gave similar results. This suggests that every possible permutation table yields a completely different generator, which understandibly is a prominent feature of the DFP as a driving oscillator in an encryption system.

11.3 The Chaotic Followers

At this point, referring back to the system scheme of Figure 11.1, we might complete our encryption system by using one DFP as a driver and a different second and third DFP as followers. Two problems would arise from this approach. First, the fact that part of the driver states, namely the bitfield $k \ldots k + 7$, is visible in the public channel, reveals some information about the driver's internal operation (note that many states coincide with a single value of this bitfield, as Figure 11.4c shows). Although in itself this need not be detrimental to our encryption scheme, it is still desirable that no information about the internal operation of the driver is derivable from what is publicly visible. To alleviate this problem, we will cascade two DFPs (called predriver and driver), via the input node shown in Figure 11.3, where the $(k, k + 7)$-bitfield of the state of predriver and driver are averaged. We note that exclusive-or-ing of these bitfields also works and is probably a bit easier to implement. The dynamical behavior of the driver is then completely perturbed by the predriver, leading to results which we will not show in the interest of saving space, but which are strikingly similar to those shown in Figure 11.5. Note the states of the predriver are not visible in the public channel.

A second problem arises from the use of DFPs as followers: it requires the need to somehow synchronize the follower-DFPs in both the receiver and transmitter so that they generate the same bitstream. From our experiments, identical driven DFPs do not synchronize when started from an arbitrary state. This difficulty could be overcome at a higher level in the communication protocol: on instantiating a transmission, both the transmitter and receiver agree to reset their follower-DFPs and then run them for an arbitrary but predetermined time with bogus input, presented by the driver, to get past the initial transient. Afterward, the message can be securely exchanged. In the next section, we present an alternative method for constructing the followers, based on generalized synchronization.

11.3.1 The Permute-Filter Block

In Figure 11.6, we show what we call a permute-filter (PF) block. Basically, this is a DFP missing the delayed self-feedback. Each block, of which many are to be used in a single follower, has a single 16-bit lossy filter. This results in a fading memory, which is a requirement for synchronization. Denoting M as the number of PF blocks in the followers used in the next experiments are parameterized as $(k, l, m, M) = (3, 4, 4, 8)$. Note that it is not necessary to have identical PF parameters in the individual blocks of a follower, nor do the bitfields of the input and output of the PF need to be identical, as is drawn in Figure 11.6. In Figure 11.7a we show how two identical followers synchronize, starting from random states. The color-intensity plot depicts the normalized Euclidian distance, over 64 samples, between corresponding PF blocks in the two followers. One

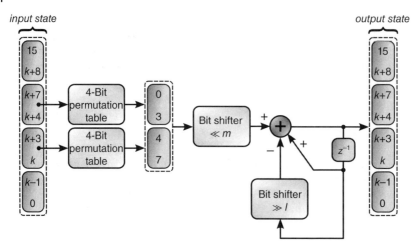

Figure 11.6 A single permutate-filter (PF) block.

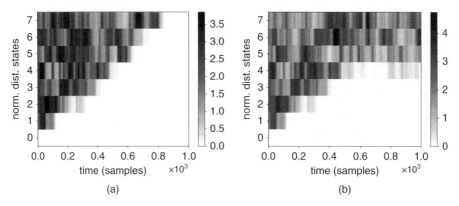

Figure 11.7 a) Synchronization of two identical followers, each having eight PF blocks, starting from a random state. b) Two almost identical followers, differing in a single permutation table swap in the fourth PF do not synchronize. The color-intensity plot depicts the normalized Euclidian distance, over 64 samples (Eq. (11.11)).

after the other, the PF blocks synchronize until both followers are entirely in sync. In contrast, if two values are swapped in one of the permutation tables, for example, in the fourth PF, then from this onward, the followers do not synchronize. This is shown in Figure 11.7b. Experimentally, we observed that the crosscorrelation between the outputs of the desynchronized followers stayed below 1.2%, which is qualitatively similar to Figure 11.5b. Note that instead of using a single 8-bit permutation table per PF block, two 4-bit permutation tables which together form a single 8-bit table are used. The reason for this is as follows. We want to maximize the effect of a difference in permutation table entries between two followers. On average, a single entry in a 8-bit permutation table in a PF would get a "hit" once every $2^8 = 256$ samples. Suppose we find ourselves in the situation where all PF blocks between two followers, except for the last one, are identical and this last PF would have only one bit error or one swap. Then by virtue of the feedforward nature of the PF many of its output samples would be correct. This might lead to parts of the message that are in between hits of the wrong value to be decoded correctly. Since for the 4-bit permutation table the hit rate is 1/16-th, there is less time for this to happen. Alternatively, one can say that the overal 8-bit permutation formed by the 2× 4-bit tables changes in 16 places for every single change in a 4-bit table. Understandibly, we pictured an extreme case, and using 8-bit permutation tables in all PF blocks of a follower will offer adequate security. A mixed approach, where internal the PF blocks benefit from an 8-bit permutation table, while the last block utilizes two 4-bit tables is also possible. The downside of this approach is that there simply exist fewer 4-bit permutation tables then 8-bit ones.

11.3.2 Brute Force Attack

Thus, one might wonder how difficult it is to break a follower consisting of a chain of eight PF blocks by brute force attack alone. Since each PF has two 4-bit tables and we exclude identical mappings $Q(x) = x$ for $x = 0 \ldots 15$, there are $(15!)^2 \approx 1.7 \times 10^{24}$ configurations for a single PF. Having eight PF blocks then delivers 7.3×10^{193} possible follower configurations. The attack might be carried out using specialized computing hardware running in parallel. Solely for the purpose of a back-of-the-envelope calculation, let's take the Cedar supercomputer (officially named GP2) at the Simon Fraser University in Canada as an example. This supercomputer, which was built in 2017, has been reported as having a theoretical peak performance of 3.6 petaflops in double precision. Let's assume that somehow this number equals the number PF configurations we can check per second. Obviously, this is a gross overestimation. To check all configurations then would take $7.3 \times 10^{193}/(3.6 \times 10^{15} \times 3600 \times 24 \times 365.5) \approx 6.4 \times 10^{170}$ years. Clearly, a brute-force attack is impractical. We stress that the strength of this method, as was the case for the system presented in [11], relies on the fact that the internal states of intermediate PF blocks are kept unobservable. A full identification and analysis of all possible attacks in a so-called threat model is only useful given a more detailed practical implementation, where it is clear how this scheme is implemented in a protocol stack. We leave such analysis for future work.

11.3.3 PF-Bitstream Generation

In Table 11.2, we show the results of the NIST randomness test suite for the bitstream derived from the LSB of the state of the final PF in a chain of eight. This follower chain was driven by a predriver-driver DFP cascade as described above. We obtained 50 million bits to derive these results. The results file states that the minimum pass rate for the random excursion (variant) test is approximately 29 for a sample size of 31 binary sequences. All tests are passed. We conclude that the bitstream generated by the followers shows no sign of deviation from randomness. In the next section, we show how the DFPs and PFs are brought together to form a complete encryption system.

Table 11.2 Results from the NIST randomness test suite, for the testing of 50 million bits obtained as the LSBs from the output of a chain of PF blocks, having parameters $(k, l, m, M) = (3, 4, 4, 8)$.

Uniformity P-value	Pass Ratio	Test
0.739918	50/50	Frequency
0.051942	50/50	BlockFrequency
0.657933	50/50	CumulativeSums (min)
0.883171	50/50	CumulativeSums (min)
0.955835	50/50	Runs
0.058984	50/50	LongestRun
0.419021	50/50	Rank
0.455937	49/50	FFT
0.020548	50/50	NonOverlappingTemplate (min)
0.991468	50/50	NonOverlappingTemplate (max)
0.998821	49/50	OverlappingTemplate
0.171867	49/50	Universal
0.983453	50/50	ApproximateEntropy
0.015963	31/31	RandomExcursions (min)
0.706149	31/31	RandomExcursions (max)
0.110952	31/31	RandomExcursionsVariant (min)
0.834308	31/31	RandomExcursionsVariant (max)
0.699313	50/50	Serial (min)
0.699313	50/50	Serial (max)
0.319084	49/50	LinearComplexity

11.4   The Complete System

Note that in a practical setup a single communication can contain both the payload data as well as an encrypted identifier of the originator, thus combining authentication and encryption. For readability, we introduce these usages separately.

11.4.1   Image Encryption Example

In Figure 11.8, we show the completed system, utilizing the DFP and PF blocks described above. The example images in this figure are actual results from our numerical experiments. The source image is serialized and encrypted by XOR-ing with the bitstream resulting from the LSB of the output state of follower 1. Follower 1 had parameters $(3, 4, 4, 8)$ in our experiments, as discussed above. Follower 1 is driven by the ouput of a predriver-driver cascade of two DFPs with parameters $(3, 4, 4, 922)$ and $(3, 4, 4, 1024)$ respectively. Clearly, the encrypted image resembles random noise. Follower 3 acts as an in-channel eavesdropper, attempting to decrypt the image. Follower 3 has the exact same parameters as those used to illustrate the desynchronization in Figure 11.7b: it is a single permutation table swap removed from being identical to followers 1 and 2.

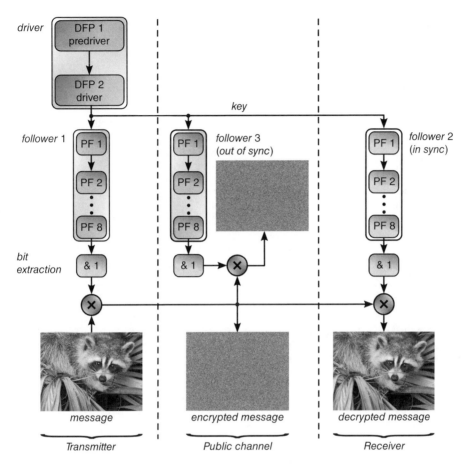

Figure 11.8 Image encryption using the proposed scheme based on generalized synchronization between two chains of permute-filter blocks, driven by delay-filter-permute blocks.

The image as decoded by follower 3 is completely unintelligable, while follower 2, in synchronization with follower 1, perfectly decrypts the image.

11.4.2 Usage for Authentication

It is clear that the same scheme can just as easily be applied for authentication, as part of the message can relay the identity of the transmitter. This is illustrated in Figure 11.9. Here, the initiator of the authentication (for example, a sensor network base station) prompts a "challenge" key, which the receiving party uses, via its PF chain, to transcribe its identity and additional payload data (for example, temperature sensor data). This encrypted message is then sent back to the initiator through the public channel. The initiator then uses his own PF chain to decode and verify the encrypted data and identity that was returned. This is possible because the initiator can construct the exact same bit sequence via its own PF follower, which is identical to that of the originator of the data. Using suitably long keys ensures that no two similar keys will ever be repeated. The follower chains in both the initiator and responder should first be synchronized, either

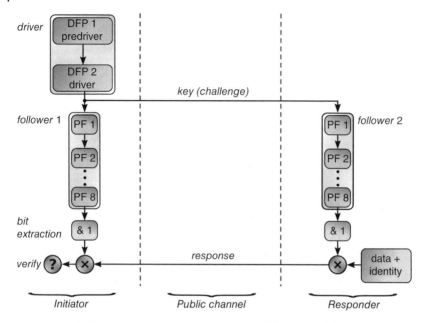

Figure 11.9 Authenticated transmission: the veracity of the origin of the response is guarantueed by having the data signed with the responder's identity. Upon decoding the response, the initiator checks the identity agains known-good responders.

by driving them long enough or by a signalling at a higher protocol level, as explained earlier.

11.5 Conclusions and Outlook

We have demonstrated an encryption and authentication scheme based on generalized synchronization of complex systems. The building blocks of these systems are straightforward to implement in FPGAs and digital ASICs or as embedded code. The strength of the system hinges critically on the unobservability of the internal states of a strongly nonlinear system having a high-dimensional phase space. The delay-filter-permute block has been shown to be an excellent random-number generator, and it has been shown that two chains of identical permute-filter blocks synchronize, while the slightest change in internal structure leads to complete decorrelation of the generated bitstreams. We emphasize that many design choices are still open to examination, as the parameters we have used in our numerical experiments are merely motivated by "what works" in a first demonstration. We have not yet systematically investigated the relation between cycle length and DFP delay memory length. This relation needs to be made clear in order to be able to optimize a practical proof-of-concept implementation. Another avenue of further research is to examine the possibility of generating several bits per transmitted key value, such that the bandwidth efficiency of the system is improved. A blunt method would be to have several PF chains working in parallel, however, this would not be very hardware efficient. As mentioned above, it is possible to generate bitstreams from other than the LSB of the final PF block in a follower chain, perhaps even from intermediate PF states. Obviously, there are many possible topologies to examine. Nevertheless, we have shown that the scheme that was introduced in [11] using analog electronics, can readily be ported to digital hardware. We conclude by remarking that complex systems are increasingly seen as alternative computational resources (for example in the field of Reservoir Computing [12], instead of mere system-dynamical curiosities to be studied in isolation. We expect all future technologies, not only IoT, to benefit from their application and have aimed show the practical viability of one such scheme.

Acknowledgements

LK and GVDS were partly supported by the Belgian Science Policy Office under Grant No IAP-7/35 "Photonics@be" and by the Science Foundation - Flanders (FWO). GVDS thanks the Research Council of the VUB.

Author Contributions Statement

The concept originated from discussions between GVDS and LK. LK performed the numerical experiments. Both authors contributed to the manuscript.

Additional Information

The authors declare no competing financial interests.

References

1 Vernam, G.S. (1926). Cipher Printing Telegraph Systems For Secret Wire and Radio Telegraphic Communications. *Journal of the IEEE* 55: 109–115.

2 Pappu, R., Recht, B., Taylor, J. and N. Gershenfeld. (2002). Physical One-Way Functions. *Science* 297 (5589): 2026–2030.

3 Gassend, B., Clarke, D., Van Dijk, M. and Devadas, S. (2002). Silicon physical random functions. *Proceedings of the 9th ACM conference on computer and communications security*, Washington, USA (18–22 November 2002). ACM.

4 Herder, C., Yu, M.D., Koushanfar, F. and Devadas, S. (2014). Physical unclonable functions and applications: a tutorial. *Proceedings of the IEEE* 102 (8): 1126–1141.

5 Rŭhrmair, U. and Holcomb, D.E. (2014). Pufs at a glance. *Proceedings of the conference on Design, Automation and Test in Europe, European Design and Automation Association*, Dresden, Germany (24–28 March 2014). IEEE.

6 Rŭhrmair, R., Sehnke, F., Sölter, et al. (2010). Modeling attacks on physical unclonable functions. *Proceedings of the 17th ACM Conference on computer and communications security*, Chicago, USA (04–08 October 2010). ACM.

7 G. T. Becker, (2015). The gap between promise and reality: On the insecurity of xor arbiter pufs. In: *Cryptographic Hardware and Embedded Systems –CHES 2015* (eds. T. Gűneysu, and II. Handschuh), 535–555. Berlin, Heidelberg: Springer.

8 Cuomo, K.M., Oppenheim, A.V. and Strogatz, S.H. (1993). Synchronization of Lorenz based chaotic circuits with applications to communications. *IEEE Transactions on Circuits and Systems II: Analog and Digital Signal Processing* 40 (10): 626–633.

9 Argyris, A., Syvridis, D., Larger, L. et al. (2005). Chaos-based communications at high bit rates using commercial fiber-optic links. *Nature* 438: 343–346.

10 Porte, X., Soriano, M.C., Brunner, D. and Fischer, I. (2016). Bidirectional private key exchange using delay-coupled semiconductor lasers. *Optics Letters* 41 (12): 2871–2874.

11 Keuninckx, L., Soriano, M.C., Fischer, I. et al. (2017). Encryption key distribution via chaos synchronization. *Scientific Reports* 7: 43428.

12 Keuninckx, L., Danckaert, J. and der Sande, G.V. (2017). Real-time Audio Processing with a Cascade of Discrete-Time Delay Line-Based Reservoir Computers. *Cognitive Computation* 9 (11): 10.1007/s12559-017-9457-5.

13 Oliveira, H.M. and Luas, L.V. (2015). Huygens synchronization of two clocks. *Scientific Reports* 5: 11548.

14 di Pellegrino, G., Fadiga, L., Fogassi, L. et al. (1992). Understanding motor events: a neurophysiological study *Experimental Brain Research* 91: 176–180.

15 Strogatz, S.H. (2001). *Nonlinear dynamics and chaos: With applications to physics, biology, chemistry, and engineering*. New York, NY: Perseus books.

16 Kato, H., Soriano, M.C., Pereda, E. et al. Limits to detection of generalized synchronization in delay-coupled chaotic oscillators. *Physical Review E* 88 (6–1): 062924.

17 Soriano, M.C., Garcia-Ojalvo, J., Mirasso, C.R. and Fischer, I. (2013). Complex photonics: dynamics and applications of delay-coupled semiconductor lasers. *Reviews of Modern Physics* 85: 421–470.

18 Ogorzalek, M.J. (1997). *Chaos and Complexity in Nonlinear Electronic Circuits*. Singapore: World Scientific Publishing.

19 Mackey, M.C. and Glass, L. (1977). Oscillation and chaos in physiological control systems. *Science* 197 (4300): 287–289.

20 Van der Sande, G., Soriano, M.C., Fischer, I. and Mirasso, C.R. (2008). Dynamics, correlation scaling, and synchronization behavior in rings of delay-coupled oscillators. *Physical Review E* 77 (5): 055202.

21 Bassham, L.E., Rukhin, A.L., Soto, J. et al. (2010). A statistical test suite for random and pseudorandom number generators for cryptographic applications. Technical Report, National Institute of Standards & Technology, Gaithersburg, MD, USA. https://ws680.nist.gov/publication/get_pdf.cfm?pub_id=906762 (accessed 25 June 2019).

Part V

IoT Use Cases and Implementations

12

IoT Use Cases and Implementations: Healthcare

Mehrnoosh Monshizadeh, Vikramajeet Khatri, Oskari Koskimies, and Mauri Honkanen

12.1 Introduction

Recently, there has been considerable global interest in exploiting digital healthcare solutions, often known as eHealth in order to improve traditional healthcare mechanisms. Currently, fewer physicians are taking care of more patients than ever; doctors do not have enough time to educate patients about their condition or care plans which causes uncertainty, stress, non-adherence to care plans, and unnecessary hospitalization. On the other hand, global healthcare spending is projected to reach $8.7 trillion by 2020 with an annual growth rate of 4.1% in 2017–2021. This growth is driven by a number of factors such as, chronic diseases (becoming more common), aging, increasing population, expansion in developing markets, new though costlier medical treatments, and rising labor costs [1, 2].

Certain chronic conditions, such as hypertension, excess weight and diabetes, are risk factors for stroke. A stroke is a condition where the flow of blood carrying oxygen and nutrients to a portion of the brain is reduced or blocked. Stroke is the second most common cause of death worldwide and ranks third most Disability Adjusted Life Years (DALY) [3].

As seen in Figure 12.1, in Finland with a population of 5.5 million, there are about 82 000 stroke patients and every year about 25 000 new people are impacted by stroke. As many as 17% of patients suffer a new stroke within a year, 10–20% develop dementia and one-third have aphasia (speech difficulties). About 50% of these patients have permanent damage of some sort while every fourth patient is of working age. Almost half of these patients require active rehabilitation following a stroke which has a relatively high impact on the economy. According to statistics, stroke is the third most expensive chronic condition in Finland with the first year of treatment costing over €20 000, lifetime cost greater than €85 000 and approximately 1.8 M in patient days caused every year. It is estimated that in Finland there will be a need for at least 100 new hospital wards by 2020 because of the increased number of stroke patients [4].

Stroke can be a lethal or a severely disabling event. Stroke patients have multiple problems with memory, normal movement, speech, and the fear of recurring strokes. One typical reason for stroke is occasional malfunction of the heart (typically in the left atrial part of the heart) that causes clotting. This clot exits the heart after physical exercise, for example, when the heart pumps heavily and ends up blocking narrow blood vessels in

IoT Security: Advances in Authentication, First Edition.
Edited by Madhusanka Liyanage, An Braeken, Pardeep Kumar, and Mika Ylianttila.
© 2020 John Wiley & Sons Ltd. Published 2020 by John Wiley & Sons Ltd.

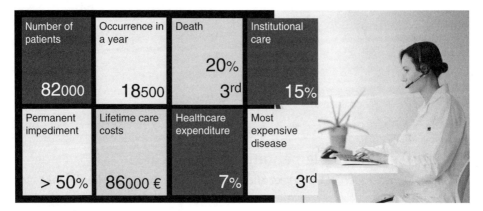

Number of patients	Occurrence in a year	Death	Institutional care
82000	18500	20% 3rd	15%
Permanent impediment	Lifetime care costs	Healthcare expenditure	Most expensive disease
> 50%	86000 €	7%	3rd

Figure 12.1 Stroke care in Finland.

the brain causing the stroke. The medical challenge is that most of these atrial fibrillation cases causing the clotting are not traceable after the event and some are not even noticed by the patient. There is, nevertheless, high risk of recurrent strokes in cases where blood flow through the heart is not functioning correctly. This creates a high interest to monitor the heart of stroke patients until normal heart functionality or dysfunctionality can be verified. Once validated, the monitoring of the impact of medication and/or functionality of a pacemaker becomes important.

There are two methods to prevent stroke: primary and secondary. Primary prevention is used for those who have not suffered from stroke previously, whereas secondary prevention applies to the patients who have already been affected by stroke. Although there are challenges such as poor adherence to medication, failed care plan and cryptogenic causes, still approximately 80% of reccurrences could be prevented with early and fast diagnosis [5].

To overcome some of the secondary prevention challenges and to expand geographical responsibility areas of healthcare providers, in this chapter, a secure digital remote patient monitoring solution is introduced. With this solution, a patient could be monitored and served remotely; health services could be provided to people far away from hospitals and by experts not necessarily available in the nearest hospital. In addition, the proposed solution applies the concept of eHealth to increase information sharing about the disease and patient awareness regarding their risk factors to improve self-care. This solution extends ElectroCardioGram (ECG) monitoring from a typical 24–48 hour holtering case to multiple 48 hour sessions across a post-discharge care plan of several months [5].

The rest of the chapter is organized as follows. Section 12.2 describes the remote patient monitoring platform in general. Section 12.3 discusses the security concerns related to digital health technologies and briefly reviews available safeguard and authentication techniques to secure a digital health system. Section 12.4 introduces a security architecture for the proposed remote patient monitoring platform and finally, and conclusions are presented in the last section.

12.2 Remote Patient Monitoring Architecture

The remote patient monitoring solution is a healthcare Internet of Things (IoT) system that can be implemented on Virtual Machines (VMs) in the cloud. The system can be

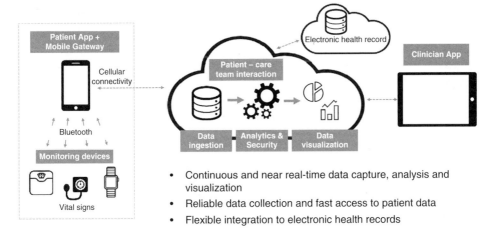

- Continuous and near real-time data capture, analysis and visualization
- Reliable data collection and fast access to patient data
- Flexible integration to electronic health records

Figure 12.2 Architecture of remote patient monitoring platform.

orchestrated to other digital services or network modules via Network Functions Virtualization (NFV). In the current use case, Docker and Amazon Application Programming Interfaces (APIs) are used for orchestration.

The remote patient monitoring platform consists of patient monitoring devices, a patient smartphone application, a cloud backend (docker containers in multiple VMs, orchestrated with Docker Swarm) and a web-based clinician application. The cloud backend could also utilize NFV for security and anomaly detection purposes.

As shown in Figure 12.2, the platform includes three main parts: patient, cloud, and hospital side.

At the patient side, wearables are provided to patients to measure parameters such as Blood Pressure (BP), weight, and ECG. The information is collected periodically, based on a care plan. Collected information is transferred via Bluetooth to a mobile device that has the Patient App. Then, information from the Patient App will be transferred to the cloud. In the cloud, there are different modules, for example, database (DB), data analytics and security and data visualization. The cloud has also an interface to electronic health records (EHRs) databases for storing patient data. The Data Analytics and Security module will classify data in predefined clusters either for anomaly detection or for future investigation by doctors. The data visualization module will transform classified data to be visualized by doctors. The doctors can log in and view the patient information through the Clinician App [5]. The Patient App and Clinician App are shown in Figure 12.3.

12.3 Security Related to eHealth

Digital health technologies may include critical and strategic services such as connected hospitals, remote surgery, patient care and preventive health plans where an anomaly in the original function can mean the difference between life and death. An attack on such devices and services could also be stealing or modifying patient health information which violates privacy and could compromise patients' health.

Wearables include personal devices such as smart watches, health monitors, and medical devices including wireless pacemakers and insulin pumps. These devices collect and transmit detailed health information about the person. If an attacker gains remote access

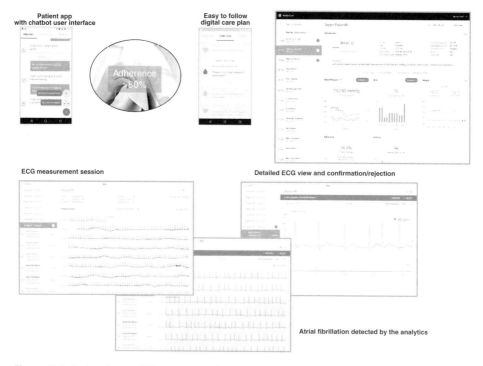

Figure 12.3 Patient App and Clinician App of remote patient monitoring platform.

or remote control of a wearable medical device, then an attacker could potentially harm or kill the person e.g. by causing a pacemaker to malfunction or an insulin pump to overdose or under-dose the insulin supplied to the person [6].

Users may prove their identity to a device using a password, Personal Identification Number (PIN) or fingerprints. Wearable devices like watches might serve as proxy devices with more natural interactions than a smartphone. As each user will perform a gesture differently, gestures can also serve as a form of authentication and therefore be used to identify a person. Furthermore, the proximity of a wearable device is helpful in identifying several contextual factors, including user location and people nearby [7].

In order to perform safe and secure data exchange between patients and their healthcare providers, different security aspects must be considered in a digital health platform:

- Technical safeguards such as encryption, authentication, and intrusion-detection methods.
- Physical safeguards such as security enclosures for workstations, media disposal procedures and asset tracking.
- Administrative safeguards such as risk analysis, access rights reviews and security training.

This chapter only concentrates on the technical safeguards. The current section surveys available authentication and cryptography methods, and the next section describes security measures in general. In addition, the next section also proposes an anomaly detection module to efficiently detect and prevent intrusion on the proposed platform.

12.3.1 IoT Authentication

To overcome the identification challenges, authentication protocols must be strong enough to be resilient against many different attacks, such as eavesdropping, replay attacks, where an attacker keeps the result of a previous authentication made by an authorized user to abuse it at a later date, man-in-the-middle attacks, dictionary attacks where an attacker tries the most probable secret keys such as names and birth dates for a password, or brute-force attacks where an attacker tries all possible password combinations [6].

To verify the identity of an IoT device, authentication can be performed in multiple ways e.g. authentication of IoT device on the cloud service and authentication of the IoT device to IoT gateway [8]. Some authentication mechanisms include: password, PIN, signature, token, voice pattern, smart card and fingerprint [9]. To transmit the authentication parameters over the network, encryption is recommended to avoid eavesdropping and credentials theft.

Authentication can be achieved using symmetric or asymmetric (public-key) cryptography. In case of symmetric cryptography authentication mechanisms, both sender and receiver share the same key, which must be delivered to IoT devices over a secure channel. Distribution of symmetric keys is often difficult to accomplish in a secure and convenient manner. One commonly used approach is to generate the keys at manufacture time and deliver them together with the device in a machine-readable form such as a QR code, so that the keys can easily be entered into the system when the IoT device is enrolled in the system.

Traditional authentication mechanisms are based on Public Key Infrastructure (PKI), where public and private keys are generated for encryption and decryption. The widely used digital signature mechanism utilizes PKI. A one-way hash of the data (or secret key of IoT device) will be created to be signed (or verified) and will then be encrypted with the private key of the IoT device. The recipient (IoT device, IoT gateway, user, cloud) uses the sent IoT device's public key to decrypt the hash and verify whether it was created with the sent IoT device's private key. This procedure performs a successful authentication [10]. A PKI-based authentication server has been proposed for authentication across multiple EHR systems [6]. However, PKI may be unfit for IoT devices as these mechanisms require significant processing, memory, storage, communication, and management overheads [11].

The following lightweight authentication algorithms are feasible for IoT devices:

A Message Queuing Telemetry Transport (MQTT) protocol identifies an IoT device by its unique device identifier, user ID, or digital certificate. Every IoT device in MQTT will have a unique client identifier based on a username and password that acts as an identifier in the authentication process. Based on client identity, the MQTT server authenticates the IoT device (client) either by using the Secure Sockets Layer (SSL) protocol or by using a password specified by the IoT device. At the transport layer, a Transport Layer Security (TLS) takes care of data encryption and avoids eavesdropping. In addition, the IoT device can utilize an X.509 certificate during the MQTT TLS handshake process for authentication [12].

The Advanced Message Queuing Protocol (AMQP) applies Simple Authentication and Security Layer (SASL) to authenticate IoT device connections to the broker. Typically, the used mechanism is PLAIN, where authentication parameters include a username and password. This PLAIN mechanism is vulnerable to eavesdropping and man-in-the-middle attacks, so it is recommended to use SSL or turn off the PLAIN mechanism [12].

Extensible Messaging and Presence Protocol (XMPP) uses SASL to offer an extensible set of authentication methods that the IoT device can use. If unsecure authentication mechanisms are offered, then the IoT device may select an unsecure authentication mechanism from offered authentication mechanisms that then allows credential theft. Therefore, it is recommended that systems are configured so that IoT devices can only use secure authentication mechanisms [12].

Message Authentication Code (MAC) algorithms are much faster than digital signatures and offer good security if a shared key is exchanged between two parties prior to the communication session establishment. Hash-based Message Authentication Code (HMAC) is an enhanced MAC consisting of a hash function and cryptographic key. HMAC typically runs fast over constrained devices such as IoT devices. However, HMAC uses the same key for encryption and decryption so all those IoT devices who have access to the secret key would be able to sign and verify the communication [12].

In [13], an IoT object-authentication framework is proposed that uses device-specific information (e.g. fingerprints), together with a transfer learning tool to detect changes in that specific information. The proposed framework includes five modules: feature extraction, fingerprint generation, similarity measure, environment estimation, and transferring knowledge. It is considered that IoT devices transmit data to a gateway, who later forwards it to a security service provider (SP) in the cloud. SP authenticates the transmitted data based on IoT device fingerprints. Considering Radio Frequency Identification (RFID) tags, possible IoT device fingerprints include wavelet-based features such as mean, variance, and skewness of electronic codes. Another example of fingerprint is environmental changes for IoT devices, such as changes in temperature, humidity, wind, physical displacement, or any physical changes. Such fingerprints make it hard to impersonate a legitimate IoT and perform successful authentication in the network.

According to [14], user authentication is divided into two categories: static authentication and continuous authentication. Static authentication is invoked at the beginning of a session. Communication begins once the user is authenticated. The user remains authenticated for a longer or indefinite time as configured at the server (i.e. session timeout). Static authentication does not offer protection against session hijacking attacks. Continuous authentication can repeatedly authenticate the legitimacy of a user during the device usage time. Continuous authentication is not a substitute for static authentication, but it complements static authentication mechanisms for enhanced security.

A lightweight continuous authentication protocol can be used for IoT devices to perform mutual authentication. This method utilizes lightweight computations such as HMAC, hash function and bitwise eXclusive-OR (XOR) functions. Initially, static authentication takes place as part of this method and an initial token for both parties is generated during the authentication period. In the continuous authentication phase, the initial authenticated token, together with IoT device data, is transferred to the IoT gateway (the receiver). The proposed protocol selects a valid authentication period and adapts the token technique according to the dynamic features of the IoT device such as remaining battery of the IoT device for enhanced security. Later, the IoT gateway checks whether the received data from the IoT device is generated in the authentication period or not. In addition, the IoT gateway verifies the Message-Digest 5 (MD5) value for integrity purposes. If all parameters are correct, then the IoT gateway sends an acknowledgement message to the IoT device [14].

In [15], a lightweight authentication scheme for eHealth application has been proposed that authenticates each object and establishes a secure channel between IoT devices and IoT gateway. HMAC ensures the integrity of communications. The proposed scheme has three phases: registration, authentication, and key establishment. Initially, a device is registered in the registration phase. In the authentication phase, mutual authentication is performed between an IoT device and IoT gateway. The IoT device generates a random value and sends a message to the IoT gateway consisting of generated value, masked identity and HMAC code. Upon receiving the message, the IoT gateway verifies the content with an associated HMAC. If verification is successful, then the IoT gateway generates a random value, and sends the message consisting of received value, generated random value and HMAC code to the IoT device. Upon receiving the message, an IoT device verifies its integrity with a supplied HMAC code; and if verification is successful then mutual authentication is carried out. After successful mutual authentication, a shared symmetric key is established for securing the communication channel. At the end, an encrypted message with a session key is sent to the IoT gateway to indicate the termination of key establishment process.

To improve the standard of life for elderly people, recently Ambient Assisted Living (AAL) systems are deployed as IoT applications. In order to protect sensitive health data generated by AAL and to establish a secure communication between medical devices and remote hosts, a proxy-based authentication with key establishment protocol has been proposed in [16]. In the proposed proxy-based approach, the resource constrained IoT medical devices, delegates the costly computational cryptographic operations to computationally rich devices located in the neighborhood of the medical sensor. The IoT sensor sends a request message (that contains identity of the source) to the responder. Upon receiving the request, the nearby computationally rich device (responder) checks the source (IoT device) of the message. If all preceding checks are successful, the key establishment session will be initiated, and data will be exchanged securely via the proxy over the TLS or IPSec (Internet Protocol Security).

To provide a secure authentication mechanism between the server and the IoT device, in [17], a multi-key (multi-password) based mutual authentication method has been proposed, as single password-based authentication mechanisms are vulnerable against side-channels and dictionary attacks. In this mechanism, the key is changed after every successful session between the server and the IoT device. The secure vault, which is a secret between the server and IoT device, is a collection of equal-sized keys. In this study, a three-way mutual authentication scheme is used; an IoT device sends a connection request to IoT gateway. Upon receiving the request, the IoT gateway sends a challenge to an IoT device; Upon receiving the challenge, the IoT device responds to the challenge from the IoT gateway and sends back an authentication challenge. If the IoT gateway verifies that the reply from the IoT device is valid, the IoT gateway responds to the challenge from the IoT device. This completes the three-way mutual authentication phase, and a session key will be produced to encrypt the exchanged communication between the IoT device and IoT gateway. Upon termination of a session, the session key will expire and later will be regenerated based on a three-way mutual authentication handshake process.

And finally, OAuth2 is a token-based authentication; where the user logs into a system and the system requests authentication in the form of a token. The user forwards the request from system to the authentication server, where the request is either rejected or accepted. If the request is allowed, the authentication server provides the token to the user and then to the requester [18].

12.4 Remote Patient Monitoring Security

The following figure illustrates the overview of security features of the remote patient monitoring platform (Figure 12.4).

12.4.1 Mobile Application Security

The mobile application (gateway service and Patient App) is secured through standard security features, namely using full disk encryption and screen lock. A different default PIN code is set for each patient and securely communicated to that patient (e.g. via text message to a personal phone or in a sealed envelope). The patient may change the PIN code to a more suitable one. Instructions must be provided to the patient on choosing a secure PIN code. The patient may also use fingerprint authentication to open the screen lock, if the phone supports it. Security credentials must be provided to the device the mobile application is running on before the device is given to the patient.

The mobile application includes a configuration user interface that is password protected. At the clinician side, the password is hardcoded into the application and only provides weak security; it is meant to prevent the patients from modifying the configuration, not to withstand a skilled attacker. It could easily be retrieved by rooting the device and decompiling the application. The cloud must treat all settings of the configuration UI as untrusted.

The application is also protected with DexGuard to protect it against modification and to detect if the patient (or someone else) roots the device. Use of DexGuard will also somewhat increase the security of configuration UI password, but not enough to trust the configuration data entered through it.

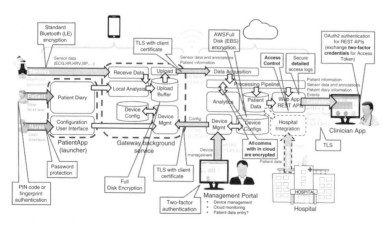

Figure 12.4 Remote patient monitoring system.

12.4.2   Communication Security

Sensor data transmitted from sensors to the patient's phone are protected by standard Bluetooth or Bluetooth Low Energy (BLE) (depending on sensor) encryption. The phones are pre-paired with the sensors.

All other communications are secured through use of TLS. A server-side TLS certificate is always used. A client-side TLS certificate is always used for the mobile application and for cloud management related HyperText Transfer Protocol (HTTP) connections. For the Clinician App, a TLS client certificate can be used if one can be installed on the devices used by the clinicians. If two-factor authentication is used for clinicians, a TLS client certificate is not required. Clinician App related connections without TLS client certificate or two-factor authentication are only accepted if they originate from the hospital network.

Administrator connections to the cloud must use two-factor authentication. This is enforced by a bastion host through which all administration connections must be routed.

Connections between cloud servers are secured through Docker Swarm, which establishes an encrypted overlay network between the containers it orchestrates.

The ID (name) of the TLS client certificate ID used by the mobile gateway application must be unique for the device, and it should be the same as the device ID of the device the mobile application is running on. The client certificate ID is used by the cloud to determine the patient ID for the data uploaded by the device, based on the patient currently assigned to that device.

12.4.3   Data Integrity

Patient-specific credentials for creating digital signatures are provisioned to the gateway during patient provisioning. The data collected by the gateway is digitally signed using these credentials. If built-in full-disk encryption cannot be used in the mobile phone, then the collected data must also be encrypted using a public key for which the corresponding private key is only available to the cloud. This guarantees that the encrypted data is safe, even if the phone is stolen and rooted.

Upon receiving data, the cloud must verify that the digital signature of the data was created by the patient to whom the device is currently allocated to. This verification prevents errors in patient allocation logic caused by data being allocated to the wrong patient. The signature and data must be retained by the cloud to make it possible to verify, at a later date, that all the data the cloud has stored or generated, based on that uploaded data, is valid.

12.4.4   Cloud Security

The two central security components are the perimeter server and authentication server. The perimeter server is a hardened server that faces the internet on one side and the cloud internal network on the other. It verifies validity of client certificates, terminates TLS connections, and sends authentication requests for authenticating incoming Representational State Transfer (REST) requests. The authentication server validates username and password credentials, grants OAuth2 access tokens based on them, and handles access token authentication requests from the perimeter server.

The authentication server will also verify two-factor authentication. If TLS client certificates are used as two-factor authentication, this means verifying that the clinician TLS client certificate matches the clinician username (for username and password credential) or the device from where an access token was originally requested (for access token credential). For mobile applications, the authentication server will map TLS client certificates to patient IDs.

12.4.5   Audit Logs

Access logs are generated when any of the REST services are used, or when data is uploaded. The access logs are stored securely on a separate server with its own set of access credentials. Additionally, on the patient side, a request identifier is generated for all incoming requests, and that identifier is added to all related requests and logs. If a service uses other services to fulfill the original request, the request identifier can be used to keep track of all logs related to an original request.

12.4.6   Intrusion Detection Module

In addition to a traditional intrusion detection system, the system also uses an experimental Hybrid Anomaly Detection Model (HADM) for intrusion detection. In HADM, different linear and learning algorithms are combined in a wide range to investigate a hybrid model for achieving high-detection performance and yet relatively low detection time, and will be deployed to detect intrusion attempts on IoT devices. In principle, it would be possible to use the data-mining mechanisms to detect different types of attacks such as corrupting and tampering data, platform configuration corruption or overloading it. In general, attacks on a remote patient monitoring platform include any kinds of malicious behaviors that cause unavailability or degradation of services, or stealing and corrupting patient information. A concrete example is the Denial of Service (DoS) attack on eHealth platform through a burst of signaling messages.

In order to secure a remote patient monitoring platform, HADM is introduced to detect and prevent previously mentioned attacks. The platform uses a combination of linear and learning algorithms combined with protocol analyzer as shown in Figure 12.5. The linear algorithms filter and extract distinctive attributes and features of the cyber-attacks while the learning algorithms use these attributes and features to identify new types of cyber-attacks [19].

As is shown in Figure 12.6, the HADM functionality is divided into three phases: protocol analyzer, dynamic machine learning and validator and database.

12.4.6.1   Protocol Analyzer

The first phase of the proposed HADM is called the protocol analyzer which filters vulnerable protocols. The protocol refers to communication protocol over which the traffic is carried on such as HTTP and Transmission Control Protocol (TCP). As is shown in Figure 12.7, a protocol analyzer consists of five modules.

i. Decision module

    This module includes a list of vulnerable protocols which are predefined and also dynamically updated based on the received feedback from log files via the database.

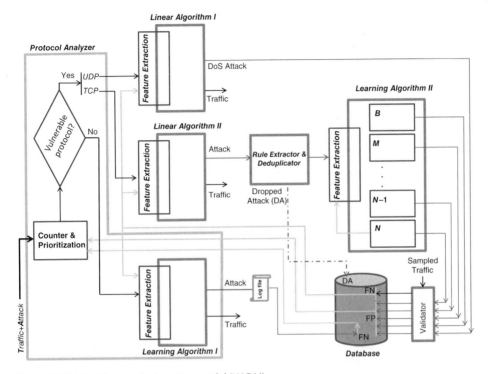

Figure 12.5 Hybrid anomaly detection model (HADM).

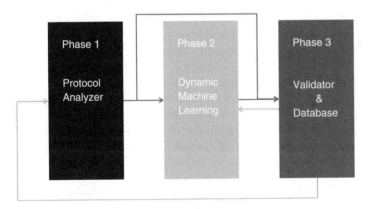

Figure 12.6 HADM on operational level in brief.

Some protocols such as HTTP and TCP are well-known vulnerable protocols while others like Real Time Streaming Protocol (RTSP) could be a safe protocol. RTSP checks whether traffic is carried on any of the listed vulnerable protocol.

ii. Counter and Prioritization module

The function of this module is based on the occurrence threshold (n) and prioritization. It means that if the vulnerable protocol carries suspected traffic for n times, then this module will forward the suspicious traffic to the next layer for detection and labeling. The idea is to cycle all possible vulnerable protocols over an agreed time

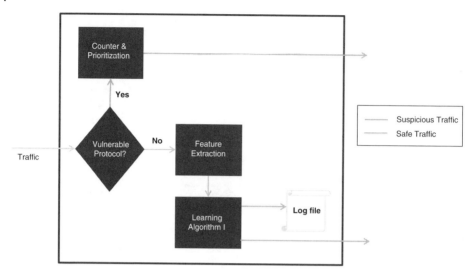

Figure 12.7 Protocol analyzer.

window (one hour, one day, etc.). The module only keeps a certain number of vulnerable protocols in the list which is based on prioritization. For example, if we already have 20 vulnerable protocols and a 21st comes up, then the counter must prioritize only 20 of these protocols. The prioritization is based on the counting occurrence over the time window. The sole purpose of this technique is to reduce the computation load of traffic analysis.

iii. Feature Extraction

This extracts the best features from the suspicious protocol. This module is utilized in second phase as well.

iv. Learning Algorithm I

If the protocol (that carries the input traffic) is not listed as vulnerable, traffic is still sent to this learning algorithm for analysis and reconfirmation. The learning algorithm I will check whether the protocol is vulnerable or not. Our proposed platform is tested with Extreme Learning Machines (ELM), Self-Organizing Map (SOM) and MultiLayer Perceptron (MLP) algorithms.

v. Log file

Every time the learning algorithm I in the protocol analyzer detects a new vulnerable protocol, it is recorded into the log file and feedback will be sent to the decision module via a database. The log file records packet features such as time stamp, packet size, Internet Protocol (IP) header and information on other layers (Ethernet, TCP, application layer).

12.4.6.2 Dynamic Machine Learning

The second phase of the proposed HADM combines linear and learning algorithms for efficient attack detection. As is illustrated in Figure 12.8, dynamic machine learning consists of the following modules, in addition to feature extraction, that was explained earlier.

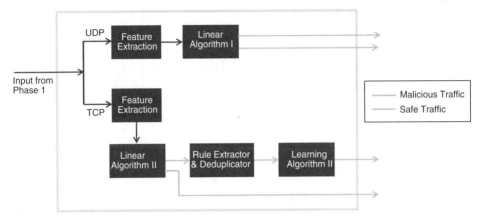

Figure 12.8 Dynamic machine learning.

i. Linear Algorithm I

This module analyzes User Datagram Protocol (UDP) traffic to detect UDP DoS attacks. Therefore, a separate algorithm such as a Decision Tree (DT) is considered for this module in order to avoid overloading the rest of the proposed hybrid model. However, the decision-tree algorithm can be replaced based on the operator demand, we have considered it because of its low processing time.

ii. Rule Extractor and Deduplicator

This module filters the known attacks that the system is already protected against, by using other deployed security mechanisms and forwards other attacks to learning algorithm II for labeling. A set of rules are extracted from those deployed security mechanisms in the network, the extracted information is compared with received attacks from linear algorithm I and is dropped if it is similar. Rules in this module are updated dynamically, based on input from the parallel security mechanisms.

iii. Learning Algorithm II

This module is the last detection layer. Initial features and clusters are defined for the algorithm during the training process in order to cluster different attacks such as Botnet attack (B) and Malicious codes (M). At first, the traffic is labeled to one of the clusters based on their similarity or distance. As the traffic that arrives to this module has been already identified as an attack, if it does not belong to any of the mentioned clusters then it is considered as new type of attack (N) and a cluster will be created for it. The features of the new type of attack (N) must be added to the algorithm accordingly. The implementation of the proposed platform with Artificial Neural Networks (ANNs) and Genetic Algorithm (GA) is already ongoing by authors and results will be presented in a future paper. Other potential unsupervised algorithms can be SOM and hierarchical clustering.

12.4.6.3 Validator and Database

The last phase of the proposed HADM validates detected attacks, stores them into the database and shares the updates to all relevant modules. It consists of the following modules as shown in Figure 12.9.

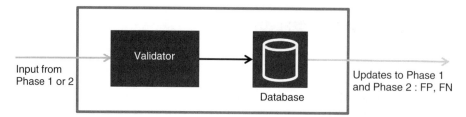

Figure 12.9 Validator and database.

The validator acts similarly to the error detection module in order to decrease False Positive (FP) and False Negative (FN) rates. If the actual result differs from the expected result, then the result is considered as error and is not registered in the database (DB). Validator should be always updated with labeled data and output from detection algorithms. The database saves all the results of detection algorithms; from each algorithm, a sample of the outcome (feedback) is sent to the database that will be used in future detection. A database contains known attacks, new attacks and dropped attacks.

12.4.7   Authentication Architecture

There are multiple types of authentication being used:

i. Browser Authentication without TLS Client Certificates
   This authentication mechanism is used for browser connections if a form of two-factor authentication is used that does not make use of personal TLS client certificates. The validation of two-factor authentication may require calling an external service.
ii. Browser Authentication with Personal TLS Client Certificates
   This form of authentication is used if personal TLS client certificates are used for two-factor authentication.
iii. Device Authentication
   This form of authentication is used for connections from the Patient App. Patient diary upload is authenticated in the same way but uses a different cloud endpoint and does not use Kafka.

12.4.7.1   Authentication Logic

Table 12.1 describes the authentication logic in more detail. Note that a specific method for implementing the validation is not mandated. One possibility is that the OAuth2 token is simply a random string and the authentication server keeps a database of valid and recently expired tokens. A token can then be validated with a simple database lookup. A session timeout can then be easily implemented with a last-access timestamp; if the timestamp is too old, the token is invalid.

If TLS-client certificates are used for clinician web app, then browser connections lacking a TLS-client certificate are rejected unless they are originate from the hospital network. In addition to the above, data encryption at rest is used in the cloud.

Table 12.1 Authentication logic.

Request type	Credentials included	Auth server validation logic	Additional validation	Side effects	Return value from Auth server
Login request from browser	• Clinician username/password • Personal TLS client certificate ID from perimeter server • Two-factor authentication credential	• Password matches the hash in user database • Two-factor authentication credentials (TLS client certificate ID) match the username	Perimeter server verifies request is originating from hospital network or has valid TLS client certificate	• Audit logging • Session timeout if access token is invalidated • Access token is bound to client certificate (TLS client certificate ID is used)	OAuth2 access token (one day validity period)
Web page request from browser	None (static web pages do not contain confidential material)	None	Perimeter server verifies request is originating from hospital network or has valid TLS client certificate	• Audit logging	None
REST API request from browser	• OAuth2 access token • Personal TLS client certificate ID from perimeter server (if client certificates are used)	• Token validity based on digital signature or database lookup • Session timeout • TLS client cert ID from perimeter server must match the bound token	Perimeter server verifies request is originating from hospital network or has valid TLS client certificate	• Audit logging • Session timeout	Internal token to specify the access rights according to request and security policy
Data upload request from patient app	• TLS client certificate ID from perimeter server • Patient-specific digital signature for data	• TLS client certificate is mapped to patient ID • TLS client certificate corresponds the generated internal token	• Perimeter server verifies client certificate authenticity • Patient ID in the data matches the patient ID returned by Auth Server • Patient ID matches the signature of the data	• Audit logging	Internal token corresponding to the patient user ID or patient ID
Authorization request from a microservice	• Internal token • Description of the requested capabilities • TLS client certificate for the requesting microservice	• Check token is valid • Grant the requested capabilities	TLS client certificate is valid	• Audit logging	User or patient details for the token if requesting microservice is entitled to them

12.4.8 Attacks on Remote Patient Monitoring Platform

Table 12.2 shows some of the potential attacks, their impact on the platform and their detection method, if applicable.

12.5 Conclusion

Replacing traditional health systems with digital health has gained global attention due to the numerous advantages for both patients and healthcare providers. Digital health reduces healthcare costs because there are fewer hospital visits for patients and home visits by nurses. It also provides patients with easy access to physicians and health centers; remote monitoring involves patients in their care plan, treatment, and self-testing via medical devices provided to them; they are more informed about their disease risk factors, their treatment progress and care plan. Remote patient monitoring also helps physicians to improve the diagnosis via continuous monitoring and for critical diseases such as stroke that require immediate detection.

In this chapter, we briefly reviewed the challenges that patient and care providers have for chronic disease in a traditional healthcare system. To overcome some of the aforementioned challenges, we introduced a remote patient monitoring platform that consists of three main parts: patient monitoring devices, cloud backend and hospitals' clinician application. The system has been implemented as a pilot project and in joint research with neurological and cardiology departments of Helsinki University Hospital (HUS). The pilot was conducted in the Uusimaa region of Finland with approximately 30 patients monitored at the same time. The duration of the pilot was three months but from an individual patient point of view it could also be shorter (approx. one week minimum, but typically might be from two weeks up to one month). The technical target of the pilot was to provide remote vital signs monitoring and patient engagement for those in post-acute stroke phase and with minimal setup or intervention required from medical personnel. This efficient and continued monitoring led to medication adjustments in 40% of the patients.

The outcome of the pilot project is summarized in Figure 12.10.

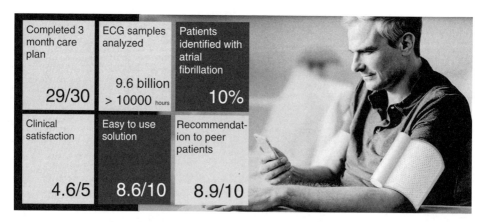

Figure 12.10 Results from pilot project.

Table 12.2 Attacks on remote patient monitoring platform.

Description	Impact	Controls
Attacker tap into connection between sensor and mobile app	• Access and modify data (sensor data, configuration data)	• Encryption
Attacker access to device	• Change the diagnosis of the other patients • Wrong treatment given for a patient • Re-configure device without admin rights, e.g. upload data to malicious server • DoS attack	• Patient-specific client certs • HADM • Blacklist client cert • Encryption with patient public key • Application to recognize rooted devices
Patient fakes data by generating ECG pulses via heart rate simulator or puts device on someone else	• Fraud (e.g. cheating on insurance or airline)	• HADM for ECG anomaly detection
Attack on mobile network	• Disclosure of patient data • Modify the system configuration • Send data somewhere else • Compromise encryption in TLS	• HADM will be installed at edge of core network to detect anomalies
Illegally create "app" that device thinks is a legal app	• Access device data • Fake data for a patient • Fake credentials • Re-configure the GW	• Application signing • If app is replaced with a fake version, it lacks the credentials for connecting to cloud
Update, reset or clear patient id	• Important data not available or to wrong person • Causing misdiagnosis	• Notification of missing data
Listen link between device and app	• Eavesdrop of data	• Bluetooth encryption
Fake or change own user ID or authorization to either app or device	• Privilege access for authorized users • Falsify access logs	• Client certificates bound to user ID • Data signing with user credentials
Buffer overflow or other vulnerability to change the SW inside device	• Tampered software	• Mobile OS security
Fake cloud, make the device to connect to a malicious server	• Access to private and sensitive data • Tamper data (e.g. remove sever alarms) • Re-configure device • DoS	• Server certificates pinned to device

(Continued)

Table 12.2 (Continued)

Description	Impact	Controls
Eavesdrop the device – cloud connection	• Access private and sensitive data	• Encryption
TLS misconfigured or broken	• Data disclosure • Loss of data integrity	• Application level encryption • Signature for all data
Malware attack on system or admin devices		• AV scans • HADM anomaly detection
Denial of service attacks on perimeter and other server in unsecure hospital environment servers	• Service unavailability	• Firewalls on every instance • Communication via TLS • Client certificates • Device certificates

And finally, we introduced some techniques to protect the proposed platform against attacks. While the majority of related studies discuss sole methods for eHealth security, our proposed digital health platform not only introduces a novel intelligent architecture to guarantee the platform and patient data security against the intrusions, but it also applies a combination of different authentication methods to protect the system from unauthorized access. The HADM applies several machine learning algorithms together with a traffic-filtering mechanism to detect anomalies on control planes such as DoS or an attack on administrative interfaces; also, to detect attacks on the user plane such as corrupting patient data (ECG and BP) and stealing patient data, etc.

References

1 Global health care sector outlook. (2017). Deloitte. https://www2.deloitte.com/content/dam/Deloitte/global/Documents/Life-Sciences-Health-Care/gx-lshc-2017-health-care-outlook-infographic.pdf (accessed 16 July 2019).

2 Global health care outlook. (2018). Deloitte. https://www2.deloitte.com/content/dam/Deloitte/global/Documents/Life-Sciences-Health-Care/gx-lshc-hc-outlook-2018.pdf (accessed 16 July 2019).

3 Hankey, G.J. (2013). The global and regional burden of stroke. *The Lancet Global Health* 1 (5): 239–240.

4 AVH in Figures. (2013). Association of brain diseases in Finland. https://www.aivoliitto.fi/files/1091/avh_lukuina2013_web.pdf (accessed 16 July 2019).

5 Ijäs, P. and Honkanen, M. (2017). Teknologia tehokkaaseen käyttöön: Stroke remote care -projekti HUS:n neurologisella osastolla, Aivosairaudet-symposium.https://www.slideshare.net/MauriHonkanen/teknologia-tehokkaaseen-kyttn-stroke-remote-care-projekti-husn-neurologisella-osastolla (accessed 16 July 2019).

6 Saleem, K., Tan, Z., and Buchanan, W. (2017). Security for cyber-physical systems in healthcare. In: *Health 4.0: How Virtualization and Big Data Are Revolutionizing Healthcare* (eds. C. Thuemmler and C. Bai), 233–251. Springer.

7 He, W., Golla, M., Padhi, R. et al. (2018). Rethinking access control and authentication for the home internet of things (IoT). In: *27th USENIX Security Symposium*. Baltimore, USA (15–17 August 2018): USENIX.

8 Chahid, I. and Marzouk, A. (2017). A secure IoT data integration in cloud storage systems using ABAC control policy. *International Journal of Advanced Engineering Research and Science* 4 (8): 34–37.

9 Maksimović, M. and Vujović, V. (2017). Internet of things based E-health systems: ideas, expectations and concerns. In: *Handbook of Large-Scale Distributed Computing in Smart Healthcare* (eds. S.U. Khan, A.Y. Zomaya and A.M. Abbas), 241–280. Springer.

10 Miller, L. (2016). Choosing the right IoT solutions. In: *IoT Security for Dummies, INSIDE Secure Edition* (ed. L. Miller), 27–37. Wiley.

11 Neto, A.L.M., Souza, A.L.F., Cunha, I. et al. (2016). AoT: authentication and access control for the entire IoT device life-cycle. In: *Proceedings of the 14th ACM Conference on Embedded Network Sensor Systems (SenSys)*. Stanford, USA (14–16 November 2016): ACM.

12 Manzoor, A. (2016). Securing device connectivity in the industrial internet of things (IoT). In: *Connectivity Frameworks for Smart Devices the Internet of Things from a Distributed Computing Perspective* (ed. M. Zaigham), 3–22. Springer.

13 Dabbagh, Y.S. and Saad, W. (2018). Authentication of Everything in the Internet of Things: Learning and Environmental Effects. http://arxiv.org/abs/1805.00969 (accessed 17 July 2019).

14 Chuang, Y., Lo, N., Yang, C., and Tang, S. (2018). A lightweight continuous authentication protocol for the internet of things. *Sensors* 18 (4): 1104.

15 Khemissa, H. and Tandjaoui, D. (2015). A lightweight authentication scheme for E-health applications in the context of internet of things. In: *9th International Conference on Next Generation Mobile Applications, Services and Technologies*, Cambridge, UK (9–11 September 2015). Cambridge, UK: IEEE.

16 Porambage, P., Braeken, A., Gurtov, A. et al. (2015). Secure end-to-end communication for constrained devices in IoT-enabled ambient assisted living systems. In: *IEEE 2nd World Forum on Internet of Things (WF-IoT)*, Milan, Italy (14–16 December). Milan, Italy: IEEE.

17 Shah, T. and Venkatesan, S. (2018). Authentication of IoT device and IoT server using secure vaults. In: *17th IEEE International Conference on Trust, Security and Privacy in Computing and Communications/12th IEEE International Conference on Big Data Science and Engineering (TrustCom/BigDataSE)*, New York, USA (1–3 August 2018). New York, NY: IEEE.

18 OAuth, 3 Common Methods of API Authentication Explained. Nordic APIs. https://nordicapis.com/3-common-methods-api-authentication-explained (accessed 17 July 2019).

19 Monshizadeh, M., Khatri, V., Atli, B., and Kantola, R. (2018). An intelligent defense and filtration platform for network traffic. In: *IFIP 16th International Conference on Wired/Wireless Internet Communications (WWIC)*, Boston, USA (18–20 June 2018). Boston: Springer.

13

Secure and Efficient Privacy-preserving Scheme in Connected Smart Grid Networks

An Braeken and Pardeep Kumar

Abstract

In a connected smart grid, several kinds of sensors have been integrated in smart meters to measure energy usage and execute the control commands and/or instructions from the utility companies. This becomes possible through the digital processing and communications organized in the smart grid (SG) networks. As, a smart grid enables the two-way communication between the electricity supplier and users, an efficient security solution should be integrated in the SG in order to guarantee the user's privacy and authenticity of the data. This chapter proposes a secure and efficient privacy preserving scheme based on elliptic curve cryptography, outperforming both in computation and communication the best currently available schemes in literature. The proposed scheme is resistant against internal attackers, provides authentication, integrity and confidentiality. In addition, it offers the possibility of pinpointing the smart meter that is providing false input. The security features of confidentiality and unforgeability are formally proven.

13.1 Introduction

The traditional power plants were concentrated systems where the energy flow was unidirectional bulk generation to domestic or industrial consumers via the traditional transmission and distribution lines. The new idea of connected smart grids is a paradigm for distributed power generation and consumption which requires a complex two-way communication infrastructure, sustaining power flows between intelligent components. SG will provide various notable features, e.g., reduce pollution, efficient use of energy [1, 2]. However, the initial steps of building a connected SG network are already in progress – many countries have started rolling out smart meters to the consumer premises. These smart meters are digital in nature and are able to collect the fine-grained consumption data within the premises. It is anticipated that the smart meter measurement data will be aggregated at regular intervals such as 15/30/45 minutes, as shown in Figure 13.1. The consumption data is then further utilized for many different purposes, for instance, better control and management, demand and response, billing, etc.

IoT Security: Advances in Authentication, First Edition.
Edited by Madhusanka Liyanage, An Braeken, Pardeep Kumar, and Mika Ylianttila.
© 2020 John Wiley & Sons Ltd. Published 2020 by John Wiley & Sons Ltd.

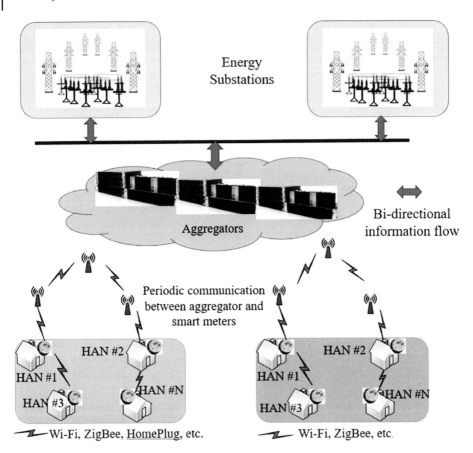

Figure 13.1 The periodic communication between smart meter and aggregator in smart grid network.

The SG network improves the efficiency, reliability, and sustainability of the traditional power-grid. However, it raises many issues [3]: (i) privacy for the individuals as the smart meter has to send consumption usages (or to be aggregated), periodically, to the power providers. It has been demonstrated that power consumption profiles at a granularity of 15/30/45 minutes may disclose whether a property is empty. Moreover, such close profiling on consumption usages may reveal the daily routines of the individuals, e.g., sleeping patterns, individual behavior patterns, appliances and multimedia used. Consequently, the power consumption usages could be used for criminal purposes. (ii) The frequent data (or consumption) aggregation issue is not only limited to the individual privacy from the home network but it can also pose security threats at the grid side network. For instance, in the SG network, the data aggregation typically relies on the digital (and open) communications and technologies. By exploiting such technologies, an attacker may counterfeit devices (e.g., unauthorised smart meters and relay nodes, e.g., puppet node [4]), to control the smart metering infrastructure in a SG network. Consequently, an attacker could lead to disruption in the smooth functioning of the energy generation, transmission, and distribution. (iii) As the energy usage value is generated by the smart meter, an energy cheater (i.e., thief) may have multiple means for data tampering. A few energy cheating methods, such as smart meter tampering, bypassing, and meter location switching, could be utilized to tamper with the consumption usage from the source of origin (i.e., smart meter). More precisely, the smart meter is only source for the data aggregation in the SG network, whether the smart meter readings are accurate or tampered, the utility company has no way to prove the authenticity and correctness of the consumption usage that is provided by the smart meter [5]. Therefore, security and privacy are major issues that need to be addressed in the SG, in order to overcome internal and external attacks and guarantee the user's privacy (i.e., confidentiality), authenticity of smart meter, and correctness of data (i.e., integrity, and detection of cheater).

13.1.1 Related Work

In the domain of the SG network, many data aggregation schemes have been proposed in recent years and each scheme has focused on attaining specific security and privacy goals. Security and privacy issues in the SG domain have been addressed mainly in two different ways, depending on who is allowed to derive the individual consumption data of each smart meter. On one hand, there are a lot of papers applying classical security techniques to protect the privacy of the individual meter data from the outsiders. Examples of schemes in the SG security apply either zero knowledge based mechanisms [6], blind signatures [7], group signatures [8, 9], or use pseudonyms [10].

On the other hand, other privacy-preserving data aggregation schemes that protect privacy not only from the outsiders, but also from insiders like the aggregator by computing the aggregated sum of the measurements. Many data aggregation schemes for the SG network security have been proposed using different types of underlying cryptographic primitives. A very large category of schemes is based on the Paillier homomorphic encryption scheme [11–15]. Also the Diffie Hellman homomorphic encryption in [16] and the El Gamal homomorphic encryption in [17] have been proposed to aggregate smart meter data securely while maintaining the privacy. However, in [18], Fan et al. pointed out that many of the earlier proposed data aggregation schemes are vulnerable to internal attacks. The first solution against the internal attackers was introduced by Fan et al. [18]. Their scheme exploited the blinding factors to protect the smart meter data. Later, Bao-Li commented that Fan et al.'s scheme can violate the data integrity [19] and therefore cannot be practical for the SG networks. In [20], and [21], the authors utilized the Boneh, Goh, and Nissim (BGN) method to aggregate the smart meters data in privacy-preserving manners enabling integrity as well. Other schemes, e.g., [23], and [24], applied bilinear pairings combined with the point to hash operations in order to collect data from the smart meters and provided security services, e.g., authentication, confidentiality and integrity.

Indeed, most of current state-of-the-art schemes (e.g., [18, 20–22]), considered the classical security properties (confidentiality, authentication and integrity). However, when the integrity check on the aggregated value is incorrect, the whole procedure needs to be repeated without knowing the originator of the resulting error. Therefore, for efficiency reasons, it is very important to identify the SM that is not providing correct information, in case of problems. Another problem with most of aforementioned mechanisms is that they are very computationally demanding, which is not evident at the side of the smart meters, those with limited computational and communicational resources. Recently in [22], a data aggregation scheme solely based on elliptic curve operations has been proposed that requires less resources at the smart meter but leave out the possibility of finding the potential cheater in case of an incorrect integrity check. We will show in this paper how to develop another ECC-based scheme, which is more performant and offers more security features at the same time.

Finally, many of the proposed schemes in the literature require the existence of a secure channel between the smart meters on one hand and the trusted third party (TTP) on the other hand during meter registration. As shown in Figure 13.2 (refer to the left hand side), the registration phase of the schemes proposed in [18, 21, 22] require a secure communication between the smart meter and the aggregator – the red lines in the communication correspond with a secure channel while the green lines represent communication over the open channel. However, this requirement is sometimes difficult to establish, especially for updating the security material (e.g., if an SM is in a remote place, then the Internet connectivity is not always good).

13.1.2 Our Contributions

From the aforementioned reasons, the main contributions of this chapter can be summarized as:

- We propose the secure and efficient privacy-preserving scheme, which is based on elliptic curve cryptography (ECC).The Elliptic Curve Qu Vanstone (ECQV) implicit certificates have been utilized in the proposed scheme. Moreover, the proposed scheme not only considers the classical security features (e.g., authentication, session key establishment, confidentiality, integrity) but can also identify the malicious smart meter.
  Unlike state-of-the-art protocols ([18, 21, 22]), no secure channel is required during the registration of the smart meter in the proposed scheme, as shown in Figure 13.2 (refer to the right hand side).
- We formally prove the security of the proposed scheme and discuss the different security features in them.

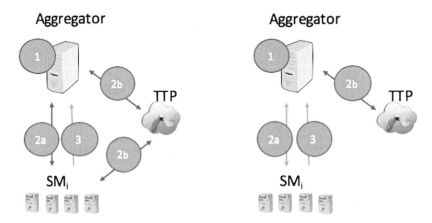

Figure 13.2 The different architectures and communication flows are presented, the one of [18, 21, 22] on the left and the one for the proposed scheme on the right.

- We compare our scheme's efficiency, addressing both computational and communication aspects, with recently proposed privacy-preserving schemes.

13.1.3 Structure of Chapter

The chapter is organized as follows. Section 13.2 deals with the preliminaries. Section 13.3 describes the proposed scheme. In Section 13.4, we discuss the security analysis of the proposed scheme. The computational complexity and communication cost are explained in Section 13.5. Finally, we end with conclusions in Section 13.6.

13.2 Preliminaries

13.2.1 System Model

We adopted the system model from the literature, e.g., [18, 21, 22] and the main participants are as follows:

- Trusted third party (TTP). This entity is considered to be a honest but curious entity. This means that it will execute the required processes, but will try to abuse information for own purposes. The main task of the TTP is the distribution of the blinding factor to the aggregator.
- The aggregator (Agg). This party is responsible for the management of the smart meters in a certain region. This includes, among others, the analysis of the power quality, the real-time maintenance and the real-time pricing. Only the sum of the meter's data is required for proper management of the SG. The Agg is assumed to be powerful and to possess sufficient tamper-proof storage for storing the secret shared keys with the SMs. The Agg generates and publishes the system parameters for the security scheme. It also publishes all the public information about the smart meters under its responsibility. The Agg is also considered to be a honest but curious party.
- Smart meter (SM). The smart meter is a resource-constrained device, which will send, at each fixed period, the real-time measurements of the electricity consumption (can also involve gas and water measurements) to the Agg. The communication to the Agg is through recent low power and long-range technology, like LoRA, Sigfox, etc. The private key is only known by the SM. We assume that the key material is stored in a tamper-resistant module of it.

13.2.2 Security Requirements

Our scheme satisfies the security features required by a privacy-preserving data aggregation scheme, as discussed in [21]. We also add the non-repudiation and public-verifiability feature. These security requirements are summarized below.

- Confidentiality. It is of utmost importance that an adversary is unable to gain any information on the usage data of any single customer throughout the communication from the smart meter to the Agg. This information could be abused by an attacker to derive information on the user's behavior, for instance, to learn if somebody is at home or not.

- Integrity. In order to avoid data being corrupted or changed by an adversary during the transmission, the data integrity should be checked.

  For some schemes in the literature (e.g., [22]), the integrity check is only possible for the global sum and not for the individual measurement data. For other schemes (e.g., schemes with underlying homomorphic encryption algorithm [21, 24], the integrity check can be organized at an individual level but requires an enormous overhead due to the high computational cost of the used cryptographic mechanisms. In any case, it is not proposed as a standard step in the schemes as it is too computationally demanding, even for the Agg.

- Authentication. This feature ensures that the transmitted data is coming from a registered smart meter. It avoids impersonation and man-in-the-middle attacks.

- Non-repudiation and public verifiability. These features enable everybody to check that the message sent by the legal smart meter cannot be denied.

The scheme should be able to resist classical attacks such as replay, modification and man-in-the middle attacks. In these types of attack, adversaries can eavesdrop on the traffic in order to replay and change messages, inject new messages, or spoof other identities in the communication messages sent between the smart meters and aggregator. Another important class of attacks are the impersonation attacks in which an adversary may impersonate one or a group of SMs and send false data on behalf of them to the Agg.

We assume the TTP and Agg to be honest but curious entities. In particular, the TTP and Agg should not be able to derive individual consumption data of the users. We do not consider that they collaborate.

13.2.3 Cryptographic Operations and Notations

The scheme is based on ECC [25], allowing lightweight public key cryptographic solutions. For instance, corresponding with an 80 bit security parameter, a field size of 160 bits for ECC is sufficient, while RSA-based solutions require 1024 bits. ECC is based on the algebraic structure of elliptic curves (ECs) over finite fields. We denote the curve in the finite field F_p by $E_{p(a,b)}$, defined by the equation $y^2 = x^3 + ax + b$ with a and b two constants in F_p and $\Delta = 4a^3 + 27b^2 \neq 0$. We denote by P the base-point generator of $E_{p(a,b)}$ of prime order q. All points on $E_{p(a,b)}$, together with the infinite point form an additive group. In [26, 27] standardized curve parameters are described.

The product $R = rP = (R_x, R_y)$ with $r \in F_q$ and $R_x, R_y \in F_p$ results in a point of the EC and represents an EC multiplication. When we send an EC point, it is sufficient to send its x coordinate. The scheme relies on two computational hard problems.

- The Elliptic Curve Discrete Logarithm Problem (ECDLP). This problem states that given two EC points R and Q of $E_{p(a,b)}$), it is computationally hard for any polynomial-time bounded algorithm to determine a parameter $x \in F_q^*$, such that $Q = xR$.

- The Elliptic Curve Diffie Hellman Problem (ECDHP). Given two EC points $R = sP, Q = yP$ with two unknown parameters $s, y \in F_q^*$, it is computationally hard for any polynomial-time bounded algorithm to determine the EC point syP.

In addition, we denote a one-way cryptographic hash function (e.g., SHA2 or SHA3) with output a number in F_q^* by $H()$. Our scheme will utilize five different hash functions

$H_i : \{0,1\}^* \to F_q^*, (i = 1, \ldots, 5)$. Also a symmetric key encryption algorithm $E_K(M)$ will be utilized to perform encryption of M using the secret shared key K. The concatenation of two messages M_1 and M_2 is denoted by $M_1 \| M_2$. We further assume that the EC parameters and the associated EC operations, together with the hash function are implemented in each entity participating the scheme.

13.3 Proposed Scheme

The proposed scheme consists of three phases, initialization phase, smart meter registration phase, and secure communication phase between smart meter and aggregator. The detailed description of each phase is as follows:

13.3.1 Initialisation Phase

Here, the Agg performs the following steps:

- It selects an EC $E_{p(a,b)}$ and its generator point P, together with the hash function $H(.)$.
- It generates a random value α as the private key. Next, the Agg computes its public key $P_\alpha = \alpha P$.
- It selects an initial timestamp T_0 and its associated time interval I. The time interval is typically around 15 minutes. Denote the timestamp $T_t = T_0 + tI, t \geq 1$.

The system parameters, $E_{p(a,b)}, P, P_\alpha, H_1(.), \ldots, H_5(.), T_0, I$, together with the public key P_{TTP} of the TTP are published. The secret key α securely stored in a tamper-resistant memory.

13.3.2 Smart Meter Registration Phase

The registration phase consists of two steps, as follows:

- Registration of a smart meter (SM) with Agg.
  This registration follows the ECQV scheme and works as follows. First each smart meter (e.g., SM_i) with identity ID_i chooses a random value $r_i \in F_q^*$ and computes $R_i = r_i P$. Denote the total number of smart meters with n, then $i \in \{1, \ldots, n\}$. The values (ID_i, R_i) are sent to the Agg. Upon receiving the registration request from the smart meter, the Agg stores the identity related information, and then it selects a random value $r_a \in F_q^*$ and computes $R_a = r_a P$. Next, it computes

$$cert_i = R_i + R_a$$
$$r = H_1(cert_i \| ID_i)r_a + \alpha$$

The values $(cert_i, r)$ are sent to the smart meter. The SM_i now computes its private key

$$d_i = H_1(cert_i \| ID_i)r_i + r$$

It accepts the registration if its public key P_i satisfies the following equality

$$P_i = d_i P = H_1(cert_i \| ID_i)cert_i + \alpha P \tag{13.1}$$

If Eq. 13.1 holds, the SM also computes the following

$$b_i = H_2(r_i)d_i$$
$$H_i = b_i P_\alpha$$
$$E_i = E_{B_{ST}}(b_i)$$
$$B_{ST} = d_i P_{TTP}, \qquad S_i = d_i P_\alpha$$

Here, b_i will be used as a blinding factor, H_i as verification for the blinding factor by the Agg, and S_i is the secret value shared with the Agg. In order to securely transmit b_i to the TTP, it will use the common shared secret key B_{ST} as symmetric encryption key in E_i. The SM stores, besides the system parameters, the following values $\{ID_i, cert_i, P_i, S_i, b_i, d_i\}$. Especially the last three values are stored in tamper-resistant memory of the SM.

In addition, a confirmation message containing the parameters $ID_i, cert_i, H_i, E_i, T_t, H_3$ $(S_i\|ID_i\|cert_i\|H_i\|E_i\| T_t)$ is sent by SM_i to the Agg, where T_t denotes the current timestamp. Upon arrival of this message, the Agg constructs P_i by using Eq. 13.1 and $ID_i, cert_i$ from the received message. Next $S_i^* = d_\alpha P_i$ is computed using its private key d_α. Finally, the Agg checks if the last part of the received message equals to $H_3(S_i^*\|ID_i\|cert_i\|H_i\|E_i\|T_t)$. If so, $S_i^* = S_i$, and the SM_i with parameters $\{ID_i, cert_i, P_i, S_i, H_i, E_i\}$ is added to the list of active SMs in its region. Note that the parameter S_i is stored in tamper-resistant memory. The last parameter E_i can be removed once the Agg has registered the SMs with the TTP.

- Construction of blinding factor by the TTP for the Agg. After the total number of SMs (say n) is registered with the Agg, the Agg sends to the TTP the list of $\{ID_i, cert_i, H_i, E_i\}_{1 \le i \le n}$. Based on this list, the TTP can compute the corresponding public keys P_i of each SM_i by Eq. 13.1. Denote the secret key of the TTP by d_{TTP}. As a consequence, the TTP can derive $B_{ST} = d_{TTP}P_i$ and b_i by decryption of E_i. Next, it can verify the validity of b_i, by checking if $b_i P_\alpha = H_i$. If so, the SM is validated and the resulting blinding factor B for the agg is then determined by

$$B = -\sum_{i=1}^{n} b_i \bmod q \tag{13.2}$$

The value B is securely sent to the Agg through the secure channel.

13.3.3 Secure Communication Between Smart Meter and Aggregator

This phase invokes at fixed timestamps $T_t = T_0 + tI$, with $t \ge 1$. During the data aggregation phase, each SM first extracts its consumption data cm_i (e.g., electricity, gas, water, etc.). This value will, in practice, be rather small (e.g., smaller than 32 bits), as it corresponds with the consumption over a short period of time. However, in order to increase the security of the proposed scheme, we will extend cm_i for the other remaining $q - 32$ bits with a random value r_{q-32}. Consequently, let us denote $m_i = (r_{q-32}\|cm_i)$. Choose a random $y_i \in F_q^*$. Next, the SM computes the following parameters:

$$Y_i = y_i P$$
$$\mu_i = H_4(ID_i\|cert_i\|S_i\|T_t\|y_i P_\alpha)(m_i + tb_i)$$
$$c_i = m_i + b_i + H_5(ID_i\|cert_i\|S_i\|\mu_i\|T_t)$$

The message (c_i, Y_i, μ_i, ID_i) is sent by SM_i to the Agg. The Agg now collects n tuples (c_i, Y_i, μ_i, ID_i) at time T_t from each registered SM_i. First, the Agg computes for each tuple $d_a Y_i$ and temporarily adds this to the stored list of identity-related information of the SMs, $\{ID_i, cert_i, P_i, S_i, H_i\}$. Next, the Agg performs the following two operations:

$$\sum_{i=1}^{n} c_i + B - \sum_{i=1}^{n} H_5(ID_i \| cert_i \| S_i \| \mu_i \| T_t) = \sum_{i=1}^{n} m_i$$

$$\sum_{i=1}^{n} H_4(ID_i \| cert_i \| S_i \| T_t \| d_a Y_i)^{-1} \mu_i P_\alpha$$

$$-t \sum_{i=1}^{n} H_i == \left(\sum_{i=1}^{n} m_i \right) P_\alpha \tag{13.3}$$

Only the last 32 bits of $\sum_{i=1}^{n} m_i$ are taken into account, which corresponds to the actual sum of the consumption data $\sum_{i=1}^{n} cm_i$. Note that the first equation in 13.3 holds due to the construction of the blinding factor of the Agg (cf. Eq. 13.2). The first equation contributes to the confidentiality and authentication feature, while the second equality in Eq. 13.3 to the integrity feature.

Moreover, suppose that if there is an issue with the integrity check on the total sum (second equality in Eq. 13.3), then the individual messages (c_i, Y_i, μ_i, ID_i) of each SM_i will be analysed in depth. For the Agg, it is sufficient to verify if m_i is incorporated in a correct way in the parameters c_i, μ_i, by checking the following equality.

$$(c_i - H_5(ID_i \| cert_i \| S_i \| \mu_i \| T_t)) P_\alpha - H_i$$
$$= H_4(ID_i \| cert_i \| S_i \| T_t \| d_a Y_i)^{-1} \mu_i P_\alpha - tH_i$$

which can be simplified by

$$(c_i - H_5(ID_i \| cert_i \| S_i \| \mu_i \| T_t)) P_\alpha + (t-1) H_i$$
$$= H_4(ID_i \| cert_i \| S_i \| T_t \| d_a Y_i)^{-1} \mu_i P_\alpha \tag{13.4}$$

If this is not the case, the tuple (c_i, Y_i, μ_i, ID_i) will be rejected.

13.4 Security Analysis

We first give a formal proof of the confidentiality and unforgeability of the proposed protocol using the security model proposed in [22], which have been based on mechanisms defined for signcryption schemes [28]. Next, the strength of the proposed scheme will be informally analyzed with respect to the different security features as discussed in Section 13.2.C.

13.4.1 Formal Proof

In the security model, two games are defined to be executed by a Challenger, denoted by $\mathscr{C}$, and an Attacker, denoted by $\mathscr{A}$. Five queries of $\mathscr{A}$ to $\mathscr{C}$ are allowed.

1) $H_i(m)$. In this query, $\mathscr{C}$ randomly selects a value $r \in F_q^*$, sends r as response to $\mathscr{C}$ and stores in L_{H_i} the tuple (m, r) with $i = 1, \ldots, 5$.

2) CreateSM(ID_i). Here, $\mathscr{C}$ generates a private key, blinding factor, and certificate, which are used to derive the public key, and the secret shared key with the Agg. The tuple $(ID_i, d_i, cert_i, P_i, b_i, S_i)$ is stored in the table L_{SM}.

3) CorruptSM(ID_i). In this query, $\mathscr{C}$ sends the secret parameters of the SM, being the private secret key, blinding factor and secret shared key with the Agg, to $\mathscr{A}$.

4) Signcrypt(ID_i, m_i). For this query, $\mathscr{C}$ creates a ciphertext c_i, Y_i, μ_i, ID_i, taking as input ID_i, m_i.

5) Unsigncrypt(c_i, Y_i, μ_i, ID_i). Here, $\mathscr{C}$ checks the validity of the ciphertext and decrypts it to get the message m_i.

We now distinguish two games and their corresponding definitions related to security.

13.4.1.1 Game 1

In this game, $\mathscr{C}$ first produces the system parameters and sends them to $\mathscr{A}$. Next, $\mathscr{A}$ selects a challenging identity ID_i^*, chooses two messages m_0 and m_1 and sends them to $\mathscr{C}$. Then, $\mathscr{C}$ picks a random value $b \in \{0,1\}$ to select one of the messages for which a signcrypted message $(c_i, Y_i, \mu_i, ID_i^*)$ is generated and sent to $\mathscr{A}$. Finally, using the five queries (except the Unsigncrypt with identity ID_i^*) defined above, $\mathscr{A}$ guesses the value of b in order to distinguish which message has been signcrypted.

Definition 1. A data aggregation scheme provides confidentiality (also called indistinguishability against adaptive chosen ciphertext attacks - IND-CCA) if an attacker is not able to win Game 1 with a non-negligible advantage, i.e., to guess b' as the correct value of b. The advantage of $\mathscr{A}$ is defined by

$$Adv_{\mathscr{A}}^{IND-CCA} = 2\|Pr(b = b') - 1\|$$

13.4.1.2 Game 2

In this game, $\mathscr{C}$ first produces the system parameters and sends them to $\mathscr{A}$. Next, $\mathscr{A}$ selects a challenging identity ID_i^* and outputs a ciphertext $(c_i, Y_i, \mu_i, ID_i^*)$ corresponding with the challenging identity ID_i^*, by using the five queries defined above (except the CorruptSM with identity ID_i^*).

Definition 2. A data aggregation scheme provides unforgeability (also called existential forgeability against adaptive chosen message attacks - EUFCMA) if no attacker is able to win Game 2 with a non-negligible advantage, i.e. if $\mathscr{A}$ is not able to make a valid ciphertext without usage of the Signcrypt query.

Theorem 13.1 The proposed scheme is able to provide confidentiality if the ECDHP is hard. Proof.

Proof: We will prove that if $\mathscr{A}$ succeeds to win the game with non-negligible advantage ϵ, then also $\mathscr{C}$ will be able to solve the ECDHP with non-negligible advantage, which is a contradiction against the hardness of the problem.

Consequently, we consider an instance of the ECDHP, being $R = sP, Q = yP$. First, $\mathscr{C}$ sets $P_\alpha = R$ and publishes it together with the system parameters $E_{p(a,b)}, P, H_1(.)$, .., $H_5(.), T_0, I, P_{TTP}$. Then, $\mathscr{A}$ randomly selects an identity ID_I, timeslot t, and two messages m_0, m_1. Next, $\mathscr{C}$ picks a random element $b = 0, 1$ to determine which message will be signcrypted. Then, it extracts the identity related information with ID_I of the table L_{SM}. The variable Y_i used in the derivation of the ciphertext is set to Q by $\mathcal{C}$. Then, $\mathcal{C}$ randomly selects the parameters c_i, μ_i and stores the corresponding values in L_{H_4}, L_{H_5}. Finally, $\mathcal{C}$ outputs (c_i, Y_i, μ_i, ID_i) to $\mathcal{A}$.

Now, $\mathcal{A}$ can proceed with the previously defined queries, which are answered by $\mathcal{C}$ as follows.

1) $H_i(m)$. For each hash function H_i, a table L_{h_i} with tuples (m, r) is stored by $\mathcal{C}$ containing the input m and output r of the hash function $H_i()$, $(i = 1, ..., 5)$. If the tuple does not exist in the current list, a new entry is generated. $\mathcal{A}$ has access to the tables.
2) CreateSM(ID_i). If ID_i is not yet in the table L_{SM}, $\mathcal{C}$ randomly selects d_i, b_i and $\alpha_{1i} = H_1(cert_i \| ID_i)$. Next, it computes $cert_i = (d_iP - P_a)(\alpha_{1i})^{-1}$. It further computes P_i and S_i. The tuple $(ID_i, d_i, cert_i, P_i, b_i, S_i)$ is added to the table L_{SM}.
3) CorruptSM(ID_i). If L_{SM} does not contain an entry corresponding with ID_i, $\mathcal{C}$ makes a CreateSM(ID_i) query.
4) Signcrypt(ID_i, m_i). For this query, $\mathcal{C}$ first checks if there is an entry in L_{SM} related to ID_i. If not, it first makes a CreateSM(ID_i) query. Using this info, $\mathcal{C}$ outputs a ciphertext c_i, Y_i, μ_i, ID_i.

After applying these queries, $\mathcal{A}$ decides on the value of b. If $\mathcal{A}$ is able to win the game with non-negligible advantage, then $\mathcal{C}$ can also solve the ECDLP as it needs to find the tuple $(R, \mu_i(m_i + tb_i)^{-1})$ from L_{H_4}, resulting in R (last part) as the solution of the considered problem.

If q_{H_4} corresponds with the number of H_4 queries, the probability that $\mathcal{C}$ can solve the ECDHP equals to $\frac{\epsilon}{q_{H_4}}$. Consequently, this leads to a contradiction and we can conclude that the proposed scheme provides confidentiality.

Theorem 13.2 The proposed scheme is able to provide unforgeability if the ECDLP is hard.

Proof: We proof with contradiction that if $\mathcal{A}$ is able to win the game, then $\mathcal{C}$ is able to solve the ECDLP. Let $Q = xP$ and the derivation of x be the instance of the ECDLP that we will consider. First, $\mathcal{C}$ randomly picks an integer α, computes P_α and determines the system parameters $E_{p(a,b)}, P, H_1(.), ..., H_5(.), T_0, I, P_{TTP}$. Then $\mathcal{A}$ chooses the challenging identity ID_I and $\mathcal{C}$ answers the queries of $\mathcal{A}$ as follows:

1) $H_i(m)$. For each hash function H_i, a table L_{h_i} with tuples (m, r) is stored by $\mathcal{C}$ containing the input m and output r of the hash function $H_i()$, $(i = 1, ..., 5)$. If the tuple does not exist in the current list, a new entry is generated. $\mathcal{A}$ has access to the tables.
2) CreateSM(ID_i). If ID_i is not yet in the table L_{SM}, we distinguish two situations. If $ID_i = ID_I$, then $P_i = Q$ and random values are taken for $cert_i, b_i, S_i$. The tuple $(ID_i, \perp, cert_i, P_i, b_i, S_i)$ is added to the table L_{SM}. In the situation $ID_i \neq ID_I$, $\mathcal{C}$ randomly selects d_i, b_i and $\alpha_{1i} = H_1(cert_i \| ID_i)$. Next, it computes $cert_i = (d_iP - P_\alpha)(\alpha_{1i})^{-1}$. It further computes P_i and S_i. The tuple $(ID_i, d_i, cert_i, P_i, b_i, S_i)$ is added to the table L_{SM}.
3) CorruptSM(ID_i). If L_{SM} does not contain an entry corresponding with ID_i, $\mathcal{C}$ makes a CreateSM(ID_i) query.
4) Signcrypt(ID_i, m_i). This query can only be made if $ID_i \neq ID_I$. In such situation, $\mathcal{C}$ first checks if there is an entry in L_{SM} related to ID_i. If not, it first makes a CreateSM(ID_i) query. Using this info, $\mathcal{C}$ outputs a ciphertext c_i, Y_i, μ_i, ID_i.

5) Unsigncrypt(c_i, Y_i, μ_i, ID_i). Here, $\mathscr{C}$ checks the validity of the ciphertext and decrypts it to get the message m_i.

As a result of the game, $\mathscr{A}$ outputs a ciphertext (c_i, Y_i, μ_i, ID_i). If $\mathscr{A}$ is able to generate a valid ciphertext, then we show that $\mathscr{C}$ will be able to solve the ECDLP. Using the forking lemma of [29], $\mathscr{C}$ is able to construct another valid ciphertext $(c_i^*, Y_i, \mu_i, ID_i)$ by choosing a different Hash function H_2. Take into account that $b_i = H_2(r_i)d_i$ This leads to the following two equations:

$$c_i P = m_i P + H_2(r_i)d_i P + H_5(ID_i \| cert_i \| S_i \| \mu_i \| T_t)P$$
$$c_i^* P = m_i P + H_2(r_i)^* d_i P + H_5(ID_i \| cert_i \| S_i \| \mu_i \| T_t)P$$

Subtracting both equations, leads to the following equality

$$(c_i - c_i^*)P = (H_2(r_i) - H_2(r_i)^*)d_i P$$

and thus $(c_i - c_i^*)(H_2(r_i) - H_2(r_i)^*)^{-1}$ as the solution of the ECDLP challenge.

Denote the size of L_{H_2} by q_{h_2}. To compute the hardness of this challenge, the probability is equal to the probability that a different hash value can be chosen $1/q_{h_2}$ times the probability ϵ that $\mathscr{A}$ is able to win the game, resulting in $\frac{\epsilon}{q_{h_2}}$. Consequently if ϵ is non-negligible, this ECDLP challenge too, which is a contradiction.

13.4.2 Informal Discussion

We also informally show how the proposed scheme addresses the security features, as discussed in 13.2.B.

13.4.2.1 Confidentiality

In our scheme, the individual consumption data m_i of the SMs cannot be revealed by any attacker. This follows from the fact that given the message tuple (c_i, Y_i, μ_i), an adversary either needs to solve the ECDLP or needs to know both b_i and S_i. The first parameter is uniquely shared with the TTP and the second parameter with the Agg and its security is based on the ECDLP.

Even if one (or a group of) SMs would collaborate, they will never find the value b_i due to construction of these parameters, as either the private key of the TTP or the involved SM should be known. Moreover, also the secret key S_i cannot be found as it would require the knowledge of the private key of the Agg or the involved SM.

Also the Agg on its own is not able to derive the value m_i as it would need to know b_i. Due to the ECQV construction of the private key of the SM, whose security is also based on the ECDLP, the Agg is not able to derive the private key of the SM and thus derive the blinding factor b_i.

Finally, the TTP is not involved in the construction of the private key of the SM. Consequently, the TTP is not able to derive the secret key S_i shared with the Agg.

13.4.2.2 Authentication

In the proposed scheme, the authentication follows from the integration of the terms b_i and $H_5(ID_i \| cert_i \| S_i \| \mu_i \| T_t)$ in the computation of c_i. Only a valid user with knowledge of the private key of the involved SM_i is able to construct both parameters b_i, S_i, due to the ECDLP. If wrong values for either b_i or S_i are used, the second equality of Eq. 13.3 does not hold anymore.

13.4.2.3 Integrity

The integrity of m_i is assured by the second parameter μ_i. This follows from the associativity property of EC addition, i.e. $\sum_{i=1}^n m_i P_\alpha = (\sum_{i=1}^n m_i) P_\alpha$.

In order to harm the message integrity in the proposed scheme, suppose an attacker (outsider, one or group of legitimated SMs) wants to add/decrease a certain amount of consumption m_i^* to/from m_i for a particular $i \in \{1, \dots, n\}$, thus changing c_i into $c_i + m_i^*$, it will lead to

$$\sum_{i=1}^n c_i + B - \sum_{i=1}^n H_5(ID_i \| cert_i \| S_i \| \mu_i \| T_t) = \sum_{i=1}^n m_i + m_i^*$$

Consequently if $\sum_{i=1}^n H_4(ID_i \| cert_i \| S_i \| T_t \| d_\alpha Y_i)^{-1} \mu_i$ will be computed, it will not be equal anymore to $(\sum_{i=1}^n m_i) P_\alpha + t \sum_{i=1}^n H_i$ as there is an additional term $m_i^* P_\alpha$ left. Thus, without being able to solve the ECDLP, it is impossible to alter the message during transmission.

13.4.2.4 Identification of the Cheater

We consider a SM to cheat if a wrong combination of (c_i, Y_i, μ_i) is sent to the Agg. Since the Agg is aware of H_i, this fact is checked through Eq. 13.4.

13.4.2.5 Resistance Against an Honest TTP and Agg

As explained before, in order to derive m_i from c_i, the combined knowledge of b_i and S_i is needed. Since S_i is only shared with SM_i and the Agg, b_i with the SM_i and the TTP, both are not able to reveal m_i on their own, without solving the ECDLP.

However, we must note that the scheme is vulnerable for a replay attack of the Agg, since he can successfully replay a message c_i at a later timestamp by replacing c_i to

$$c_i - H_5(ID_i \| certi_i \| S_i \| \mu_i \| T_t) + H_5(ID_i \| certi_i \| S_i \| \mu_i \| T_{t'})$$

This replacement will have no impact on the other values of the transmitted message (μ_i, Y_i, ID_i), as long as only Eq. 13.3 is verified. However, such an attack does not make sense in the currently proposed architecture, as it would fool himself. In case the Agg needs to further forward the information to another party for verification, it could hide in some of the remaining $q - 32$ bits of m some information of the blinding factor instead of a random value.

Table 13.1 compares the security features of the proposed scheme with the most relevant and recent privacy preserving schemes described in literature [18, 21, 22]. Note that the schemes of [18, 21, 22] do not offer the possibility to derive the identity of the SM sending false input. Except [18] provides the possibility to derive the identity of the SM sending false input, however requiring a significant higher amount of computations. Moreover, this functionality is not included in the description of the scheme.

13.4.2.6 Final Note

We also want to mention that with the cost of one additional parameter e_i in the report of the SM (c_i, Y_i, μ_i, ID_i), non repudiation and public verifiability can be obtained. These

Table 13.1 Comparison of security features

Feature	[18]	[21]	[22]	Proposed Scheme
Confidentiality	Yes	Yes	Yes	Yes
Authentication	No	Yes	Yes	Yes
Integrity	No	Yes	Yes	Yes
Identificiation of the cheater	No	No	No	Yes
No need for secure channel	No	No	No	Yes

features avoid that a smart meter can deny any performed malicious actions. We also need two other Hash functions H_6, H_7. This parameter can then be defined as

$$e_i = d_i H_6(ID_i \| cert_i \| c_i \| Y_i \| \mu_i \| T_t) + c_i y_i H_7(ID_i \| cert_i \| \mu_i \| T_t)$$

Only public key or known values are required to perform the check:

$$e_i P = H_6(ID_i \| cert_i \| c_i \| Y_i \| \mu_i \| T_t) P_i + c_i H_7(ID_i \| cert_i \| \mu_i \| T_t) Y_i$$

13.5 Performance Analysis

In this section, we analyse the performance of our scheme and compare it with the most relevant and recent privacy preserving schemes described in literature [18, 21, 22]. Note that [21] uses the BGN encryption, and [18] uses the bilinear pairing operation, whereas [22] is limited to elliptic curve operations like in our scheme. We will discuss both the computational and communication costs.

13.5.1 Computation Costs

For a fair comparison, we compare the schemes under the same security level, corresponding to a security parameter of 80 as proposed in [22]. This means that for the BGN encryption scheme of [21], two 512 bit prime numbers are applied, for the bilinear paring scheme of [18] a Type A elliptic curve with 512 and 160 bits is used for the computations, and for the ECC operations an EC with 160 bits is assumed.

Table 3 in [22] summarises the runtime results of the operations involved in the different schemes. They are obtained by using the cryptographic library MIRACL and implemented on an Intel I5-3210 2.50GHz Center Processor Unit. Note that the numbers are representative for estimating the timings at the side of the Agg. However, the timings at the side of the SM are smaller, due to the constrained nature of these devices. Nevertheless, as we are mainly interested in the comparison, we directly adopted these numbers for the complexity analysis at the SM's side. We compare the schemes of [18, 21, 22] with our proposed scheme.

Table 13.2 Comparison of the performance, runtime in μs

Scheme	Runtime SM	Runtime Agg
Fan et al. [18]	35.254	$32.909n + 28.225$
He et al. [21]	20.145	$6.264n + 48.249$
He et al. [22]	1.974	$2.969n + 1.972$
Proposed scheme	1.974	Best: $(0.992)n + 2.958$
		Worst: $3.954n + 2.958$

The output of the SM in the data aggregation phase equals to (c_i, Y_i, μ_i, ID_i). To construct this message by the SM, 2 EC multiplication and 2 hash operations need to be executed. The processing of this message by the Agg requires in the best case, meaning that the integrity check of the aggregated sum is positive, $n + 3$ EC multiplications, n EC additions and $2n$ hashes. However, if the integrity check is negative and the SM providing false input should be detected, 3 additional EC multiplications and 1 EC addition are needed for each SM, resulting in a total of $4n + 3$ EC multiplications $2n$ EC additions and $2n$ hashes. This corresponds with the worst case scenario.

Note that for the other ECC based scheme [22], at the side of the SM a total of 2 EC multiplications and hashes is required, while the Agg needs to execute $3n + 2$ EC multiplications, $2n$ EC additions and $3n$ hashes.

Table 13.2 compares the performance at the side of SM and Agg of the proposed scheme with the ones of [18, 21, 22]. As can be concluded, the schemes applying EC operations (the proposed scheme and [22]) highly outperform the other schemes. This follows from the fact that BGN encryption and bilinear pairing operations have a considerable higher computational complexity. The complexity of the proposed and the scheme proposed by [22] is similar at the side of the SM. However, at the side of the Agg in the best case, there is an improvement of factor 3 compared to [22] for a large number of SMs. Even in the worst case, where each individual message of the SM should be checked, our scheme is still in similar range as [22].

13.5.2 Communication Costs

This subsection analyses the number of messages and their length in total, sent in the aggregation phase by the SM. We assume that the length of the identity equals to 32 bits and for the matter of equivalence that in all schemes the timestamp is known and should not be explicitly send anymore. Table 13.3 describes the comparison of the communication costs for the different schemes. In [18], the SM sends three messages with a total of 260 bytes, and provides less security services. He et al.'s scheme [21] needs four messages in the communication, corresponding to 280 bytes, which is still more expensive than [18] and provides only confidentiality, authentication and integrity. Whereas the scheme proposed in [22] requires five messages to be transmitted with a total of 84 bytes and provides similar security services as in [21]. Finally, the proposed scheme needs four messages, resulting to 64 bytes, as shown in Table 13.3 and provides more security services than [18, 21, 22].

Table 13.3 Comparison of the communication costs

Scheme	Messages	Size (in Bytes)
Fan et al. [18]	3	260
He et al. [21]	4	280
He et al. [22]	5	84
Proposed Scheme	4	64

This difference in communication cost for one individual message has a huge impact in the total traffic volume gain (i.e., overhead) at the aggregator. For the sake of example purposes, consider a virtual smart village, where each aggregator serves N number of consumers (i.e., SM). Let each SM generate a message (i.e., power consumption report) every 15 minutes and send to the Agg. The total volume of messages that requires to be verified every 15 minutes by the Agg will be significantly high. If the packet size equals to p then the communication overhead at Agg is $N \times p$. Consequently, the proposed scheme is able to decrease this overhead with $24\%, 75\%$ and 77%, compared to [21, 22], and [18] respectively.

13.6 Conclusions

The new idea of connected smart grids is a paradigm for distributed power generation and consumption which requires a complex two-way communication infrastructure, sustaining power flows between intelligent components. However, this bi-directional communication raises many security and privacy challenges This chapter presents the secure and efficient privacy-preserving scheme that can be applied in a connected smart grid context, offering the best results from a computation and communication point of view.

Our proposed scheme also adds additional security features compared to the other state of the art schemes from the literature, like the possibility to detect the malicious SM contributing to the aggregated sum and the avoidance of a secure channel between SM and TTP.

References

1 Statistical office of the European Union (2016). Shedding light on energy in the EU: A guided tour of energy statistics, Technical Reports. http://ec.europa.eu/eurostat/cache/infographs/energy/ (accessed 25 June 2019).

2 Act of Congress, Energy Independence and Security Act of 2007, Pub.L. 110–140. https://www.gpo.gov/fdsys/pkg/PLAW-110publ140/html/PLAW-110publ140.htm (accessed 25 June 2019).

3 Ferrag, M.A., Maglaras, L.A., Janicke, H. and Jiang, J. (2016). A survey on privacy-preserving schemes for smart grid communications, *ARxIV:1611.07722V1* 1–32.

4 Yi, P., Zhu, T., Zhang, Q. et al. (2016). Puppet attack. *Journal of Networking and Computer Applications* 59: 325–332.

5 Zhifeng, X., Xiao, Y. and Du, D.H.-C. (2013). Non-repudiation in neighborhood area networks for smart grid. *IEEE Communications Magazine* 51 (1): 18–26.

6 Jawurek, M., Johns, M., and Kerschbaum, F. (2011). Plug-in privacy for smart metering billing. *Proceedings in Privacy Enhancing Technologies: 11th International Symposium, PETS*, Waterloo, Canada (27–29 July 2011). Springer.

7 Chim, T.W., Yiu, S., Hui, L.C.K. et al. (2012). Selling power back to the grid in a secure and privacy-preserving manner. *Proceedings Information and Communications Security: –The International Conference on Information and Communication Systems*, Baghdad, Iraq (9–12 April 2012). IEEE.

8 Bohli, J.-M., Sorge, C. and Ugus, O. (2010). Privacy preserving via group signature in smart grid. *IEEE International Conference on Communications, 2010*, Capetown, South Africa (23–27 May 2010). IEEE.

9 Zargar, S.H.M. and Yaghmaee, M.H. (2013). Privacy preserving via group signature in smart grid. *Proceedings of the First Congress of Electronic Industry Automation*, Mashhad, Iran (13 February 2013).

10 Cheung, J.C.L., Chim, T.W., Yiu, S. et al. (2011). Credential-based privacy-preserving power request scheme for smart grid network. *Proceedings of the Global Communications Conference, GLOBECOM*, Texas, USA (5–9 December 2011). IEEE.

11 Garcia, F.D. and Jacobs, B. (2010). Privacy-friendly energy-metering via homomorphic encryption. *6th International Workshop on Security and Trust Management: STM*, Athens, Greece (23–24 September 2010). Springer.

12 Li, F., Luo, B., and Liu, P. (2010). Secure information aggregation for smart grids 470 using homomorphic encryption. *First IEEE International Conference on Smart Grid Communications SmartGrid-Comm*, Maryland, USA (4–6 October 2010). IEEE.

13 Ruj, S. and Nayak, A. (2013). A decentralized security framework for data aggregation and access control in smart grids. *IEEE Transactions on Smart Grid* 4 (1): 196–205.

14 Li, H., Lin, X., Yang, H. et al. (2013). EPPDR: an efficient privacy-preserving demand response scheme with adaptive key evolution in smart grid. *IEEE Transactions on Parallel and Distributed Systems* 5 (8): 1–11.

15 Lu, R., Liang, X., Li, X. et al. (2012). EPPA: An efficient and privacy-preserving aggregation scheme for secure smart grid communications. *IEEE Transactions on Parallel and Distributed Systems* 23 (9): 1621–1631.

16 Shi, E., Chan, T.H.H. and Rieffel, E. (2011). Privacy-preserving aggregation of time-series data. *Proceedings of the Annual Network and Distributed System Security Symposium –NDSS Symposium*, San Diego, USA (6–9 February 2011). The Internet Society.

17 Bae M., Kim, K. and Kim, H. (2016). Preserving privacy and efficiency in data communication and aggregation for AMI. *Journal of Network and Computer Applications* 59: 333–334.

18 Fan, C.-I., Huang, S.-Y. and Lai, Y.-L. (2014). Privacy-enhanced data aggregation scheme against internal attackers in smart grid. *IEEE Transactions on Industrial Informatics* 10 (1): 666–675.

19 Bao, H. and Rongxing, L. (2016). Comment on a privacy-enhanced data aggregation scheme against internal attackers in smart grid. *IEEE Transactions on Industrial Informatics* 12 (1): 2–5.

20 Chen, L., Lu, R., Cao, Z. et al. (2015). MuDA: Multifunctional data aggregation in privacy-preserving smart grid communications. *Peer-to-Peer Networking and Applications* 8 (5): 777–792.

21 He, D., Kumar, N., Zeadally, S. et al. (2017). Efficient and privacy-preserving data aggregation scheme for smart grid against internal adversaries. *IEEE Transactions on Smart Grid* 8 (5): 2411–2450.

22 He, D., Zeadally, S., Wang, S. and Liu, Q. (2017). Lightweight data aggregation scheme against internal attackers in smart grid using elliptic curve cryptography. *Wireless Communications and Mobile Computing* (ID 3194845): 1–11.

23 Vahedi, E., Bayat, M., Pakravan, M.R. and Aref, M.R. (2017). A secure ECC-based privacy preserving data aggregation scheme for smart grids. *Computer Networks* 129(Part 1): 28–36.

24 He, D., Kumar, N. and Lee, J.-H. (2016). Privacy-preserving data aggregation scheme against internal attackers in smart grids. *Wireless Networks* 22 (2): 491–502.

25 Hankerson, D., Menezes, A.J. and Vanstone, S. (2003). *Guide to Elliptic Curve Cryptography*. New York: Springer-Verlag New York, Inc.

26 SEC 2: Recommended elliptic curve domain parameters (2000). Certicom Research, Standards for Efficient Cryptography Version 1.0. https://www.secg.org/SEC2-Ver-1.0.pdf (accessed 25 June 2019).

27 Recommended elliptic curves for Federal Government use. (2015). National Institute of Standards and Technology. https://csrc.nist.gov/news/2015/fips-186-4-rfc-nist-recommended-elliptic-curves (accessed 25 June 2019).

28 Barbosa, M. and Farshim, P. (2008). Certificateless signcryption. *Proceedings of the ACM Symposium on Information, Computer and Communications Security (ASIACCS ?08)*, Tokyo, Japan (18–20 March 2008). ACM.

29 Pointcheval, P. and Stern, J. (2000). Security arguments for digital signatures and blind signatures. *Journal of Cryptology* 13 (3): 361–396.

14

Blockchain-Based Cyber Physical Trust Systems

Arnold Beckmann, Alex Milne, Jean-Jose Razafindrakoto, Pardeep Kumar, Michael Breach, and Norbert Preining

Abstract

Cyber Physical Trust Systems (CPTS) are Cyber Physical Systems and Internet of Things (IoT) enriched with trust as an explicit, measurable, testable system component. In this chapter, we propose to use blockchain technology as the trust-enabling system component for CPTS. Our proposed approach shows that a blockchain-based CPTS achieves the security properties of data authenticity, identity, and integrity. We describe results of a testbed which implements a blockchain-based CPTS for physical asset management.

Keywords *cyber physical systems; internet of things; data authenticity; identity; integrity; asset management*

14.1 Introduction

Cyber Physical Systems (CPS) integrate computation, networking, and physical processes [1]. As CPS and Internet of Things (IoT) are overlap quite a bit, the distinction to IoT is blurred, with CPS serving as IoT devices, and IoT devices being components of CPS. Advances enabled by CPS are vast, including electric power generation and delivery, personalized healthcare, traffic flow management, and emergency response, as well as in many other areas just now being envisioned. As shown in Figure 14.1, many millions of connected CPS devices will be communicating over the public network and will provide services to their respective applications.

For many applications of CPS, the identity of devices and data generated by devices form an important part of the overall ecosystem they are integrated in. Often, there are a number of actors, which may be devices or humans, that participate in such ecosystems, and who in general do not trust each other. While some actors may interact with devices directly, they often share virtual representation of device identities and their data. The challenge in such a situation is how actors can gain trust in the integrity of identities and data in an explicit, measurable, testable way. Trust can be defined as *reliance on the character, ability, strength, or truth of someone or something; one in which confidence is placed* [2], or as *the firm belief in the reliability, truth, or ability of someone or something* [3].

IoT Security: Advances in Authentication, First Edition.
Edited by Madhusanka Liyanage, An Braeken, Pardeep Kumar, and Mika Ylianttila.
© 2020 John Wiley & Sons Ltd. Published 2020 by John Wiley & Sons Ltd.

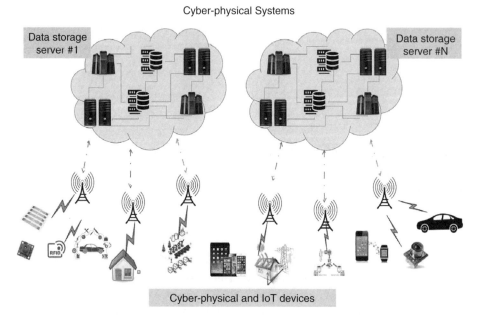

Figure 14.1 Connected cyber physical system.

We define *Cyber Physical Trust Systems* (CPTS) as CPS which have explicit mechanisms for gaining trust on integrity of identities and data built into them. This has to be contrasted with trustworthiness of CPS, which is the combination of security, privacy, safety, reliability, and resilience [4]. Trustworthiness is a property which is implicit to CPS, often established as a form of certificate. It cannot be tested on a CPS system level but exists externally to it.

Definition (Cyber Physical Trust Systems [CPTS]) *A* Cyber Physical Trust System *integrates computation, networking, physical processes, and explicit mechanisms for gaining trust in integrity of data about processes.*

In this chapter, we propose to use blockchain technology as a way of establishing explicit mechanisms for gaining trust. A blockchain (or ledger), as its name suggests, is a growing chain of blocks that contain transaction data of various types – such as financial transactions related to exchange of assets – and linked together using cryptography. On a blockchain, transactions are recorded chronologically, forming an immutable chain – hence, making its data verifiable and auditable. The ledger is distributed across all participants in the network. And because of the immutability property of the blockchain, and a clever mix of cryptography and game theory, everyone in the network agrees with a single copy of the blockchain. Figure 14.2 shows a pictorial high-level view of a blockchain.

In addition to being a system of record, a blockchain can also be a platform for smart contracts. Basically, a smart contract is an autonomous agent stored on the blockchain and is encoded as part of a special transaction, which introduces the contract to the blockchain. One can also view a smart contract as a state machine with its current state somehow represented on the blockchain. Any transaction invoking a smart contract stored on the blockchain will trigger its execution. Once execution is finished, all

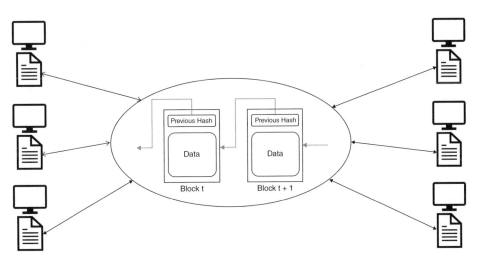

Figure 14.2 A pictorial representation of a blockchain.

relevant actors in the network will unanimously agree on the new state of the smart contract and record that state on the blockchain.

There are different types of blockchains. The most widely used type of blockchains are of public type. In a public blockchain, such as Bitcoin [5] or Ethereum [6], anyone can participate without permission. However, one can also have permissioned blockchains. Such blockchains are built such that they grant special permissions to each participant for specific functions to be performed – such as read, write, and access information on the blockchain. Here, we are mainly focused on permissioned blockchains; and in particular, the Hyperledger Sawtooth, which is an open source project originally developed by Intel [7] and now under the Hyperledger umbrella. Among the consensus options in Sawtooth, there is a novel consensus protocol known as "Proof of Elapsed Time," a lottery-design consensus protocol that optionally builds on trusted execution environments provided by Intel's Software Guard Extensions (SGX) [7].

As an application of CPTS, we will consider traceable assembly systems. Assets as assembly systems are ubiquitous in our modern world. For various societal and economic challenges, it is essential to provide identities to components of assembly system, and to enable a group of untrusted economic players to create trust about identity and usage of such components. Use cases can be found in circular economy (trust in usage of components enables reuse and supports recycling), subscription models (instead of consumer owning assets, e.g. cars, ownership is retained with manufacturer and consumer subscribe to asset pools of various quality), preventing fake parts in the automobile industry [8], and various other models of refined distributed ownership of assembly systems.

Traditional identity management is not able to provide the required level of trust in the identity and usage of assembly systems and other real-world CPS. We have taken a different approach based on blockchain, employing blockchain's data immutability and provenance, and consensus mechanisms. We obtain a blockchain-based CPTS that provides trust in the integrity of identities in assembly systems and their usage data, using the following basic idea: We assume that physical components have digitally represented physical identities – there are solutions available already via security tags in the form of enhanced radiofrequency identification (RFID) tags which cannot be removed from physical objects without being destroyed, and which are enabled with suitable cryptographic primitives for signing data. We also assume that part of the assembly systems are IoT devices for recording usage data, and that those IoT devices are enhanced by cryptographic primitives for signing data. Based on those assumptions, we have achieved a demonstrator which implements a blockchain-based CPTS for establishing trust in the integrity of identities and usage data for assembly systems. The demonstrator provides a decentralized application (DApp) consisting of a permissioned blockchain built on the Hyperledger Sawtooth framework, integrated into a wider business logic, which interacts with physical objects via simulated security tags.

The rest of the chapter is organized as follows: Section 14.2 discusses the state-of-the-art work where blockchain is used as a security service. Section 14.3 presents an overview of use-cases and security design goals. Section 14.4 describes the details of the proposed technique, and Section 14.5 discusses the testbed results and security analysis. Finally, conclusions are drawn in Section 14.6.

14.2 Related Work

Recently, blockchain as a security-service has attracted more and more attention from both academia and industry and it is spanning across several domains, including supply chain systems, banking, healthcare, asset management, etc. This section presents the state-of-the-art work on the blockchain-based security services (such as authentication, trust, integrity, etc.) in IoT, wireless sensor networks and other domains.

In blockchain, the transaction-recording and non-duplicability services make it a good technological choice for several applications. More precisely, these services demonstrate the blockchains' suitability for public key infrastructure (PKI). Blockchain-based PKI solutions are distributed and have no centralized point of failure. As a result, certificate-based PKI can be used to realize authentication in blockchain [9, 10]. However, public-key certificates have their own shortcomings and issues. In order to solve certificate issues, Lin et al. propose an identity-based linearly homomorphic signature scheme and its application in blockchain [11]. In an identity-based scheme, a node's ID can be the node's name or any arbitrary string that can be used as a public key. The encryption approach consists of four phases: setup, extract, encrypt, and decrypt. In addition, the scheme is proven to be secure against existential forgery on adaptively chosen messages and identity attack under the random oracle model.

Lewison-Corella propose a blockchain-based distributed database to store data securely [12]. The main idea of that paper is to allow the certificate authority (CA) to publish an unsigned certificate. The blockchain stores the hash value of the certificate and that stored value is controlled by entities, such as banks or governments. These entities make use of two blockchain databases, one for the issued certificates and another for the revoked certificates. When verifying the certificate, an entity first assures the corresponding data is stored to the blockchain. If the certificate's hash value is found in the database, then the certificate is a valid certificate. Otherwise, it is not a valid certificate and it will then be revoked from the blockchain. This idea is simple and provides several advantages such as an easy verification with low-delay guarantees.

However, the implementation and evaluation results are missing, therefore the viability of this approach is a big question.

Lin et al. propose a blockchain-based secure mutual authentication and access control system for Industry 4.0 [13]. They claim to provide various security services, including anonymous authentication, auditability, and confidentiality and privacy. The authors utilized attribute-based signatures to achieve anonymous authentication and fine-grained access control. Lin et al. adopted consensus procedure, which is based on the practical byzantine fault tolerance (PBFT) approach. However, PBFT suffers from the scalability issues as discussed in [14].

As the number of IoT devices is exploding, it is almost impossible to create an efficient centralized authentication system. Hammi et al. propose a decentralized blockchain-based authentication system for IoT [15]. To achieve their goals, the proposed scheme relies on the security advantages provided by blockchains, and serves to create secure virtual zones (bubbles) where things can identify and trust each other.

Another research focuses on blockchain-based digital identity management also known as *"BIDaaS: Blockchain based ID as a Service"* [16]. This research mainly targets identity management in mobile telecommunication networks. Three entities are being involved: user (e.g. mobile user), BIDaaS provider (e.g. telecommunication company), and partner of the BIDaaS provider (e.g. partner of the telecommunication company). The basic idea of the scheme is that a mutual authentication is performed between the user and the partner. The scheme did not utilize any pre-shared information or security credential shared among them. More detailed survey papers on security services using blockchain can be found in [17, 18].

14.3   Overview of Use-Cases and Security Goals

14.3.1   Use-Cases

Blockchain-based approaches are popular in many real-world applications. We describe two examples which are relevant to our approach.

14.3.1.1   Asset Management

Asset assemblies can have thousands of tracked components, e.g. for aviation assets. It is, therefore, crucial to have suitable asset management in place to handle the organization, identification, and value creation of asset assemblies. Asset management systems also need to support the creation of subassemblies and virtual representation of them in order to reduce the underlying complexity. In the context of the need for our economies to become circular, reusing components within assets becomes a necessity. Thus, such asset management systems should also manage the accurate recording of usage against assembly systems and their components, as well as their history, to support value creation from reused components.

14.3.1.2   Traffic Management Using a Smart Road Radar

A smart road radar is a road radar that can be controlled and set up remotely without human intervention to avoid road accidents [15]. The main responsibility of a road radar is to measure the speed of vehicles on the road. If a vehicle is violating the speed limit then the radar can detect the speed violator, and send a message including a photograph of the vehicle, license plate and the measured speed to the blockchain-based traffic management systems. In such a system, car authenticity, identity, and data integrity is of high importance.

14.3.2   Security Goals

Following the literature survey, we have identified that a blockchain-based CPTS must fulfill a number of security requirements in order to attain the sustainability and resiliency in the CPS. Therefore, this subsection describes the main security goals, as follows.

14.3.2.1   Data Authentication [19]

In the real-world CPS, message authentication is an important goal. As a malicious user can easily inject fake data to a CPS, the blockchain-based systems must ensure data authenticity and check whether the data originates from the trusted or claimed node. In general, data authentication allows a receiver entity to check the legitimacy of data and that the data really was sent by the claimed entity.

14.3.2.2   Data Integrity [18]

Data integrity ensures the receiver that the received data/transaction is not altered by an adversary.

14.3.2.3   Secure Identity Management [20]

A massive number of devices will be deployed in a CPS network, and each device will have its own identity. However, identity management can play a major role in real-world CPS networks to track and trace the information/status of the devices. Therefore, secure identity management is an important requirement for blockchain-based CPTS.

14.4   Proposed Approach

The basic design idea of blockchain-based CPTS is that CPS are linked to a blockchain ledger which is distributed among the actors of the ecosystem. The key blockchain features, as discussed before, will ascertain that the data stored on the blockchain and smart contracts executed by the blockchain are trusted amongst the actors.

In the following, we describe a scheme in which data from a CPS device will be recorded on the blockchain. For our scheme, we assume that the recording is requested by one of the actors who has an interest in the data being documented at this point in time. We assume that the actors also want to gain trust in the time when the data was recorded, in addition to gaining trust in the data itself. The data is stored on the blockchain in an associative array, where the keys are given by the IDs of CPS devices, and the values are the process data, and the block number within the blockchain where the recording took place is a timestamp.

Our scheme has three components:

- A CPS device, that, in addition to being able to compute and to communicate, has a cryptographic identity through an asymmetric pair of keys. It is able to communicate its identity in the form of its public key, to communicate data related to its processes, to receive additional data, and to sign data (i.e. its process data or received data).
- A client representing one of the actors who aims to record the current process data of the CPS device on the blockchain. All actors are registered with the blockchain. Thus, the client can interact with the blockchain by sending transactions which will be executed by the blockchain. The client can also communicate with the CPS device, and will have computing capabilities, e.g. to form transactions.
- A permissioned blockchain system that stores and executes smart contracts. The actors are permissioned to interact with the blockchain, thus transactions sent by them will be executed by the blockchain system.

The scheme then operates using the following six steps and the flow of the proposed approach is depicted in Figure 14.3:

1. Client wants to record process data of the CPS device on blockchain. He sends a transaction to the blockchain that requests a nonce.
2. Blockchain processes the transaction and sends nonce back to client.
3. Client requests the CPS device to sign its current process data together with the nonce.
4. CPS device sends the signed process data and nonce.
5. Client builds the transaction for the blockchain that contains the CPS device ID, signed data and nonce as its payload, the action is to record the data against the device ID on the blockchain.
6. Blockchain executes a smart contract to check authenticity of identity, data, and time with the following steps:
   o Verify that the signed process data was signed by the claimed CPS device.
   o Check that the data was signed with the correct nonce and that the nonce has not timed out.
   o If all checks are true then save the data against the ID on the blockchain.

We claim that the scheme satisfies our security requirements as described above, under a set of regularity assumptions:

Claim. Assuming that the blockchain system is able to produce an unpredictable nonce, and that the cryptographic primitives are secure, data, and timestamps against an ID as recorded on the blockchain are identical to the data produced by the device ID at the corresponding time.

Thus, this scheme achieves the security requirements of data authenticity and integrity, and secure identity management. This realizes a CPTS.

14.5 Evaluation Results

This section discusses the security features and testbed results for the proposed approach.

14.5.1 Security Features

14.5.1.1 Data Authentication and Data Integrity
In the proposed ecosystem, each transaction of an entity uses a public key cryptography (PKC) based signature, which is generated by the private key of the entity. The inherent features of the PKC-based signature (i.e. *sign[usage data, nonce], usage data*) ensures that the proposed approach can achieve data authentication and integrity.

14.5.1.2 Device Identification
Each device/entity owns an identity, which is a unique identity. Here, the identity is issued at the time of registration of devices utilizing its public key. Note that entity registration is out of the scope of this chapter. However, the trustworthiness of the issued identity is assured by the signature. Each transaction of an entity is computed over its private key, and the key is only associated to the device identity. Hence, the approach can easily identify the device.

14.5.1.3 Non-repudiation
As shown in Figure 14.3, in each transaction, the data is signed using the private key (i.e. *sing[usage data, nonce], usage data)*, which is possessed by its owner entity. More precisely, this is the only owner who can generate and use the transaction. Therefore, it cannot deny the fact of signing a message.

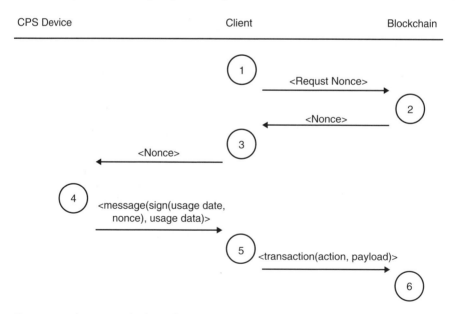

Figure 14.3 The proposed scheme for recording data from a CPS device on the blockchain.

14.5.1.4 Secure Against Replay Attack

A replay attack is the most common threat, where an ill-intentioned adversary can replay old messages. However, the proposed approach utilizes the random nonce to avoid the replay attack. Thus, an adversary cannot replay the old messages in the proposed approach.

14.5.1.5 Protection from Spoofing Attack

Indeed, an attacker can spoof the identity of the object from the open communication channels. However, it cannot verify the spoofed identity, as he/she does not have the knowledge of the private key of the real entity. Therefore, spoofing identity does not help him/her.

14.5.2 Testbed Results

In the following we describe our testbed results. The testbed has been developed as part of a CHERISH-DE[1] funded Escalator project *Blockchain for Subscription Models*.

14.5.2.1 Testbed Overview

The setup we are using is to have a blockchain system communicating through a client with a security tag/IoT device that is attached to an asset, see Figure 14.4. The blockchain system we are using is Hyperledger Sawtooth v1.05. We use a client to interact with the security tag that is ingrained in the asset in order to sign data.

14.5.2.2 Hyperledger Sawtooth

Hyperledger Sawtooth is like other blockchain system in that its main purpose is to ensure many different parties agree on a common set of data. Sawtooth stores this data in addresses in a Merkle tree. With each change in the data, the root hash of the Merkle

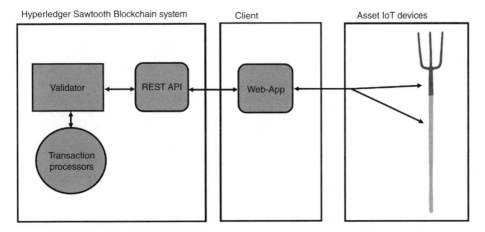

Figure 14.4 A simple overview of the implemented cyber physical trust system, with a gardening fork as an example asset of two components.

---

1 http://cherish-de.uk The CHERISH Digital Economy Centre is a multidisciplinary research centre at Swansea University addressing the impact of the digital economy on humans, society, and industry.

tree changes. The current Merkle root is stored in each block as the current state of the system. Each party can verify a block by performing the given transactions from the block on their own data and then making sure that the Merkle root of the data is identical to the Merkle root in the block. This ensures that all parties are running code with an identical effect in their transaction processors because if they were not, they would produce a different Merkle root.

14.5.2.3   IoT Devices

In the testbed implementations, our IoT devices/tags are attached to the asset in a way that they are not removable without destroying the device. This makes them digital representations of the physical assets. Each asset that wants to be represented on the blockchain will need to have a (digital) tag attached. These tags store a public and private key pair that are generated securely on the creation of the device and are never changeable, with the private key never been accessible outside the chip. A tag's identity is its public key and each tag has the ability to sign data given to it to prove its identity.

Some of these IoT devices have extra functionality; the ability to track usage of the device. They store this data securely and can sign it to prove that it was, in fact, the usage of this asset. These tags with extra functionality will prove their identity by signing their stored usage value along with a nonce. This is of course because allowing them to sign arbitrary data would result in an attacker being able to fake a usage not from the tag by asking the tag to sign it.

During our project we used RFID tags to emulate the functionality of these security tags. However, there are products available that realize such security tags.[2]

14.5.2.4   Client

The client is the interface that allows users to interact with the blockchain and the asset security tags. Users can query the tags to get the public key, the usage and get the tag to sign data.

The client can build transactions to send data to the blockchain to update the assets' state. Some transactions are transactions to; update the usage; update the usage within a timeframe; create assets' digital representation on the blockchain, assemble assets into assemblies of assets, etc.

14.5.2.5   Demonstrator

The assets in our demonstrator are forks and their security tag enhanced components. The action of starting a leasing of a fork using the web-app client can been seen in Figure 14.5. Here, the client will obtain the usage and signed usage from the tag, seen in Figure 14.6, and send this in a transaction to the blockchain. As the usage is used when determining the cost of the leasing in our model, it needs to be accurately updated at both the start and end of a lease.

14.5.2.6   Protocol for Updating the Usage

The protocol for updating the usage is an instantiation of the general scheme as described before, see Figure 14.2. The timing requirement is implemented by using

---

2 The NXP Mifare DESFire provides highly secure microcontroller-based ICs which can be used for provide security tags, see https://www.nxp.com/products/identification-and-security/mifare-ics/mifare-desfire.

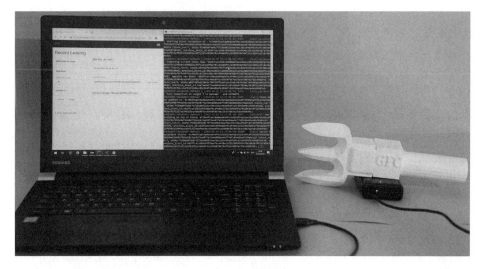

Figure 14.5 The demonstrator system reading and signing the security tags stored usage of 200 and then sending this in a transaction to the blockchain where it is accepted and updated.

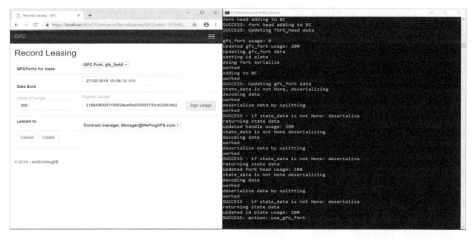

Figure 14.6 The running demonstrator system reading and signing, using the security tag in the fork head, the stored usage data.

the Merkle tree root hash from the blockchain as a nonce. We can do that while making the assumption that the blockchain system is in regular use (blocks are added frequently), and that not all of the transactions changing the blockchain Merkle tree can be predicted. Making those assumptions, the Merkle root hash will be unpredictable. We note that it is a general issue with blockchain systems to generate random numbers: As everyone must agree on the random number deterministically for there to be consensus, it cannot be random.

Applying our results from the previous section, we can say that the demonstrator implements a CPTS. Hence, the system provides trust in the data on usage and identity of assets. With regard to time stamps it obtains a guarantee that the usage was read out from the tag within the interval between the block containing the update of the usage and the block containing the Merkle tree root that served as a nonce.

14.6 Conclusion

Data authenticity and integrity, and identity security are big security issues for an ever-growing number of CPS and IoT devices. We have introduced blockchain-based CPTS as CPS enhanced with blockchain as an explicit, measurable, testable system component for providing trust in data authenticity and integrity, and identity security. We have proposed a PKC-based approach for data exchange between devices and blockchain, and argued that it achieves the security requirements of data authenticity and integrity, and identity security. We presented results from a testbed that implemented a CPTS for asset management.

In future work, we will conduct in-depth formal and informal security analysis of our proposed scheme. We will also extend the testbed into a generic application for supporting CPTS, and conduct in-depth performance analysis ranging from theoretical testbeds based on theoretical performance assumptions of blockchain technology, to practical testbeds in relation to an enhanced testbed implementation. We will also explore other application domains, in which CPTS can be applied.

References

1 Lee, E.A. and Seshia, S.A. (2017). *Introduction to Embedded Systems – A Cyber-Physical Systems Approach*, 2e. MIT Press.

2 Trust. (2019). Merriam-Webster.com. https://www.merriam-webster.com/dictionary/trust (accessed 17 July 2019).

3 Trust. (2019). Oxford Online Dictionary. https://en.oxforddictionaries.com/definition/trust (accessed 17 July 2019).

4 Framework for Cyber-Physical Systems, Release 1.0. (2016). Cyber Physical Systems Public Working Group, National Institute of Standards and Technology.

5 Nakamoto, S. (2009). Bitcoin: A peer-to-peer electronic system. http://Bitcoin.org (accessed 17 July 2019).

6 Ethereum Foundation. (2014). Ethereum's white paper. https://github.com/ethereum/wiki/wiki/White-Paper(accessed 17 July 2019).

7 Bucci, D. (2019). Blockchain and its Emerging Role in Health IT and Health-related research. U.S. Department of Health and Human Services, Office of the National Coordinator for Health Information Technology.

8 Fake Vehicle parts are on the rise. https://www.gov.uk/government/news/fake-vehicle-parts-are-on-the-rise (accessed 17 July 2019).

9 Matsumoto, S. and Reischuk, R.M. (2017). IKP: turning a PKI around with decentralized automated incentives. In: *2017 IEEE Symposium on Security and Privacy (SP)*, San Jose, USA (25 May 2017). San Jose, CA: IEEE.

10 Moinet, A., Darties, B. and Baril, J.-L. (2017). Blockchain-based trust & authentication for decentralized sensor networks. https://arxiv.org/abs/1706.01730 (accessed 17 July 2019).

11 Lin, Q., Yan, H., Huang, Z. et al. (2018). An ID-based linearly homomorphic signature scheme and its application in blockchain. *IEEE Access* 6 (99): 1.

12 Lewison, K. and Corella, F. 2016 Backing Rich Credential with Blockchain PKI. Technical Report.

13 Lin, C., He, D., Huang, X. et al. (2018). BSeIn: a blockchain based secure mutual authentication with fine-grained access control system for industry 4.0. *Journal of Network and Computer Applications* 116: 42–52.

14 Vukolić, M. (2015). The quest for scalable blockchain fabric: proof-of-work vs. BFT replication. In: *International Workshop on Open Problems in Network Security*. Kolkata, India (16–20 December 2015): IEEE.

15 Hammi, M.T., Hammi, B., Bellot, P., and Serhrouchni, A. (2018). Bubbles of trust: a decentralized blockchain-based authentication system for IoT. *Computer & Security* 78: 126–142.

16 Lee, J.H. (2017). BIDaaS: blockchain based ID as a service. *IEEE Access* 6 (99): 1.

17 Lin, I.C. and Liao, T.C. (2017). A survey of blockchain security issues and challenges. *International Journal on Network Security* 19 (5): 653–659.

18 Salman, T., Zolanvari, M., Erbad, A. et al. (2018). Security services using blockchains: a state of the art survey. *IEEE Communications Surveys, & Tutorials* 21: 858–880.

19 Khan, M.A. and Salah, K. (2018). IoT security: review, blockchain solutions, and open challenges. *Future Generation Computer Systems* 82: 395–411.

20 Preuveneers, D., Joosen, W., and Zudor, E.I. (2017). Identity management for cyber-physical production workflow and individualized manufacturing in industry 4.0. In: *Proceedings of the Symposium on Applied Computing*. Marrakech, Morocco (3–7 April 2017): ACM.

Index

IoT Security: Advances in Authentication, First Edition.
Edited by Madhusanka Liyanage, An Braeken, Pardeep Kumar, and Mika Ylianttila.
© 2020 John Wiley & Sons Ltd. Published 2020 by John Wiley & Sons Ltd.